AF697329

Progress in Physics
Vol. 10

Edited by
A. Jaffe, G. Parisi,
and D. Ruelle

Birkhäuser
Boston · Basel · Stuttgart

Statistical Physics and Dynamical Systems

Rigorous Results

J. Fritz,
A. Jaffe, and
D. Szász, editors

1985

Birkhäuser
Boston · Basel · Stuttgart

UNIVERSITY OF TOLEDO LIBRARIES

Editors

J. Fritz
MTA MKI
Realtanoda u. 13–15
H–1053 Budapest
(Hungary)

A. Jaffe
Dept. of Physics
Lyman Laboratory
Harvard University
Cambridge, MA 02138 (USA)

D. Szász
MTA MKI
Realtanoda u. 13–15
H–1053 Budapest
(Hungary)

Library of Congress Cataloging in Publication Data
Main entry under title:

Statistical physics and dynamical systems.
(Progress in physics ; vol. 10)
»Contains most of the invited papers of the Second Colloquium and Workshop on ›Random Fields: Rigorous Results in Statistical Mechanics‹ held in Köszeg, Hungary between August 26 and September 1, 1984« -- CIP pref.
Includes bibliographies.
1. Random fields -- Congresses. 2. Statistical mechanics -- Congresses. 3. Quantum field theory -- Congresses. I. Fritz, J. II. Jaffe, Arthur, 1937– III. Szász, D. IV. Colloquium and Workshop on »Random Fields : Rigorous Results in Statistical Mechanics« (2nd : 1984 : Köszeg, Hungary) V. Series: Progress in physics (Boston, Mass.) ; v. 10.
QC174.85.R36S73 1985 530.1'3 85–1235
ISBN 0-8176-3300-6

CIP-Kurztitelaufnahme der Deutschen Bibliothek

Statistical physics and dynamical systems : rigorous results / J. Fritz . . ., ed. – Boston ; Basel ; Stuttgart : Birkhäuser, 1985.
(Progress in physics ; Vol. 10)
ISBN 3-7643-3300-6 (Basel . . .)
ISBN 0-8176-3300-6 (Boston)
NE: Fritz, Jozsef [Hrsg.]; GT

All rights reserved.
No part of this publication may be reproduced, stored in a retrieval system, or transmitted, in any form or by any means, electronic, mechanical, photocopying, recording or otherwise, without the prior permission of the copyright owner.

© 1985 Birkhäuser Boston, Inc.
Printed in Germany
ISBN 0-8176-3300-6
ISBN 3-7643-3300-6

QC 174.85 .R36S73 1985

CONTENTS

PREFACE

This volume contains most of the invited papers of the Second Colloquium and Workshop on "Random Fields": Rigorous Results in Statistical Mechanics" held in Kőszeg, Hungary between August 26 and September 1, 1984. Invited papers whose authors could finally not attend the Colloquium are also included.

The Colloquium was organized by the generous sponsorship of the International Union for Pure and Applied Physics, the International Workshop Committee for Theoretical Physics, the International Association of Mathematical Physics, the Hungarian Academy of Sciences and the János Bolyai Mathematical Society.

Members of the International Program Committee were R. L. Dobrushin, A. Jaffe, J. L. Lebowitz, D. Ruelle, Ya. G. Sinai. The Organizing Committee consisted of J. Fritz, D. Szász (co-chairmen), D. Petz (secretary), A. Krámli, G. Lippner, P. Lukács, P. Major, A. Sütő, N. Simányi, A. Vetier.

There were 112 participants from 21 countries representing all the six continents. There were altogether 22 forty-five minute invited talks and 63 ten minute contributed papers. The Workshop was organized on the last two days of the Colloquium. It mainly concerned topics most interesting for Hungarian physicists and, moreover, its program ensured additional space for discussions.

We express our sincere gratitude to Dénes Petz for his careful work as the technical editor of this volume and to Zsuzsa Erő for her excellent and rapid retyping of several manuscripts.

The Editors

P R O G R A M

Monday

10.00	Opening Ceremony
10.20 - 11.05	R. L. Dobrushin - S. B. Shlosman: Constructive criteria of uniqueness and analiticity in statistical mechanics
11.10 - 11.55	J. L. Lebowitz: Mathematical and physical ideas in nonequilibrium statistical mechanics
3.00 - 3.45	E. Lieb: Various estimates for the eigenvalues of the Laplacian
3.50 - 4.35	H. Araki: C*-algebra approach to the ground states of the X-Y model
coffee break	
5.00 - 5.45	M. Aizenmann: Rigorous studies of critical behaviour
5.50 - 6.35	D. Kazhdan: ε-representations
7.30	Welcome party

Tuesday

9.00 - 9.45	A. Kupiainen: Non-trivial continuum limit for negative coupling ϕ^4_4
9.50 - 10.35	G. Papanicolau: Bulk diffusion and self diffusion for interacting Brownian motions
coffee break	
11.00 - 11.45	O. E. Lanford: Renormalization group methods for mappings with golden-ratio rotation number
10.50 - 12.00	H. Rost: Equilibrium fluctuations for a onedimensional nearest neighbour model
12.02 - 12.12	A. Verbeure: States stationary for the detailed balance of reversible processes
12.14 - 12.24	M. F. Chen: Jump Markov processes and interacting particle systems

12.26 - 12.36	Ra. Siegmund-Schulze: On existence of non-equilibrium dynamics of multidimensional infinite particle systems: the translation invariant case
12.38 - 12.48	J. Fritz: Interacting Brownian particles: existence and self-adjointness
3.00 - 3.10	A. Krámli - D. Szász: The problem of recurrence for Lorentz processes
3.12 - 3.22	D. Dawson: Ensemble and multilevel models of critical behaviour
3.24 - 3.34	K. Fleischman: Occupation time processes at a critical point
3.36 - 3.46	J. A. Galves - E. Olivieri - M. E. Vares: Metastable behaviour of stochastic systems: a pathwise approach
3.48 - 3.58	J. R. Klauder: Langevin equations for statistical computations
4.00 - 4.10	N. Ianiro: Stationary Boltzman-equation
4.12 - 4.22	G. Jetschke: On stochastic nonlinear parabolic differential equations
4.24 - 4.34	P. Calderoni: On the Smoluchowski limit for simple particle systems
4.36 - 4.46	H. Cranel: Stochastic systems on manifolds
coffee break	
5.14 - 5.24	C. A. Hurst: C*-algebra approach to the Pfaffian method for the Ising model
5.26 - 5.36	J. T. Lewis - J. V. Pulè: Phase transitions and the weak law of large numbers
5.38 - 5.48	K. Kuroda: The Pirogov-Sinai theory of phase transitions for continuum systems
5.50 - 6.00	Y. Higuchi: A weak version of Russo-Seymour-Welsh theorem for the two-dimensional Ising model
6.02 - 6.12	L. Laanait - A. Messager - J. Ruiz: Phases coexistence and surface tension for the Potts model
6.14 - 6.24	B. Tóth: A lower bound for the critical probability of square-lattice site percolation
6.26 - 6.36	D. Merlini: On the Temperley conjecture for the two-dimensional Ising model
6.38 - 6.48	V. Warstat: A uniqueness theorem for systems of interacting polimers at low temperature
6.50 - 7.00	M. Arató: The distribution of stochastic integrals

Wednesday

9.00 - 9.45	E. Presutti - Ya. G. Sinai - M. Soloviechik: Hyperbolicity and Möller-morphism for a model of classical statistical mechanics
9.50 - 10.35	Yu. M. Suhov - A. G. Shuhov: Linear and related models of time evolution in quantum statistical mechanics
coffee break	
11.00 - 11.45	R. Caflisch: Thermal layers for the Boltzman equation
11.50 - 12.00	J. K. Percus: Evaluation of a class of functional integral
12.02 - 12.12	J. Bricmont - J. Fröhlich: Random walks and the particle structure of lattice guage theories
12.14 - 12.24	D. Surgailis: On continuous contour-models and Arak fields
12.16 - 12.36	G. F. Lawler: Intersection properties of simple random walks
12.38 - 12.48	P. Major: Renormalization of Dyson's hierarchical vector-valued model at low temperatures

Thursday

9.00 - 9.45	L. Accardi: Quantum probability
9.50 - 10.35	H. Spohn: Equilibrium fluctuations for some stochastic particle systems
coffee break	
11.00 - 11.45	C. Marchioro: Some problems in vortex theory
11.50 - 12.00	C. Kipnis: Asymptotics for the motion of a tagged particle in the simple exclusion model
12.02 - 12.12	J. R. Fontaine - Ph. A. Martin: Equilibrium equations and Ward identities for Coulomb systems
12.14 - 12.24	Ch. Gruber: On the invariance of charged systems with respect to external fields
12.26 - 12.36	J. Jedrzejewski: Phase transitions in models of itinerant electrons
12.38 - 12.48	G. Schlijper: Rigorous results for approximate variational principles
3.00 - 3.10	D. Dürr: On Harris' collision model
3.12 - 3.22	H. Rodehausen: Diffusive behaviour for a class of Ornstein-Uhlenbeck processes
3.24 - 3.34	J. Gärtner: On long-time fluctuations of weakly interacting diffusions

3.36 - 3.46	A. Krámli - N. Simányi - D. Szász: Transport phenomena and random walks with internal states
3.48 - 3.58	Y. Elskens - H. L. Frisch: Annihilation dynamics in one dimension
4.00 - 4.10	P. Ferrari - E. Presutti - M. E. Vares: Hydrodynamical properties of a zero-range model
4.12 - 4.22	D. Szász - B. Tóth: One-dimensional persistent random walks in random environment
4.24 - 4.34	W. A. Majewski: On ergodic properties of dynamical semigroups
4.36 - 4.46	A. Wakolbinger: Time-reversal co-dimensional diffusions
coffee break	
5.14 - 5.24	S. Pogosian: Cluster property of classical spin systems
5.26 - 5.36	B. Nahapetian: Limit theorems for weakly dependent random variables
5.38 - 5.48	D. Petz: Quasi entropies for finite quantum systems
5.50 - 6.00	H. Baumgärtel: A class of nontrivial weakly local massive Wightman fields with interpolation properties
6.02 - 6.12	I. Daubechies - J. R. Klauder: Wiener measures for exp(-itH)
6.14 - 6.24	E. Brüning: On the construction of random probability measures of infinite dimensional spaces
6.26 - 6.36	K. H. Fichtner - G. Jetschke: A probabilistic model of a quantum mechanical infinite particle system
6.38 - 6.48	E. Orlandi - R. Figari: Gaussian approximation for the Green's functions of Laplacian in a domain with random holes
6.50 - 7.00	V. Schaffenberger: Borel summability in the disorder parameter of the averaged Green's function for Gaussian disorder
coffee break	
7.20 - 7.30	A. de Masi - P. Ferrari: Diffusion in percolation regime I.
7.32 - 7.42	P. Ferrari - A. de Masi: Diffusion in percolation regime II.

7.44 - 7.54	R. Kotecky: On residual entropy models
7.56 - 8.06	S. Olla: Large deviations and variational principles
8.08 - 8.18	G. Royer: De Fortret-Mourier distance and log-concave functions

Friday

9.00 - 9.45	A. Katok: Random perturbations of dynamical systems motivation, conjectures, rigorous results
9.50 - 10.35	M. Misiurewicz: Convergence of images of certain measures

coffee break

11.00 - 13.00	Discussions
3.00 - 4.00	Discussions
4.00 - 4.10	H. O. Georgii: On the critical temperature of disordered ferromagnets near the percolation threshold
4.12 - 4.22	J. L. van Hemmen: Statistical mechanics of spin glasses
4.24 - 4.34	H. Englisch - M. Endrulis: Random alloys and special energies

coffee break

5.00 - 5.10	F. Przytycki: Riemann maps and holomorphic dynamics
5.12 - 5.22	S. Pirogov: Automata systems with defects
5.24 - 5.34	A. Vetier: Ergodic properties of the Sinai billiard in an external field
5.36 - 5.46	J. Kotus: Ω-stability of vector fields
5.48 - 5.58	K. Ziemian: An almost sure invariance principle for some maps of an interval

Saturday

9.00 - 9.45	P. Collet: Phase transitions on diamond lattices
9.50 - 10.35	L. Pastur: On the spectral theory of random and almost periodic operators
10.40 - 11.25	B. Souillard: Transitions from pure point to continuous spectrum for random Schrödinger operators. Some examples
11.30 - 12.00	Closing ceremony

LIST OF PARTICIPANTS

L. Accardi
Dip. di Matematica
Università di Roma II
Via Orazio Raimondo
I-00173 Roma
(Italy)

Á. Ág
MTA KFKI
PF. 49
H-1525 Budapest 114
(Hungary)

M. Aizenman
Dept. of Mathematics
Rutgers University
Busch Campus
New Brunswick, NJ 08903
(USA)

H. Araki
230-42 Iwakura-Nagatanicho
Sakyoku
Kyoto 606
(Japan)

M. Arató
Fehérvári út 129
H-1119 Budapest
(Hungary)

H. Baumgärtel
Inst. f. Mathematik
Mohrenstrasse 39
1086 Berlin
(DDR)

G. Benfatto
Via A. Fraccaroli, 7
I-00157 Roma
(Italy)

C. Boldrighini
Istituto Matematico
Università di Camerino
I-62032 Camerino
(Italy)

J. Bricmont
FYMA 2
Chemin du Cyclotron
B-1348 Louvain-La-Neuve
(Belgium)

R. Caflisch
Courant Institute
251 Mercer St.
New York, NY 10012
(USA)

P. Calderoni
ZIF Bielefeld Universität
Wellenberg 1
D-4800 Bielefeld
(FRG)

M. Campanino
Via Bisagno 14
I-00199 Roma
(Italy)

S. Caprino
Via Armando di Tullio 27
I-00136 Roma
(Italy)

Mu Fa Chen
Dept. of Mathematics
Beijing Normal University
Beijing
(Peoples' Republic of China)

G.S. Chobanov
Math. Inst. of the
Bulg. Acad. of Sci.
P.O.B. 373
1090 Sofia
(Bulgaria)

J.S. Cohen
Prinsengracht 1055 A-2
NL-1017 JE Amsterdam
(The Netherlands)

P. Collet
26, rue Vergmiaud
F-75013 Paris
(France)

I. Daubechies
Wolfshaegen 33
B-3053 Huldenberg
(Belgium)

D. Dawson
2155 Delmar Dr.
Ottawa K1H 5P6
(Canada)

R. L. Dobrushin
Inst. of Problems of Transmission
of Information
Avia Motornaja 8
Moscow E - 24
(USSR)

D. Dürr
Ruhr-Universität Bochum
Inst. für Mathematik
Postfach 102148
D-4630 Bochum 1
(FRG)

Y. Elskens
Fac. de Sci.
Univ. Libre de Bruxelles
C.P. 231, Bvd. du Triomphe
B-1050 Bruxelles
(Belgium)

H. English
Steinstrasse 16
DDR-7030 Leipzig
(DDR)

R. Eposito
Piazzale Montesquieu 28 IA
I-00137 Roma
(Italy)

P.A. Ferrari
IHES
F-91440 Bures-Sur-Yvette
(France)

A. Fialowski
Villányi út 103
H-1118 Budapest
(Hungary)

K. Fleischmann
AdW d DDR
Institut für Mathematik
Mohrenstrasse 39
1086 Berlin
(DDR)

J.-R. Fontaine
3A Ch. des Esserts
CH-1024 Ecublens
(Switzerland)

J. Fritz
MTA MKI
Reáltanoda u. 13-15
H-1053 Budapest
(Hungary)

J.A. Galves
R. Riachuelo 296
13130 Sousas - S.P.
(Brasil)

J. Gärtner
AdW d DDR
Institut für Mathematik
Mohrenstrasse 39
1086 Berlin (DDR)

H.-O. Georgii
Sperlingweg 7
D-8031 Eichenau (FRG)

V. Gorini
Dip. di Fisica
Sez. Fis Teor.
Via Celoria 16
I-20133 Milano
(Italy)

Ch. Gruber
Inst. Phys. Théor.
EPL-Lausanne
CH-1015 Lausanne
(Switzerland)

Y. Higuchi
Dept. of Mathematics
Kobe University
Rokko
Kobe 657
(Japan)

C.A. Hurst
99 Fifth Avenue
Joslin, South Australia 5070
(Australia)

N. Ianiro
Via Urbana 143
I-00100 Roma
(Italy)

K.R. Ito
ZiF, Univ. Bielefeld
Wellenberg 1
D-4800 Bielefeld 1 (FRG)

J. Jedrzejewski
Budziszyńska 135/5
Wroclaw
(Poland)

G. Jetschke
Friedrich-Schiller Universität
Sektion Mathematik
DDR-6900 Jena (DDR)

A. Katok
1080 Spruce Str.
Berkeley, Calif. (USA)

D. Kazhdan
Harvard University
Mathematics Department
1 Oxford Street
Cambridge, MA 02138 (USA)

C. Kipnis
17 Rue Mathis
F-75019 Paris
(France)

J.R. Klauder
AT and T Bell Labs
Murray Hill, NJ 07974 (USA)

R. Kotecký
Dept. Math. and Phys.
V. Holesovickách 2
Praha 8
(Czechoslovakia)

J. Kotus
Szczęślinicka 29 m 29
P-02-353 Warsaw
(Poland)

A. Krámli
MTA SZTAKI
Victor Hugo u. 18-22
H-1132 Budapest
(Hungary)

R. Kuik
Sloep 299
NL-9732 CT Groningen
(The Netherlands)

A. Kupiainen
Abrahamink. 17 C59
Helsinki 18
(Finland)

K. Kuroda
Dept. of Math. Keio University
Hiyoski 3-14-1
Kohoku-ku
Yokohama 223
(Japan)

L. Laanait
C.P.T. CNRS
Lumniy Case 907
F-13288 Marseille Cedex 9
(France)

O.E. Lanford
Inst. des Hautes Etudes Sci.
35, Route de Chartres
F-91440 Bures-Sur-Yvette
(France)

G.F. Lawler
311 S. Lasalle St. Apt. 30 A
Durham, NC 27705 (USA)

J.L. Lebowitz
52 Locust Lane
Princeton, NJ 08546 (USA)

G. Leha
Bavariastrasse 2
D-8551 Pinzberg (FRG)

J.T. Lewis
San Clemente, Vico Rd.
Dalkey, County Dublin
(Ireland)

E. Lieb
Jadwin Hall
P.O.B. 708
Princeton, NJ 08544 (USA)

G. Lippner
BME Villamosmérnöki Kar Mat. Tsz.
Müegyetem rkp. 3-9
H-1111 Budapest
(Hungary)

P. Lukács
MTA SZTAKI
Victor Hugo u. 18-22
H-1132 Budapest
(Hungary)

W.A. Majewski
Inst. Theor. Phys. Astrophys.
Univ. Gdańsk
Wita Stwosza 75
P-80-952 Gdańsk
(Poland)

P. Major
MTA MKI
Reáltanoda u. 13-15
H-1053 Budapest
(Hungary)

C. Marchioro
Dip. di Matematica
Università di Trento
I-38050 Trento
(Italy)

E. Orlandi
Dep. Mat. "G. Castelnuovo"
Cittá Universitária
Piazzale A. Moro
I-00100 Roma
(Italy)

R. Marra
Piazzale Montesquieu 28/A
I-00137 Roma
(Italy)

G. Papanicolaou
Currant Institute
251 Mercer Str.
New York, NY 10012 (USA)

A. de Masi
Ist. Matematica Univ. dell'Aquila
Aquila
(Italy)

L. Pastur
Pr. Lenina 47
Harkov (USSR)

D. Merlini
Mathematisches Institut
Ruhr-Universität
D-4630 Bochum (FRG)

J.K. Percus
340 Riverside Dr.
New York, NY 10025 (USA)

M. Misiurewicz
Asfaltowa 7/5
Warsaw
(Poland)

D. Petz
MTA MKI
Reáltanoda u. 13-15
H-1053 Budapest
(Hungary)

B.S. Nahapetian
Inst. Math. Armenian SSR
ul. Barekamutian 24
Yerevan (USSR)

S.A. Pirogov
IPPI AN USSR
Ermolovoi str. 19
101447 Moscow GSP-4 (USSR)

S. Olla
Via Bandello 52
I-09100 Cagliari
(Italy)

S.K. Pogosian
Inst. Mat. Armenian SSR
ul. Barekamutian 24
Yerevan (USSR)

E. Omerti
Dip. di Matemática
Univ. di Trento
I-38050 Povo (Trento)
(Italy)

E. Presutti
Dip. di Matematica
Univ. di Roma
I-00185 Roma
(Italy)

F. Przytycki
ul. Puszczyka 17 m 80
P-02777 Warszawa
(Poland)

M. Pulvirenti
Via A. di Tullio 27
I-00136 Roma
(Italy)

M. Rattner
Dept. of Math.
University of California
Berkeley, CA 94720 (USA)

M. Rédei
Rákó u. 47/a
H-1112 Budapest
(Hungary)

H. Rodenhausen
Schlierbacher Lanstr. 154
D-6900 Heidelberg (FRG)

H. Rost
Seidenweg 7
D-6907 Nussloch (FRG)

G. Royer
13 rue Jules Cuillerier
F-94140 Alfortville
(France)

J. Ruiz
CPT CNRS
Lumniy Case 907
F-13288 Marseille Cedex 9
(France)

E. Scaciatelli
Dip. Matematica
Univ. di Roma
Piazzale Aldo Moro 5
I-00100 Roma
(Italy)

U. Scharfenberger
Eckenheimer Landstr. 72
D-6000 Frankfurt (FRG)

A.G. Schlijper
Clavecimbellaan 399
NL-2287 VP Rijswijk
(The Netherlands)

R. Siegmund-Schulze
Elsa Brandström-Str. 38
1100 Berlin (DDR)

N. Simányi
Rózsa F. u. 5
H-2440 Százhalombatta
(Hungary)

B. Souillard
4 et 6 Rue Saint Nicolas
F-75012 Paris
(France)

H. Spohn
Theoretische Physik
Theresienstr. 37
D-8000 München 2 (FRG)

A.K. Stepanov
Inst. Fiz. AN SSSR
Moskovskoe Obl.
Černogolovka (USSR)

Yu.M. Suchov
IPPI AN SSSR
Ermolovoi St. 19
101447 Moscow GSP-4 (USSR)

D. Surgailis
Inst. of Math. and Cyb.
232600 Vilnius
Pozelos 54 (USSR)

A. Sütö
MTA KFKI
Pf. 49
H-1525 Budapest 114
(Hungary)

D. Szász
MTA MKI
Reáltanoda u. 13-15
H-1053 Budapest
(Hungary)

K. Szlachányi
MTA KFKI
Pf. 49
H-1525 Budapest 114
(Hungary)

B. Tóth
MTA MKI
Reáltanoda u. 13-15
H-1053 Budapest
(Hungary)

L. Triolo
Univ. degli Studii di Roma
Dip. Mat.
Via A. Scarpa
I-00161 Roma
(Italy)

M. Urbański
ul. Gagarina 45 HA-2
P-87-100 Toruń
(Poland)

M. Vares
Rua Sorocaba 484, Apt. 301 F
Botafogo
Rio de Janeiro
(Brasil)

P. Vecsernyés
MTA KFKI
Pf. 49
H-1525 Budapest 114
(Hungary)

A. Verbeure
Leuvense Straat 5
B-3050 Oud-Heverlee
(Belgium)

A. Vetier
BME Villamosmérnöki Kar
Matematika Tanszéke
Müegyetem rkp. 3-9
H-1111 Budapest
(Hungary)

A. Wakolbinger
Reinprechtenstr. 2
A-4040 Puchenau
(Austria)

J. Westwater
Dept. of Mathematics
Univ. of Washington
Seattle, WA 98195 (USA)

V. Warstat
Schönitzstr. 19
DDR-4020 Halle (DDR)

W.D. Wick
Phys. Dept.
Princeton Univ.
Princeton, NJ 08540 (USA)

G. Winkler
Alranstr. 17
D-8000 München 70 (FRG)

M. Winnink
Inst. for Theor. Physics
P.O.B. 800
Groningen
(The Netherlands)

K. Ziemian
ul. Na Uboczu 24 m 64
P-02802 Warszawa
(Poland)

THE ALGEBRA OF MANY-PARTICLE OPERATORS

R. A. Minlos

There exists a large class of operators which contains practically all operators of many- (or infinite-) particle physics. These operators were called in [1] cluster operators. They were investigated from a different point of view in the papers [1], [2], [3], [4]. Here we consider some pure algebraic aspects of the theory of cluster operators. These operators together with the space, where they act, naturally form a category. It is convenient to describe some constructions used in the theory of cluster operators on this language.

1. The cluster categories

We introduce here a specific category of such kind. We shall discuss the generalizations of this scheme at the end of this section.

Let $H_{n,h}$ be a Hilbert space of the form

$$H_{n,h} = \ell_2((Z^\nu)^n, h) \tag{1}$$

(Z^ν is the ν-dimensional lattice, n is an integer, h is a Hilbert space), i.e. the space of the h-valued functions $f = \{f(T),\ T=(t_1,\dots,t_n) \in (Z^\nu)^n\}$

$$\|f\| = \left\{ \sum_{T\in(Z^\nu)^n} \|f(T)\|_h^2 \right\}^{1/2} . \tag{2}$$

The objects of our category (we denote it by N) are the direct sums of the spaces $H_{n,h}$:

$$H_\Gamma = \bigoplus_{(n,h)} H_{n,h} ,$$

where $\Gamma = \{(n,h)\}$ is a finite our countable collection of pairs (n,h). We shall describe now the morphisms of the category N which were called cluster operators. Evidently, every bounded linear operator

$$B : H_\Gamma \to H_{\Gamma'}$$

where $\Gamma = \{(n,h)\}$, $\Gamma' = \{(n',h')\}$ are two collections of pairs, generates a matrix of linear operators $\{B_{(n',h'),(n,h)}\}$

$$B_{(n',h'),(n,h)} : H_{n,h} \to H_{n',h'} \ .$$

Furthermore, a linear operator

$$B: H_{n,h} \to H_{n',h'}$$

generates the matrix of operators

$$\{B_{T',T} \ , \ T' \in (Z^\nu)^{n'} \ , \ T \in (Z^\nu)^n\}$$

$$B_{T',T} : h \to h' \ .$$

We denote by $N_{n',n} = N_{n'} \cup N_n$ the disjunctive union of the sets $N_n = \{1,\ldots,n\}$ and $N_{n'} = \{1',\ldots,n'\}$ and by $A_{n',n}$ the lattice of all partitions

$$\gamma = (\alpha_1,\ldots,\alpha_s)$$

of the set $N_{n',n}$ such that

$$\alpha_i \cap N_n \neq 0 \ , \ \alpha_i \cap N_{n'} \neq 0 \ , \ i=1,\ldots,s \ .$$

It is convenient to assume that the blocks of the partition γ are enumerated so that the numbers of the smallest elements of the sets $\alpha_i \cap N_n$ $i=1,\ldots,s$ increase. The order $\gamma \leq \gamma'$ in $A_{n',n}$ means that every block of γ' contains some block of γ. For each $n \geq 1$, $n' \geq 1$, each partition $\gamma = (\alpha_1,\ldots,\alpha_s) \in A_{n'n}$ and each collection $(\tau_1,\ldots,\tau_s) \in \in (Z^\nu)^s$ we denote by $T^n_\gamma \in (Z^\nu)^4$ and $T^{h'}_\gamma \in (Z^\nu)^n$ the vectors $T^n_\gamma = \{t_1,\ldots,t_n\}$, $T^{n'}_\gamma = \{t'_1,\ldots,t'_{n'}\}$ so that $t_j = = t'_{j'} = s_i$ if $j \in N_n \cap \alpha_i$, $j \in N_{n'} \cap \alpha_i$.

We say that the operator

$$A = A_\gamma : H_{n,h} \to H_{n',h'}$$

is connected with respect to $\gamma = (\alpha_1,\dots,\alpha_s) \in \mathbb{A}_{n'n}$, if its matrix blocks $A_{T'T}: h \to h'$ satisfy the following conditions:

1) for every collection $(\tau_1,\dots,\tau_s)$

$$A_{T'+T^{n'}_\gamma, T+T^n_\gamma} = A_{T',T}$$

2) the series

$$\sum_{T,T'}^{\gamma} \|A_{T',T}\|_{h,h'} < \infty$$

converges. Here the sum $\sum^\gamma$ means that the summation is taken over $T \in (Z^\nu)^n$ and T so that

$$t_{j_i^{min}} = 0 \ , \ i=1,\dots,s \ , \ j_i^{min} = \min_{j \in \alpha_i \cap N_n} j$$

($\|\cdot\|_{h,h'}$ is the operator norm).

The operator

$$B: H_{n,h} \to H_{n',h'}$$

is called a cluster operator if there exists a representation

$$B = \sum_{\gamma \in \mathbb{A}_{n',n}} A_\gamma \tag{3}$$

where every operator A_γ is connected with respect to the partition $\gamma \in \mathbb{A}_{n',n}$.

Lemma 1. Cluster operators are bounded and their representations (3) are unique.

Proof. The first assertion follows from the inequality

$$\|B\| \le \frac{1}{2}\left(\sup_T \sum_{T'} \|B_{T',T}\|_{h,h'} + \sup_{T'} \sum_T \|B_{T'T}\|_{h,h'}\right) . \tag{4}$$

Conditions 1) and 2) imply that the right side of (4) is finite. Again, 1), 2) and (4) imply for any $\gamma \in \mathbb{A}_{n',n}$ that

$$\lim_{\substack{\tau_i - \tau_j \to \infty \\ i \neq j, i,j=1,\dots,s}} B_{T'+T_\gamma^{n'}, T+T_\gamma^{n}} = \sum_{\tilde{\gamma} \geq \gamma} (A_{\tilde{\gamma}})_{T',T} = (D_\gamma)_{T',T}$$

(in the sense of uniform convergence of operators). Then

$$(A_\gamma)_{T',T} = \sum_{\tilde{\gamma} \geq \gamma} \mu_{\mathbb{A}_{n',n}} (\gamma, \tilde{\gamma}) (D_{\tilde{\gamma}})_{T',T} ,$$

where $\mu_{\mathbb{A}_{n',n}}(\cdot,\cdot)$ is the Möbius function of the lattice $\mathbb{A}_{n',n}$. The representation is called the cluster expansion of the operator B, and the operators A_γ are its connected components.

More generally, the operator

$$B : H_\Gamma \to H_{\Gamma'}$$

is called a cluster one if all its blocks $B_{(n',h')(n,h)}$ are cluster operators.

The set $\mathrm{Mor}(H_\Gamma, H_{\Gamma'}) \equiv \mathrm{Mor}(\Gamma, \Gamma')$ of the morphisms from H_Γ to $H_{\Gamma'}$ in the category consists of the cluster operators $B: H_\Gamma \to H_{\Gamma'}$. The correctness of this definition follows from the lemma.

Lemma 2. 1) For every Γ the unit operator $E_\Gamma: H_\Gamma \to H_\Gamma$ is a cluster one. 2) Let $B_1: H_\Gamma \to H_{\Gamma'}$ and $B_2: H_{\Gamma'} \to H_{\Gamma''}$ be cluster operators. Then their composition $B = B_2 B_1 : H_\Gamma \to H_{\Gamma''}$ is also a cluster operator.

The first assertion of the lemma is evident. The second one follows from the next lemma.

Lemma 3. Let the operators $A^{(1)}_{\gamma_1}: H_{n,h} \to H_{n',h'}$ and $A^{(2)}_{\gamma_2}: H_{n',h'} \to H_{n'',h''}$ be connected with respect to the partitions $\gamma_1 \in \mathbb{A}_{n',n}$ and $\gamma_2 \in \mathbb{A}_{n'',n'}$ correspondingly. Then the composition $A^{(2)}_{\gamma_2} A^{(1)}_{\gamma_1} = A \equiv A_{\bar{\gamma}}$ is connected with respect to the partition $\bar{\gamma} \in \mathbb{A}_{n'',n}$ of set $N_{n'',n}$ which is defined in the following way. We denote by $\tilde{\gamma}_1$ and $\tilde{\gamma}_2$ the partitions of the set $N_n \cup N_{n'} \cup N_{n''} = N_{n'',n',n}$ which are obtained from γ_1 and γ_2 with the help of additional partitions of the sets N_n and $N_{n''}$ into one-point blocks. Then

$$\bar{\gamma} = \tilde{\gamma}_1 \vee \tilde{\gamma}_2 |_{N_{n'',n}} \tag{5}$$

where $\tilde{\gamma}_1 \vee \tilde{\gamma}_2$ is the greatest lower bound of $\tilde{\gamma}_1$ and $\tilde{\gamma}_2$ and $\tilde{\gamma}_1 \vee \tilde{\gamma}_2 |_{N_{n'',n}} \in A_{n'',n}$ is the partition of $N_{n'',n}$ generated by the partition $\tilde{\gamma}_1 \vee \tilde{\gamma}_2$ of $N_{n'',n',n}$.

The proof follows easily from conditions 1) and 2).

Let $s(\gamma)$ be the number of the blocks of the partition γ . Then, as easily follows from the definition (5),

$$s(\bar{\gamma}) \leq \min(s(\gamma_1), s(\gamma_2)) .$$

Let B be a cluster operator $B: H_{n,h} \to H_{n',h'}$ and A_γ its connected components. The largest value $s(\gamma)$ for the partitions $\gamma \in A_{n',n}$ such that

$$A_\gamma = 0$$

is called the rang of the cluster operator B . If $B: H_\Gamma \to H_{\Gamma'}$ is a cluster operator we put

$$r(B) = \max r(B_{(n',h'),(n,h)})$$

where $B_{(n',h'),(n,h)}$ are the matrix blocks of B .

Corollary of Lemma 3. Let Mor_s be the set of morphisms of the category M which have rang $r \leq s$. Then Mor_s is a two-sided ideal of the category N .

Let $B^*: H_{\Gamma'} \to H_\Gamma$ be the operator conjugate to the cluster operator $B: H_\Gamma \to H_{\Gamma'}$. Then B^* is a cluster operator and the map $*: B \to B^*$ generates a contravariant functor from N to itself.

The set $M(\Gamma,\Gamma)$ of cluster operators from H_Γ to itself is an algebra with unity and involution.

Let $\{U_t^{(n,h)}, t \in Z^\nu\}$ be unitary representation of the group Z in the space $H_{n,h}$

$$(U_t^{(n,h)} f)(T) = f(T+t) .$$

The direct sum of these representations give unitary representation

$$U_t^{(\Gamma)} = \bigoplus_{(n,h)\in\Gamma} U_t^{(n,h)}$$

in the space H_Γ. Evidently, every cluster operator $B: H_\Gamma \to H_{\Gamma'}$ satisfies the relation

$$U_t^{(\Gamma')}B = BU_t^{(\Gamma)} \qquad t\in Z^\nu , \tag{6}$$

Remarks

1. The previous constructions permit the following generalizations:

a) instead of the group Z^ν we can consider another infinite group G;

b) instead of the spaces H_Γ we can consider a more general class of spaces and representations of the groups $(Z^\nu)^n$ (or G^n) which act in these spaces;

c) we can replace condition 2) of the definition of the operator A_γ connected with respect to the partition γ by some other condition providing "connectedness" of the variables "belonging to the same block of the partition γ".

2. We shall indicate certain subcategories of the category N:

a) subcategory N^{fin}; its objects are the direct sums of spaces $H_{n,n}$ corresponding to finite-dimensional spaces h;

b) subcategory N^p $p>0$; connected components A_γ of morphisms of this category satisfy the condition:

$$\sum_{T,T'}^{\gamma} |T|^{\beta}|T'|^{\beta'} \|(A_\gamma)_{T',T}\|_{h,h'} < \infty$$
$$\text{for } |\beta|, |\beta'| \le p$$

where $\beta = (\beta_1,\ldots,\beta_n)$, $\beta' = (\beta'_1,\ldots,\beta'_{n'})$ are multiindexes, $|\beta| = \beta_1+\ldots+\beta_n$, $|\beta'| = \beta'_1+\ldots+\beta'_{n'}$, $T^\beta = t_1^{\beta_1}\ldots t_n^{\beta_n}$, $T^{\beta'} = t_1^{\beta'_{1'}}\ldots t_n^{\beta'_{n'}}$; ;

c) subcategory $N^{\text{h.f.}} \subset N^{\text{fin}}$ (hyperfinite), its objects are finite direct sums of the spaces $H_{n,h}$, where h is finite dimensional.

2. The category $\tilde{N}$ and the Fourier functor

The category N constructed above corresponds to the "space"-representation (x -representation) of states and observables in quantum mechanics. We shall introduce now another category $\tilde{N}$ which corresponds to the "moment"-representation (p -representation).

The objects of the category $\tilde{N}$ are the spaces of the form

$$\tilde{H}_\Gamma = \bigoplus_{(n,h)\in\Gamma} \tilde{H}_{n,h}$$

where $\tilde{H}_{n,h} = L_2((T^\nu)^n, h)$ is the space of the h -valued functions

$$\tilde{f} = \{\tilde{f}(\Lambda),\ \Lambda=(\lambda_1,\ldots,\lambda_n) \in (T^\nu)^n\}$$

(T^ν : ν -dimensional torus) with the norm

$$\|f\| = \left(\int_{(T^\nu)^n} \|f(\Lambda)\|_h^2 d\Lambda \right)^{1/2}$$

($d\Lambda = d\lambda_1 \ldots d\lambda_n$, $d\lambda$: the Haar measure on T^ν).

The operator

$$B:\ \tilde{H}_{n,h} \to \tilde{H}_{n',h'}$$

is called cluster operator (or Λ -cluster) if it can be represented in the form

$$(B\tilde{f})(\Lambda') = \int_{(T^\nu)^n} K(\Lambda',\Lambda)\tilde{f}(\Lambda)d\Lambda \tag{7}$$

(for a dense set of the functions $\tilde{f}$) and the kernel $K(\Lambda',\Lambda)$ is the operator-valued distribution of the form

$$K(\Lambda',\Lambda) = \sum_{\gamma\in \mathfrak{A}_{n',n}} a_\gamma(\Lambda',\Lambda)\delta_\gamma , \tag{8}$$

where $\Lambda=(\lambda_1,\ldots,\lambda_n)$, $\Lambda'=(\lambda'_1,\ldots,\lambda'_{n'})$, and for the partition $\gamma = (\alpha_1,\ldots,\alpha_s)$

$$\delta_\gamma = \prod_{i=1}^{s} \delta(\pi_i - \pi'_i)$$

$\pi_i = \sum_{j\in\alpha_i\cap N_n} \lambda_j$, $\pi_i' = \sum_{j'\in\alpha_i\cap N_{n'}} \lambda_{j'}'$, with $\delta(\cdot)$ a δ -function on the torus T^ν and a_γ continuous operator-valued functions on the manifolds

$$D_\gamma = \{\Lambda,\Lambda' : \pi_i = \pi_i' \, , \; i=1,\ldots,s\}$$

called the cluster functions of the operator B . The integral in (7) is understood as Bochner integral. More generally, the bounded operator

$$B: \tilde{H}_\Gamma \to \tilde{H}_{\Gamma'}$$

is called Λ -cluster operator if all its blocks $B_{(n',h'),(n,h)}: \tilde{H}_{n,h} \to \tilde{H}_{n',h'}$ are the same. Assertions similar to Lemmas 1 and 2 are true.

Lemma 4. 1) An operator of the form (7), (8) is bounded.

2) The representation of the kernel $K(\Lambda',\Lambda)$ in the form (8) is unique.

3) The composition $B = B_2 B_1$ of the Λ -cluster operators $B_1: H_\Gamma \to H_{\Gamma'}$ and $B_2: H_{\Gamma'} \to H_{\Gamma''}$ is a Λ -cluster operator.

4) The unit operator $E_\Gamma: H_\Gamma \to H_\Gamma$ is a Λ -cluster operator.

The set $\widetilde{\mathrm{Mor}}(H_\Gamma, H_\Gamma) \equiv \widetilde{\mathrm{Mor}}(\Gamma,\Gamma')$ of the morphisms of the category $\tilde{N}$ consists of the cluster operators. As it follows from Lemma 4 this definition is correct. We introduce two important subcategories of the category $\tilde{N}$ similar to the corresponding subcategories of N .

1. Subcategory $\tilde{N}^{fin}$: its objects are the direct sums of the spaces $\tilde{H}_{n,h}$, where h are finite-dimensional spaces.

2. Subcategory $\tilde{N}_p$: its morphisms are Λ -cluster operators with p -smooth cluster functions a_γ .

3. Hyperfinite category $\tilde{N}^{h.f.} \subset \tilde{N}^{fin}$: its objects are finite sums of the spaces $H_{n,h}$ where h is finite dimensional.

Let $F_{n,h}: H_{n,h} \to \tilde{H}_{n,h}$ be the unitary transformation (Fourier transformation)

$$(F_{n,h}f)(\Lambda) \equiv \tilde{f}(\Lambda) = \frac{1}{(2\pi)^{n\nu/2}} \sum_{T\in(Z^\nu)^n} \exp\{i\sum(\lambda_j,t_j)\}f(T)$$

where $\Lambda = \{\lambda_1,\ldots,\lambda_n\} \in (T^\nu)^n$, $T = (t_1,\ldots,t_n) \in (Z^\nu)^n$.

Let us assume also that

$$F_\Gamma = \sum_{(n,h)\in\Gamma} F_{n,h}: H_\Gamma \to \tilde{H}_\Gamma .$$

The transformation F_Γ generates the covariant functor $\mathcal{F}: \mathcal{N} \to \tilde{\mathcal{N}}$ (the Fourier functor) from the category $\mathcal{N}$ to the category $\tilde{\mathcal{N}}$:

$$\mathcal{F}: H_\Gamma \to \tilde{H}_\Gamma \tag{10}$$

and for $B \in \mathrm{Mor}(\Gamma,\Gamma')$

$$\mathcal{F}B \equiv \tilde{B} = F_{\Gamma'}BF_\Gamma^{-1}: \tilde{H}_\Gamma \to \tilde{H}_{\Gamma'} . \tag{10^a}$$

Lemma 5. Definitions (10), (10^a) are correct i.e. $\mathcal{F}B \in \mathrm{Mor}(\Gamma,\Gamma')$ and in fact the map $\mathcal{F}$ is a functor.

Remarks

1. The image $\mathcal{F}\,\mathrm{Mor}(\Gamma,\Gamma') \neq \widetilde{\mathrm{Mor}}(\Gamma,\Gamma')$, i.e. the class of Λ -cluster operators is wider than the class of Fourier images (10^a) of cluster operators.

2. The functor $\mathcal{F}$ transforms the subcategory $\mathcal{N}_p$ into the subcategory $\tilde{\mathcal{N}}_p$ but $\mathcal{F}\,\mathrm{Mor}_p(\Gamma,\Gamma') \neq \widetilde{\mathrm{Mor}}_p(\Gamma,\Gamma')$ where $\mathrm{Mor}_p(\Gamma,\Gamma')$ and $\mathrm{Mor}_p(\Gamma,\Gamma')$ denote the set of morphisms of the categories $\mathcal{N}_p$ and $\tilde{\mathcal{N}}_p$ respectively.

3. The representation $\{U_t^\Gamma,\ t\in Z^\nu\}$ of the group Z^ν in the space H_Γ is transformed by the Fourier transform F into the representation $\{\tilde{U}_t^\Gamma,\ t\in Z^\nu\}$, which in the spaces $H_{n,h} \subseteq H_\Gamma$ has the form

$$(\tilde{U}_t^\Gamma\tilde{f})(\Lambda) = \exp\{i(t,\lambda_1+\ldots+\lambda_n)\}\tilde{f}(\Lambda)$$

$$\Lambda = (\lambda_1,\ldots,\lambda_n) \in (T^\nu)^n .$$

3. The s-superstructures and s-functors

We give some very useful constructions for the investigation of the cluster operators.

Let s and $n\geq s$ be integers and $\Xi(s,n)$ the set of pairs (b,σ) where b is a partition of the set N_n consisting of s blocks and $\sigma\in\Pi_s$ is a permutation of the set $N_s=\{1,\ldots,s\}$. We denote by $H^s_{n,h}$ the Hilbert space

$$H^s_{n,h} = \ell_2((Z^\nu)^n,\ h\otimes\ell_2(\Xi(s,n)))\ .$$

We can consider its elements as functions $f=\{f_{(b,\sigma)},\ (b,\sigma)\in\Xi(s,n)\}$ with the values in $\ell_2((Z^\nu)^n,h)$. For every $\Gamma=\{(n,h)\}$ we call the space

$$H^s_\Gamma = \oplus\ H^s_{n,h}$$

the s-superstructure of the H_Γ. Evidently, the spaces H^s_Γ are the objects of the category N. Every operator $B\colon H_{n,h}\to H^s_{n',h'}$ generates a matrix of operators

$$B(b',\sigma'),(b,\sigma)\colon H_{n,h}\to H_{n',h'}\ (b',\sigma')\in\Xi(s,n')\ ,$$
$$(b,\sigma)\in\Xi(s,n).$$

We consider the cluster operators $B\colon H^s_{n,h}\to H^s_{n',h'}$ which satisfy the following conditions:

a) the operators $B_{(b',\sigma'),(b,\sigma)}$ have the form

$$B_{(b',\sigma'),(b,\sigma)} = M_{b',b,\sigma',\sigma^{-1}}$$

where $M_{b',b}$ is a cluster operator acting from $H_{n,h}$ to $H_{n',h'}$;

b) the cluster expansion of $M_{b',b,\sigma}$ has the form

$$M_{b',b,\sigma} = \sum_{\gamma\geq\bar\gamma} A^{\bar\gamma}_\gamma \tag{11}$$

where the partition $\bar\gamma=\bar\gamma(b',b,\sigma)=(\alpha_1,\ldots,\alpha_s)$ is defined in the following way: $\alpha_i=\beta_i\vee\beta'_{\sigma^{-1}(i)}$ $i=1,\ldots,s$, $b=(\beta_1,\ldots,\beta_s)$ $b'=\{\beta'_1,\ldots,\beta'_s\}$ (the enumeration of the blocks of b and b' is similar to the enumeration of the blocks of the partition $\gamma \in A_{n',n}$). We denote by $Mor^s(\Gamma,\Gamma') \subseteq Mor(H^s_\Gamma, H^s_{\Gamma'})$ the set of cluster operators

$$B: H^s_\Gamma \to H^s_{\Gamma'}$$

such that their blocks $B_{(n',h'),(n,h)}: H^s_{(n,s)} \to H^s_{(n',h')}$ satisfy conditions a) and b).

Lemma 6. The sets of the objects $\{H^s_\Gamma\}$ and the morphisms $Mor^s(\Gamma,\Gamma')$ form a subcategory of the category N ; we denote it by N^s and call it the s-superstructure of N.

Proof. The unit operator $E^s_\Gamma: H^s_\Gamma \to H^s_\Gamma$ belongs for every Γ to $Mor^s(\Gamma,\Gamma)$. Further, the expansions (11) of the blocks $B^{(i)}_{(b',\sigma')(b,\sigma)}$ $i=1,2$ of the operators $B^{(1)}: H^s_{n,h} \to H^s_{n',h'}$ and $B^{(2)}: H^s_{n',h'} \to H^s_{n'',h''}$ imply again (with the help of Lemma 3) the expansion of the blocks of operator $B = B^{(2)}B^{(1)}$ in the sums (11).

Similarly we can define the categories $\tilde{N}^s$ $s=1,2,\ldots$ as the superstructures of the category $\tilde{N}$. The objects of $\tilde{N}^s$ are the direct sums of the spaces $\tilde{H}^s_{n,h}$ consisting of the functions $\{\tilde{f}_{(b,\sigma)}, (b,\sigma) \in \Xi(s,n)\}$ with values in $L_2((T^\nu)^n,h)$. The morphisms of $\tilde{N}^s$, $B: H^s_{n,h} \to H^s_{n',h'}$ consist of the blocks $B_{(b',\sigma'),(b,\sigma)}$ defined by the kernels of the forms

$$K_{(b',\sigma'),(b,\sigma)}(\Lambda',\Lambda) = \sum_{\gamma \geq \bar{\gamma}} a^{\bar{\gamma}}_\gamma(\Lambda',\Lambda)\,\delta_\gamma .$$

(The partition $\bar{\gamma} = \bar{\gamma}(b',b,\sigma',\sigma^{-1})$ is defined above.) For every space $H^s_{n,h}$ we define a representation of the group $(Z^\nu)^s$ by the formula

$$[U^s_{n,h}(\tau)f]_{(b,\sigma)}(T) = f_{(b,\sigma)}(T+\tilde{T}), \quad f \in H^s_{n,h}$$

$\tau=(\tau_1,\ldots,\tau_s) \in (Z^\nu)^s$, $b=(\beta_1,\ldots,\beta_s)$, $\tilde{T}=\tilde{T}(\tau,(b,\sigma))=(\tilde{t}_1,\ldots,\tilde{t}_n)$,

where $\tilde{t}_j = \tau_{\sigma^{-1}(i)}$ if $j \in \beta_i$.

We define the representations $\{U_\Gamma^s(\tau),\ \tau\ (Z^\nu)^s\}$ of the group $(Z^\nu)^s$ as the direct sums of representations (12). In addition for every $B \in \mathrm{Mor}^s(\Gamma,\Gamma')$

$$U_{\Gamma'}^s B = BU_\Gamma^s .$$

From here it follows that the spaces H_Γ^s, the operators U_Γ^s and the morphisms $B \in \mathrm{Mor}^s(\Gamma,\Gamma')$ can be decomposed in the direct integrals:

$$H_\Gamma^s = \int_{(T^\nu)^s} H_\Gamma^s(\pi_1,\dots,\pi_s) \prod_{i=1}^s d\pi_i .$$

$$H_\Gamma^s = \int_{(T^\nu)^s} \exp\{i\sum(t_i,\pi_i)\}E^s(\pi_1,\dots,\pi_s) \prod_{i=1}^s d\pi_i \tag{13}$$

$$B = \int_{(T^\nu)^s} B(\pi_1,\dots,\pi_s) \prod_{i=1}^s d\pi_i .$$

Here $\{H_\Gamma^s(\pi_1,\dots,\pi_s)(\pi_1,\dots,\pi_s) \in (T^\nu)^s\}$ is the family of spaces, $E_\Gamma^s(\pi_1,\dots,\pi_s)$ is the unit operator in $H_\Gamma(\pi_1,\dots,\pi_s)$ and $B(\pi_1,\dots,\pi_s)$ is an operator from $H_\Gamma^s(\pi_1,\dots,\pi_s)$ into $H_{\Gamma'}^s(\pi_1,\dots,\pi_s)$. In addition for every collection $(\pi_1,\dots,\pi_s)$ the spaces $\{H_\Gamma^s(\pi_1,\dots,\pi_s)\}$ and the operators $\{B(\pi_1,\dots,\pi_s)\}$ form the category $N^s(\pi_1,\dots,\pi_s)$. The decomposition (13) are called the canonical ones of the category N^s. Similarly we can introduce the canonical decomposition of the category $\tilde{N}^s$, that has a very clear structure. Namely, for each partition b of N_n we can introduce new variables on the torus $(T^\nu)^n$ with the help of the decomposition $(T^\nu)^n = (T^\nu)^s \times L_b : (\lambda_1,\dots,\lambda_n) \to (\pi_1,\dots,\pi_s,\mu)$ $\pi_i \in T^\nu$ $i=1,\dots,s$, $\mu \in L_b$, where $L_b \subseteq (T^\nu)^n$ is the subgroup of $(T^\nu)^n$:

$$L_b\{(\lambda_1,\dots,\lambda_n):\ \hat{\pi}_i = \sum_{j\in\beta_i} \lambda_j = 0 \quad (i=1,\dots,s),\ b \neq (\beta_1,\dots,\beta_s) .$$

Thus each function $f(\lambda_1,\dots,\lambda_n)$ can be written as $f(\pi_1,\dots,\pi_s,\mu)$. The space $H_{n,h}^s(\pi_1^0,\dots,\pi_s^0)$ consists of the functions $\{f_{(b,\sigma)}(\mu),(b,\sigma) \in \Xi(s,n),\ \mu \in L_b\}$ and the decomposition (13) in this case means

$$\{f_{(b,\sigma)}(\lambda_1,\ldots,\lambda_s,\mu)=f_{(b,\sigma)}(\pi_1,\ldots,\pi_s,\mu)\} \to \{\hat{f}_{(b,\sigma)}^{\pi_1^0,\ldots,\pi_s^0}(\mu)\},$$

where $\hat{f}^{\pi_1^0,\ldots,\pi_s^0}(\mu) = f_{(b,\sigma)}(\pi^0_{\sigma^{-1}(1)},\ldots,\pi^0_{\sigma^{-1}(s)},\mu)$. The operator $B(\pi_1^0,\ldots,\pi_s^0)$ is represented with the help of the kernels of the form

$$K^{\pi_1^0,\ldots,\pi_s^0}_{(b',\sigma'),(b,\sigma)}(\mu',\mu) = \sum_{\gamma\geq\bar{\gamma}} a_\gamma^{\bar{\gamma}}(\mu',\mu)\tilde{\delta}_\gamma(\mu',\mu)$$

where $a_\gamma^{\bar{\gamma}}(\mu',\mu)$ are continuous functions on the manifolds

$$\tilde{D}_\gamma = D_\gamma \cap L_{b'}\times L_b ,$$

$\bar{\gamma} = \bar{\gamma}(b',b,\sigma'\bar{\sigma}^1)$ and $\tilde{\delta}_\gamma(\mu',\mu)$ is δ-function concentrated on this manifold. We define now the functor $T^s\colon N \to N^s$; $T^s\colon H_\Gamma \to H_\Gamma^s$ and for

$$\begin{aligned} B &= \sum_\gamma A_\gamma \in \mathrm{Mor}(H_{n,h},H_{n',h'}) \\ (T^sB)_{(b',\sigma'),(b,\sigma)} &= \sum_{\gamma\geq\bar{\gamma}} A_\gamma \end{aligned} \tag{14}$$

where $\bar{\gamma} = \bar{\gamma}(b',b,\sigma^1\sigma^{-1})$ is defined above. The morphism $T^sB \in \mathrm{Mor}^s(\Gamma,\Gamma')$ is defined by (14) for every morphism $B \in \mathrm{Mor}(\Gamma,\Gamma')$ of the category N .

Lemma 7. The map T^s $N \to N^s$ is a covariant functor and $\mathrm{Ker}\ T^s$ coincides with the ideal $\mathrm{Mor}_s \subset \mathrm{Mor}\, N$.

The proof is simple and we omit it. Similarly we can introduce the functors $\tilde{T}^s$ acting from $\tilde{N}$ to $\tilde{N}^s$. The decompositions (13) and its analogous counterpart generate the decompositions of the functors T^s and $\tilde{T}^s$ into the family of the functors $\{T^s(\pi_1,\ldots,\pi_s)\}$ and $\{\tilde{T}^s(\pi_1,\ldots,\pi_s)\}$.

n4. The hyperfinite category $\tilde{N}^{h.f.}$

Theorem 8. Let $B\colon H_\Gamma \to H_\Gamma$ be an invertible cluster operator acting in the finite sum of the spaces $H_{n,h}$.

Then the inverse operator B^{-1} is a cluster operator.

We give a short outline of the proof of this theorem. If we assume that B^{-1} is a cluster operator with the cluster function $\{b_\gamma\}$ we obtain equations for them after a substitution of the cluster expansions for B and B^{-1} in the equalities

$$BB^{-1} = B^{-1}B = E_\Gamma \ .$$

After the application of the functors $T^s_{(\pi_1,\dots,\pi_s)}$ we can obtain the system of equations

$$B^s(\pi_1,\dots,\pi_s)(B^{-1})^s(\pi_1,\dots,\pi_s) = (B^{-1})^s(\pi_1,\dots,\pi_s)B^s(\pi_1,\dots,\pi_s) =$$

$$= E^s_\Gamma(\pi_1,\dots,\pi_s) \quad s=1,2,\dots, \quad s_0 = \max_{(n,h)\in\Gamma} n$$

where $B^s(\pi_1,\dots,\pi_s) = T^s(\pi_1,\dots,\pi_s)B$. This system of equations have a hierarchical structure: the first of them for $s=s_0$ contains only the cluster functions $\{b_\gamma\}$ of B^{-1} with $s(\gamma)=s_0$ and for every fixed collection $(\pi_1,\dots,\pi_{s_0})$ has the form of an algebraic linear equation. The matrix of this equation is non-degenerate as it easily follows from the existence of the inverse operator B^{-1} . Thus the cluster functions $\{b_\gamma\}$ exist with $s(\gamma)=s_0$. Then, the next equation for $s=s_0-1$ is reduced with the help of the solution of first equation to an equation of Fredholm type. This equation has the cluster functions b_γ of B^{-1} with $s(\gamma)=s_0-1$, as its solution. By repeating this arguments we get all cluster functions of B^{-1} .

In the case when B has p -smooth cluster functions the inverse operator B^{-1} has the same property.

The notations and the constructions developed above are very useful for an investigation of the spectrum and the scattering of a self-adjoint cluster operator (see [4]).

References

[1] Abdulla-Zadeh F. H., Minlos R. A., Pogosian S. K. Multicomponent random systems. Ed. Dobrushin R. L., Sinai Ya. G. Marcel Dekker Inc. N. Y. - Basel, pp. 1-37.

[2] Malyshev V. A., Minlos R. A.,
I. J. Stat. Phys. 21, N 3, 231-242 (1979).
II. Commun. Math. Phys. 82, 211-226 (1981).

[3] Малышев В. А., Минлос Р. А. Труды семинара им. Петровского. МГУ, 9, 63-80 (1983).

[4] Minlos R. A., Mogilner A. I. (in preparation).

C*-ALGEBRA APPROACH TO GROUND STATES OF THE XY-MODEL

Huzihiro ARAKI and Taku MATSUI
Research Institute for Mathematical Sciences
Kyoto University, Kyoto 606, JAPAN

1. Introduction

C*-algebras are known to provide a proper mathematical framework for quantum statistical mechanics of spin systems on an infinitely extended lattice in the same way as Hilbert spaces for quantum mechanics. In the present article, we try to show that theory of C*-algebras is useful not only for a general formulation but also for a concrete computation by illustrating how all ground states of the one-dimensional XY-model in an external transversal field can be determined through C*-algebra methods.

The XY-model on a finite one-dimensional lattice has been "exactly" solved through the Jordan-Wigner transformation [6]. In the case of equilibrium state at non-zero temperature, the infinite volume limit of the finite volume Gibbs canonical ensemble provides the unique (KMS) equilibrium state for the infinite system. On the other hand, a (kink) soliton state for the Ising model on an infinite one-dimensional lattice is a ground state in the sense that any local perturbation of the state produces states of equal or higher energy (i.e. it is a state of the lowest energy) and is not obtained as the limit of the ground state of a finite system. We formulate the whole problem directly for an infinite system and determine all such states, if they exist. In particular, soliton states are shown not to exist as ground states for the XY-model, except for the Ising model case (which corresponds to specific choices of parameters of our model).

In the present approach, C*-algebra methods play a crucial role in two different places. One is with the Jordan-Wigner transformation, which changes Pauli spin observables to anticommuting Fermion creation and annihilation operators. Each creation or annihilation operator at a lattice site, in its definition in terms of Pauli spin operators, contains a product of spin operators on all sites to the left of the lattice point (a "string" connecting the left end of the lattice to the lattice site of the "Fermion") and this product does not converge in an infinite lattice. Our remedy taken from an earlier work [3] is to introduce this "string" operator as a new operator T outside of the given C*-algebra A of Pauli spin operators on all lattice sites, define a larger C*-algebra $\hat{A}$ generated by A and T and define Fermion creation and annihilation operators (by the Jordan-Wigner transformation involving T) which generate a subalgebra A^{CAR} of $\hat{A}$ which is different from A. As far as A^{CAR} is concerned, we can follow usual "exact" computation and find various information about it including its ground state which is unique for all values of the parameter. We then try to determine ground states of A from information about A^{CAR}. This is the second place where C*-algebra methods provide a powerful tool. In particular, the criterion for equivalence or disjointness of restrictions of two Fock states to the even part A_+^{CAR} of A^{CAR} (even relative to Fermion numbers) in terms of the Hilbert-Schmidt norm (of the difference of proejection operators defining the two Fock states) and the Z_2-index introduced in [2] plays a decisive role in finding the number of pure ground states.

For large values of the external transversal field, the ground states turn out to be unique, while for smaller values of the external transversal field there exist two and only two ground states breaking the symmetry of 180° degree rotation of all spins around the z-axis, with exceptions of (i) the case of XY-symmetry and (ii) the case of Ising model. At the critical value of the external transversal field (i.e. at the boundary of the above two regions) as well as in the XY-symmetric case with subcritical external field strength, correlation functions, which are otherwise real analytic in two paramters for external field and XY-asymmetry, become singular. Ground states are unique also for those critical values of parameters. For the case of Ising model, there exist two pure ground states, which analytically continue to two pure ground states for nearby values of parameters, plus two sets of infinitely many pure ground states corresponding to kink soliton solutions which are not translation invariant (a reason for the infinite number of them), the two sets being distinguished by non-equivalence of associated representations of the algebra of spin operators, and again breaking the symmetry under 180° rotation of all spins around the z-axis.

The two phases with unique and non-unique ground states are also distinguished by its stability under local perturbation. The unique ground state will return to itself under a local perturbation but, in the region of non-unique ground states, not only mixture but also pure ground states do not return to itself under a local perturbation in general.

While the main emphasis of this work is on the application of an abstract technique to determination of number of pure ground states, the present analysis yields also an explicit formula for correlation functions. However, we shall give only a brief sketch of C*-algebra arguments for determination of number of ground states and detailed analysis is referred to forthcoming publication [4].

2. Model

At each site j of a one-dimensional infinite lattice $\mathbb{Z}$, observables for 1/2 spin are elements of the algebra of Pauli spin operators

$$\sigma_x^{(j)} \approx \begin{pmatrix} 0 & 1 \\ 1 & 0 \end{pmatrix}, \quad \sigma_y^{(j)} \approx \begin{pmatrix} 0 & -1 \\ i & 0 \end{pmatrix}, \quad \sigma_z^{(j)} \approx \begin{pmatrix} 1 & 0 \\ 0 & -1 \end{pmatrix},$$

which commute with Pauli spin operators at other lattice sites. The C*-algebra (of observables) generated by all Pauli spin operators at all lattice sites is denoted by $\mathcal{A}$ below.

The Hamiltonian of an XY-model for a finite interval $[-N,N]$ $(N = 1,2,\ldots)$ is as follows:

$$H_N = -J\Big[\sum_{j=-N}^{N-1} \{(1+\gamma)\sigma_x^{(j)}\sigma_x^{(j+1)} + (1-\gamma)\sigma_y^{(j)}\sigma_y^{(j+1)}\} + 2\lambda \sum_{j=-N}^{N} \sigma_z^{(j)}\Big]. \tag{2.1}$$

Here J, γ, and λ are real parameters and we assume $J > 0$. For the infinite system, the above Hamiltonian defines the dynamics as follows:

$$\alpha_t(A) = \lim_{N\to\infty} e^{iH_N t} A \, e^{-iH_N t}. \tag{2.2}$$

Here A is any element of the observable algebra $\mathcal{A}$, $\alpha_t(A)(-\infty < t < \infty)$

is its time evolution and the limit in the defining equation (2.2) is known to exist by a general theory on spin lattice systems [5].

Signs of each of three real constants J (total strength), γ (x-y asymmetry) and λ (strength of transversal external magnetic field relative to nearest neighbour interactions of x and y spins) can be changed by the following symmetry transformations, which are automorphisms of the C*-algebra $\mathcal{A}$:

(i) The sign of J is reversed by 180° rotation of spins at all even sites around the x-axis and 180° rotation of spins at all odd sites around the y-axis, which are obtained by the $N \to \infty$ limit of the unitary transformation uAu^* $(A \in \mathcal{A})$ with

$$u = \prod_{j=-N}^{N} (\sigma_x^{(2j)} \sigma_y^{(2j+1)}).$$

(ii) The sign of γ is reversed by 180° rotation of spins at all even sites around the z-axis, which is obtained by the $N \to \infty$ limit of the unitary transformation uAu^* $(A \in \mathcal{A})$ with

$$u = \sum_{j=-N}^{N} \sigma_z^{(2j)} .$$

(iii) The sign of λ is reversed by 180° rotation of spins at all sites around the x-axis, which is obtained by the $N \to \infty$ limit of the unitary transformation uAu^* $(A \in \mathcal{A})$ with

$$u = \prod_{j=-N}^{N} \sigma_x^{(j)} .$$

The Hamiltonian (2.1) is invariant under 180° rotation of spins at all sites around the z-axis given by

$$\Theta(A) \equiv \lim_{N\to\infty} \left(\sum_{j=-N}^{N} \sigma_z^{(j)} \right) A \left(\sum_{j=-N}^{N} \sigma_z^{(j)} \right) . \tag{2.3}$$

(More concretely, $\Theta(\sigma_x^{(j)}) = -\sigma_x^{(j)}$, $\Theta(\sigma_y^{(j)}) = -\sigma_y^{(j)}$, $\Theta(\sigma_z^{(j)}) = \sigma_z^{(j)}$.) Consequently, for all $A \in \mathcal{A}$

$$\Theta(\alpha_t(A)) = \alpha_t(\Theta(A)). \tag{2.4}$$

3. Ground States

A state φ of the C*-algebra $\mathcal{A}$, as an expectation functional $\varphi(A)$ for $A \in \mathcal{A}$, is given as a vector state

$$\varphi(A) = (\Omega_\varphi, \pi_\varphi(A)\Omega_\varphi) \tag{3.1}$$

by a unit vector Ω_φ in a Hilbert space $\mathcal{H}_\varphi$, which carries a representation $\pi_\varphi(A)(A \in \mathcal{A})$ of the C*-algebra $\mathcal{A}$ and which is minimal in the sense that the vectors $\pi_\varphi(A)\Omega_\varphi$, $A \in \mathcal{A}$, are dense in $\mathcal{H}_\varphi$. (The GNS-triplet.)

If a state φ is stationary in the sense that $\varphi(\alpha_t(A)) = \varphi(A)$ for all $A \in \mathcal{A}$ and for all real t, then there is a continuous one-parameter group of unitary operators $U_\varphi(t)$ determined by

$$U_\varphi(t)\ \pi_\varphi(A)\Omega_\varphi = \pi_\varphi(\alpha_t(A))\Omega_\varphi \qquad (A \in \mathcal{A}) \tag{3.2}$$

and they implement the time evolution automorphism α_t :

$$U_\varphi(t)\ \pi_\varphi(A)\ U_\varphi(t)^* = \pi_\varphi(\alpha_t(A)). \tag{3.3}$$

We shall denote the generator of $U_\varphi(t)$ by H_φ : $U_\varphi(t) = \exp itH_\varphi$. As a special case of (3.2) with $A = 1$, Ω_φ is invariant under $U_\varphi(t)$ and hence is an eigenvector of the selfadjoint operator H_φ belonging to an eigenvalue 0 .

Interpreting H_φ as an energy operator in the space $\mathcal{H}_\varphi$ under consideration, $H_\varphi \geq 0$ is the same as saying that the state φ given by the vector Ω_φ has the lowest energy at least among states given by vectors in $\mathcal{H}_\varphi$. (Since the vectors $\pi_\varphi(A)\Omega_\varphi$ with A a polynomial of a finite number of Pauli spins are already dense in $\mathcal{H}_\varphi$, these states may be regarded as a result of more or less local perturbation on φ.) If this is the case (namely if φ is stationary and $H_\varphi \geq 0$), φ is called a <u>ground state</u>.

Let f be a complex-valued function of class $\mathcal{S}$ of a real variable and define

$$\tilde{f}(t) = \int e^{-ipt} f(p)dp/(2\pi), \tag{3.4}$$

$$A(\tilde{f}) = \int \alpha_t(A)\ \tilde{f}(t)dt. \tag{3.5}$$

Then

$$\pi_\varphi(A(\tilde{f}))\Omega_\varphi = \int \pi_\varphi(\alpha_t(A))\Omega_\varphi \quad \tilde{f}(t)dt$$
$$= \int U_\varphi(t)\ \tilde{f}(t)dt \quad \pi_\varphi(A)\Omega_\varphi$$
$$= f(H_\varphi)\ \pi_\varphi(A)\Omega_\varphi\ .$$

Therefore, if the support of f is contained in the open interval $(-\infty,0)$, then $f(H_\varphi) = 0$ follows for a ground state φ, i.e.

$$\varphi(A(\tilde{f})^*A(\tilde{f}))\ (= ||\ \pi_\varphi(A(\tilde{f}))\Omega_\varphi||^2\ ) = 0. \tag{3.6}$$

Conversely, if $\varphi(A(\tilde{f})^*A(\tilde{f}))$ vanishes whenever the support of f is in $(-\infty,0)$, then φ is shown to be stationary and $H_\varphi \geq 0$, i.e. φ is shown to be a ground state.

The characterization of a ground state by vanishing of $\varphi(B_f)$ for a certain set of positive B_f $(= A(\tilde{f})^*A(\tilde{f})) \in \mathcal{A}$ (i.e. by (3.6)) immediately implies that the set of all ground states is a "face" of the (compact) convex set of all states, namely

(1) any mixture

$$\varphi(A) = \lambda\varphi_1(A) + (1-\lambda)\varphi_2(A) \quad (0 < \lambda < 1) \tag{3.7}$$

of ground states φ_1 and φ_2 is also a ground state;

(2) any limit

$\varphi(A) = \lim \varphi_\nu(A) \quad (A \in \mathcal{A})$

of ground states φ_ν is also a ground state;

(3) any decomposition (3.7) of a ground state φ into states φ_1 and φ_2 (with $0 < \lambda < 1$) yield ground states φ_1 and φ_2 due to

$$0 = \varphi(B_f) \geq \lambda\varphi_1(B_f) \geq 0$$

for $B_f = A(\tilde{f})^*A(\tilde{f}) \geq 0$ with $\operatorname{supp} f \subset (-\infty,0)$.

If a ground state φ does not have a decomposition (3.7) into other ground states φ_1 and φ_2, then φ is said to be a pure ground state. By (3), it is then a pure state, i.e. it does not have a decomposition (3.7) to any other states φ_1 and φ_2. It is known that a state φ is pure if and only if the representation π_φ is irreducible, i.e. H_φ does not have any non-trivial $\pi_\varphi(A)$-invariant subspaces.

It also follows from (1) and (2) that mixtures of pure ground states are dense in the set of all ground states. Thus we shall mostly discuss pure ground states.

Sometimes, H_φ may have many eigenvectors of the eigenvalue 0. In that case there are different pure ground states, for which the corresponding (GNS) representations π_φ are equivalent. We call these states *equivalent*. If the corresponding GNS representations are disjoint, then states are called *disjoint*, which is the same as *not equivalent* in the case of pure states.

4. Main Results

We now describe our results about qualitative features of pure ground states of the XY-model. The following 4 cases for values of parameters λ and γ are to be distinguished. (Under the assumption $J > 0$, the parameter J does not play any role for ground states.)

(1) *Θ-symmetry preserving region*: $|g| > 1$.

(2) *Critical points*: (i) $|\lambda| = 1, \quad \gamma \neq 0.$
(ii) $|\lambda| < 1, \quad \gamma = 0.$
(iii) $|\lambda| = 1, \quad \gamma = 0.$

(3) *Region of broken Θ-symmetry*: $|\lambda| < 1,\ \gamma \neq 0,\ (\lambda,\gamma) \neq (0,\pm 1)$

(4) *Ising model*: $\lambda = 0, \quad |\gamma| = 1.$

The *number of pure ground states* is as follows:

1 in cases (1) and (2).
2 in the case (3).
∞ in the case (4).

The *number of non-equivalent representations* associated with ground states is as follows:

1 in cases (1) and (2).
2 in the case (3).
4 in the case (4).

About the question of whether a pure ground state φ <u>returns to itself under local perturbation</u> in the sense that

$$\lim_{t\to\infty} \psi(\alpha_t(A)) = \varphi(A) \qquad (A \in \mathcal{A}) \tag{4.1}$$

for all $\psi(A) = (\Psi, \pi_\varphi(A)\Psi)$, $\Psi \in H_\varphi$, $\|\Psi\| = 1$, the answer is

<u>yes</u> in cases (1) and (2).

<u>no</u> in cases (3) and (4).

About the question of whether <u>Θ-symmtry is broken</u> in the sense that the state $(\varphi \circ \Theta)$ defined by $(\varphi \circ \Theta)(A) = \varphi(\Theta(A))$ is disjoint from φ, the answer is

<u>Θ-symmetric</u> $(\varphi = \varphi \circ \Theta)$ in cases (1) and (2),

<u>broken Θ-symmetry</u> for any pure ground state in cases (3) and (4).

For a fixed polynomial A of Pauli spin operators (strictly local observables), dependence of $\varphi(A)$ on parameters λ and γ is

<u>analytic</u> in regions (1) as well as in the union of regions (3) and (4) (for an appropriate choice of two pure ground states $\varphi_\pm$),

<u>singular</u> at points of (2).

Pure ground states of the Ising model has been determined in [5] and can be explicitly described as follows.

For $\lambda = 0$ and $\gamma = 1$, the Hamiltonian is

$$H_N = -2J \sum_{j=-N}^{N-1} \sigma_x^{(j)} \sigma_x^{(j+1)} . \tag{4.2}$$

The pure ground states $\varphi_\pm$, mentioned above in connection with analytic dependence of $\varphi(A)$ on (λ, γ), are the product of eigenstates $\xi_\pm^{(j)}$ of $\sigma_x^{(j)}$ belonging to eigenvalues ± 1 at each lattice site:

$$\varphi_\pm = \prod_j \xi_\pm^{(j)} . \tag{4.3}$$

A class of kink ground states are given by

$$\left(\prod_{j<k} \xi_-^{(j)}\right)\left(\prod_{j\geq k} \xi_+^{(j)}\right), \quad k = 0, \pm 1, \pm 2, \ldots \tag{4.4}$$

They are associated with a single irreducible representation of A and hence their superposition yields a pure ground state, too. Although the eigenvalue of H_N for (4.4) is larger than its eigenvalue for (4.3) for all finite N, any local perturbation of (4.4) can not eliminate the kink due to the different asymptotic behavior of the state at large distance in the right and the left (asymptotically φ_+ for $j \to +\infty$ and φ_- for $j \to -\infty$) and yields states of the same or higher energy.

Another class of pure ground states are the image of (4.4) by Θ-symmetry and are given by

$$\left(\prod_{j<k} \xi_+^{(j)}\right)\left(\prod_{j\geq k} \xi_+^{(j)}\right), \quad k = 0, \pm 1, \pm 2, \ldots \tag{4.5}$$

and their superpositions. These 4 classes exhaust all pure ground states in case (4).

For $\lambda = 0$ and $\gamma = -1$, the Hamiltonian reduces to

$$H_N = -2J \sum_{j=-N}^{N-1} \sigma_y^{(j)} \sigma_y^{(j+1)}$$

and ground states are obtained by exchanging x and y in the above.

5. Introduction of the Operator T

We need an operator T which can play the role of the non-existing limit of

$$T_N = \sigma_z^{(-N)} \ldots \sigma_z^{(-1)} \sigma_z^{(o)} \tag{5.1}$$

as $N \to +\infty$. Due to $(\sigma_z^{(j)})^* = \sigma_z^{(j)}$ and $(\sigma_z^{(j)})^2 = 1$, T_N is a self-adjoint unitary operator:

$$T_N^* = T_N, \quad T_N^2 = 1. \tag{5.2}$$

As in the case of Hamiltonian H_N, for which $N \to \infty$ limit of $e^{iH_N t}$ does not exist but (2.2) exists, we consider the $N \to \infty$ limit of the automorphism of $\mathcal{A}$ induced by the (selfadjoint) unitary operator T_N:

$$\Theta_-(A) \equiv \lim_{N\to\infty} T_N A T_N \qquad (A \in \mathcal{A}). \tag{5.3}$$

This limit exists and can be computed explicitly for Pauli spin operators:

$$\Theta_-(\sigma_x^{(j)}) = s(j)\sigma_x^{(j)},$$
$$\Theta_-(\sigma_y^{(j)}) = s(j)\sigma_y^{(j)},$$
$$\Theta_-(\sigma_z^{(j)}) = \sigma_z^{(i)}.$$

Here $s(j) = 1$ for $j \geq 1$ and $s(j) = -1$ for $j \leq 0$.

From the explicit form of Θ_- above, it immediately follows that

$$\Theta_-^2 = \text{id.} \quad (\text{i.e. } \Theta_-(\Theta_-(A)) = A) \tag{5.4}$$

first on Pauli spin operators and hence for all elements of $\mathcal{A}$. We can say that the group Z_2 (the additive group of integers n modulo 2) is acting on A as a group of automorphisms Θ^n ($= \Theta$ for odd n and $=$ id. for even n). According to a general theory of crossed product, there exists a C*-algebra $\mathcal{A}$ (written as $\mathcal{A} \times_\Theta Z_2$) which consists of elements

$$A + BT \tag{5.5}$$

with A and B arbitrary elements of $\mathcal{A}$, satisfying

$$c_1(A_1 + B_1T) + c_2(A_2 + B_2T) = (c_1A_1 + c_2A_2) + (c_1B_1 + c_2B_2)T,$$
$$(A_1 + B_1T)(A_2 + B_2T) = (A_1A_2 + B_1\Theta_-(B_2)) + (A_1B_2 + B_1\Theta_-(A_2))T,$$
$$(A + BT)^* = A^* + \Theta_-(B^*)T.$$

In effect, we introduce a new element T satisfying

$$T^* = T, \quad T^2 = 1, \tag{5.6}$$

$$T\,A\,T = \Theta_-(A) \tag{5.7}$$

and generate a new algebra $\hat{A}$ out of A and T. (The above formulas follow from (5.6), (5.7) and general rules for algebraic manipulation.)

We note that no elements of A can satisfy (5.7), i.e. Θ is an outer automorphism, which can be shown, for example, as follows: Let τ_n be the lattice translation automorphism of A satisfying $\tau_n(\sigma_\alpha^{(j)}) = \sigma_\alpha^{(j+n)}$ $(\alpha = x,y,z)$ for all j. We have the following asymptotic abelian property:

$$\lim_{n\to\infty} [\tau_n(B),A] = 0, \tag{5.8}$$

This is immediately checked when A and B are Pauli spin operators (they start commuting for large n), which generates A, and hence (5.8) holds for arbitrary A and B in A. In particular, if A is invertible, we have

$$\lim \tau_n(B)A\ \tau_n(B^{-1}) = A. \tag{5.9}$$

If T were in A, then (5.9) should hold for $B = T$, where $B^{-1} = B$ because (5.7) with $A = 1$ implies $T^2 = 1$ i.e. $T = T^{-1}$. On the other hand, (5.7) implies

$$\tau_n(T)A\ \tau_n(T) = \tau_n(T\tau_{-n}(A)T) = \tau_n\Theta_-\tau_{-n}(A)$$

which tends to $\Theta(A)$ as $n \to +\infty$ and contradicts with (5.9). Therefore Θ_- is an outer automorphism.

6. CAR Algebra

We can now write down the Jordan-Wigner transformation for an infinite system and introduce Fermion creation and annihilation operators c_j^* and c_j (at the lattice site j) as follows:

$$c_j^* = T^{(j)}\ (\sigma_x^{(j)} + i\sigma_y^{(j)})/2, \tag{6.1}$$

$$c_j = T^{(j)}\ (\sigma_x^{(j)} - i\sigma_y^{(j)})/2. \tag{6.2}$$

Here $T^{(j)}$ plays the role of the $N \to +\infty$ limit of $\sigma_z^{(-N)} \ldots \sigma_z^{(j-1)}$ and is given in terms of T by

$$T^{(j)} = T\,\sigma_z^{(1)} \ldots \sigma_z^{(j-1)} \qquad \text{if } j \geq 2, \tag{6.3a}$$

$$T^{(1)} = T, \tag{6.3b}$$

$$T^{(j)} = T\,\sigma_z^{(o)} \ldots \sigma_z^{(j)} \qquad \text{if } j \leq 0. \tag{6.3c}$$

For $j < k$, $T^{(i)}$ commutes with $T^{(k)}$, $\sigma_x^{(k)}$ and $\sigma_y^{(k)}$, while $T^{(k)}$ anticommutes with $\sigma_x^{(j)}$ and $\sigma_y^{(j)}$. Therefore

$$B_j\, B_k = -B_k\, B_j \qquad (j \neq k)$$

where B_i is either c_i^* or c_i. For $j = k$, the operator $T^{(j)}$ disappears from c_j^2, $c_j c_j^*$ and $(c_j^*)^2$ due to $(T^{(j)})^2 = 1$ and the matrix representation of σ's in Section 2 can be used to find commutation relations. As a result, we obtain the following canonical anticommutation relations (CARs):

$$[c_j, c_k]_+ = [c_j^*, c_k^*]_+ = 0, \tag{6.4a}$$

$$[c_j, c_k^*]_+ = \delta_{jk}\, \mathbb{1}, \tag{6.4b}$$

where $[A,B]_+ = AB + BA$. We shall denote a C*-subalgebra of $\hat{A}$ generated by c_j and c_j^*, $j \in \mathbb{Z}$, as A^{CAR}.

Pauli spin operators can be expressed in terms of T, c_j and c_j^* as follows: First we have

$$2c_j^* c_j - 1 = (\sigma_x^{(j)} + i\sigma_y^{(j)})(\sigma_x^{(j)} - i\sigma_y^{(j)})/2 - 1 = \sigma_z^{(j)}\,. \tag{6.5a}$$

Now $T^{(j)}$ is expressed by T and $\sigma_z^{(j)}$, $j \in \mathbb{Z}$ by (6.3). By (6.1) and (6.2), we then obtain

$$\sigma_x^{(j)} = T^{(j)}(c_j + c_j^*), \quad \sigma_y^{(j)} = T^{(j)}\, i\,(c_j - c_j^*). \tag{6.5b}$$

The automorphism Θ of A defined by (2.3) can be extended to an automorphism of $\hat{A}$ satisfying $\Theta(T) = T$. Then

$$\Theta(c_j^*) = -c_j^*, \quad \Theta(c_j) = -c_j. \tag{6.6}$$

Because of $\Theta^2 = \text{id.}$, any element $A \in \hat{A}$ can be decomposed as a sum of Θ-even and Θ-odd parts:

$$A = A_+ + A_- , \quad A_\pm = (A + \Theta(A))/2 \in \hat{A}_\pm , \tag{6.7}$$

$$\hat{A}_\pm \equiv \{A \in \hat{A};\ \Theta(A) = \pm A\}. \tag{6.8}$$

Since A and A^{CAR} are setwise Θ-invariant, we have

$$A = A_+ + A_- , \quad A_\pm = A \cap \hat{A}_\pm , \tag{6.9}$$

$$A^{CAR} = A_+^{CAR} + A_-^{CAR}, \quad A_\pm^{CAR} = A^{CAR} \cap A_\pm . \tag{6.10}$$

From (6.6) and (6.7), it follows that A_+^{CAR} is generated by even polynomials of c's and c^*'s. By (6.1) and (6.2), it is contained in A_+. (T cancels out.) On the other hand, A_+ is generated by polynomials of σ's which have even total degree in σ_x's and σ_y's. Such polynomials are even polynomials of c's and c^*'s by (6.5). Therefore,

$$A_+ = A_+^{CAR} . \tag{6.11}$$

Similar argument leads to

$$A_- = A_-^{CAR}\, T , \quad A_-^{CAR} = A_-\, T . \tag{6.12}$$

By (5.5), we have

$$\begin{aligned} \hat{A} &= A_+ + A_- + A_+ T + A_- T \\ &= A_+^{CAR} + A_-^{CAR} T + A_+^{CAR} T + A_-^{CAR} . \end{aligned} \tag{6.13}$$

We define the time evolution $\alpha_t(A)$ for $A \in \hat{A}$ also by (2.2). To show the existence of the limit, it is enough to show it for $A = T$:

$$\begin{aligned} \alpha_t(T) &= \lim_{N\to\infty} e^{iH_N t}\, T\, e^{-iH_N t} \quad (T \cdot T) \\ &= \lim_{N\to\infty} e^{iH_N t}\, \Theta(e^{-iH_N t}) T = V_t T , \end{aligned} \tag{6.14}$$

$$V_t \equiv \lim_{N\to\infty} e^{iH_N t}\, e^{-i\Theta(H_N) t} \quad (\in A_+), \tag{6.15}$$

where we have used $T^2 = 1$ in the first equality, (5.7) in the second equality and the automorphism property of Θ in writing V_t.
To see the convergence of (6.15), we note that $\Theta(H_N)$ differs from H_N by an N-independent bounded operator:

$$\Delta \equiv \Theta(H_N) - H_N = 2J[(1+\gamma)\sigma_x^{(o)}\sigma_x^{(1)} + (1-\gamma)\sigma_y^{(o)}\sigma_y^{(1)}] .$$

Therefore, $e^{H_N t} e^{-i\Theta(H_N)t}$ has an absolutely convergent perturbation expansion which converges as $N \to +\infty$ to

$$V_t = \sum_{n=o}^{\infty} (-i)^n \int_o^t dt_1 \cdots \int_o^{t_{n-1}} dt_n \, \alpha_{t_1}(\Delta) \cdots \alpha_{t_n}(\Delta). \tag{6.16}$$

Actually, we do not use an explicit expression (6.16) for V_t.

Since $H_N \in A_+ = A_+^{CAR}$, α_t commutes with Θ (by $\Theta(H_N) = H_N$) and leaves A^{CAR} (and obviously A) setwise invariant.

7. Exact Solution for A^{CAR}

As far as A^{CAR} is concerned, we can follow "exact" computation for a finite system ([6]). We describe here only the result of such a computation.

First, to describe the time evolution which turns out to be linear transformations of c's and c*'s (mixed together), it is convenient to introduce a notation which unifies c's and c*'s [1]. We define

$$B(h) = \Sigma f_i c_i^* + \Sigma g_i c_i \tag{7.1}$$

where $f = (f_i) \in \ell_2(\mathbb{Z})$, $g = (g_i) \in \ell_2(\mathbb{Z})$, $h = \binom{f}{g}$ and the sum is shown to converge (in norm). Then

$$\alpha_t(B(h)) = B(e^{itK}h) , \tag{7.2}$$

$$K = 2J \begin{pmatrix} U + U^* - 2\lambda & \gamma(U - U^*) \\ -\gamma(U - U^*) & -(U + U^* - 2\lambda) \end{pmatrix} . \tag{7.3}$$

Here U and U^* are shift operators on $\ell^2(\mathbb{Z})$ defined by

$$(Uf)_j = f_{j+1} , \quad (U^*f)_j = f_{j-1} . \tag{7.4}$$

The merit of the Jordan-Wigner transformation (6.1) and (6.2) is that α_t can be explicitly computed in the form of (7.2).

The operator K can be further analyzed by the Fourier series. Let

$$\tilde{f}(\theta) = \Sigma e^{in\theta} f_n , \quad f_n = (2\pi)^{-1} \int e^{-in\theta} \tilde{f}(\theta) d\theta . \tag{7.5}$$

Then

$$(Kh)^{\sim}(\theta) = K(\theta)\tilde{h}(\theta), \tag{7.6}$$

$$K(\theta) = 4J \begin{pmatrix} \cos\theta - \lambda & -i\gamma\sin\theta \\ i\gamma \sin\theta & \lambda - \cos\theta \end{pmatrix} . \tag{7.7}$$

The eigenvalue of $K(\theta)$ is $\pm 4Jk(\theta;\lambda,\gamma)$ where

$$k(\theta) = k(\theta;\lambda,\gamma) = [(\lambda - \cos\theta)^2 + \gamma^2 \sin^2\theta]^{1/2}. \tag{7.8}$$

We immediately see that K has an absolutely continuous spectrum if $(\lambda,\gamma) \neq (0,\pm 1)$, while K is $4J$ times a selfadjoint unitary operator with its spectrum concentrated on two points $\pm 4J$ if $(\lambda,\gamma) = (0,\pm 1)$ (Ising model).

A slight digression on operators B's is in order. The CAR's for c's are equivalent to the following relations for B's:

$$[B(h_1)^*, B(h_2)]_+ = (h_1,h_2) \, \mathbb{1} . \tag{7.9}$$

(Here $(h_1,h_2) = (f_1,f_2) + (g_1,g_2)$, $(f_1,f_2) = \Sigma \bar{f}_{1i} f_{2i}$.) In addition

$$B(h)^* = B(\Gamma \bar{h}) \tag{7.10}$$

where $\Gamma\binom{f}{g} = \binom{g}{f}$ and $(\bar{f})_i = \bar{f}_i$ is the complex conjugate of f_i. B's generate A^{CAR} and the relations (7.9) and (7.10) completely characterize the algebra of B's just as CAR's characterize the algebra of c's and c^*'s.

Any unitary operator W (acting on $h \in \ell_2 \oplus \ell_2$) commuting with Γ preserves the relations (7.9) and (7.10) and hence induces an automorphism ω of A^{CAR} satisfying $\omega(B(h)) = B(Wh)$. (It is called a Bogoliubov automorphism.) e^{iKt} appearing in (7.2) is such an operator. The commutativity with Γ is equivalent to

$$\Gamma K \Gamma = -K. \tag{7.11}$$

We now discuss the relation of c and c^* to B. To go back from B to c and c^*, one has to specify a projection operator E $(E = E^* = E^2)$ satisfying

$$\Gamma E \Gamma = 1 - E. \tag{7.12}$$

We call such a projection a basis projection. Whenever a basis projection E is given, we may interprete $B(h)$ as annihilation operator (c's) if $Eh = 0$. For a general h, we have $h = \Gamma f + g$ with $g = (1-E)h$, $f = \Gamma E h = (1-E)\Gamma h$ (by (7.12)), and $B(h) = B(f)^* + B(g)$. The CAR's for $B(g)$ follow from (7.9) and (7.10). In particular we obtain original c's and c^*'s by taking $E = \binom{1}{0}$. Immediately below, we will use a basis projection E for which annihilation operators $B(g)$ is a linear combination of the original c's and c^*'s.

We now want to find ground states of A^{CAR} (relative to the time evolution α_t), although this is not exactly the problem we are solving. For this purpose, we need a digression on Fock states.

Let E be a basis projection and φ_E be a state of A^{CAR} satisfying

$$\varphi_E(B(h)^*B(h)) = (h,Eh). \tag{7.13}$$

We can easily see that there is a unique state φ_E satisfying (7.13), called a Fock state. The uniqueness is shown as follows:

By (7.13) and the Schwarz inequality, we obtain

$$\varphi_E(A\, B(h)) = \varphi_E(B(h)^*A) = 0 \tag{7.14}$$

whenever $Eh = 0$ i.e. $B(h)$ is an (E-)annihilation operator. (We are no longer talking about the original c's.) In any polynomial of B's, a general $B(h)$ is decomposed as a sum of $B(f)^*$ and $B(g)$ with $Ef = Eg = 0$, and then all $B(f)^*$ can be brought to the left of all $B(g)$ using anticommutation relations (possibly with additonal terms). Then all terms containing B's (after such a reduction) contain either $B(f)^*$ at the extreme left or $B(g)$ at the extreme right and gives a vanishing contribution for φ_E due to (7.14). Thus only the constant term (a multiple of the identity) remains and the value of φ_E on any polynomial of B's is uniquely determined. Hence φ_E is unique.

For any basis projection E, a Fock state can be explicitly given (showing its existence) and is known to be a pure state.

Let E_+ be the spectral projection of K for $(0,+\infty)$:

$$(E_+h)^{\sim}(\theta) = E_+(\theta)\tilde{h}(\theta), \tag{7.15}$$

$$E_+(\theta) = \{1 + (4Jk(\theta))^{-1}K(\theta)\}/2. \tag{7.16}$$

By (7.11), $\Gamma E_+\Gamma$ is the spectral projection of $\Gamma K \Gamma = -K$ for $(0,+\infty)$, i.e. that of K for $(-\infty,0)$. Therefore

$$E_+ + \Gamma E_+\Gamma = 1$$

because K <u>does not have an eigenvalue 0</u>. This shows that E_+ is a basis projection. We now show that <u>a ground state of A^{CAR} is necessarily the Fock state</u> φ_E and hence is unique.

We choose $A = B(h)$ in the characterization (3.6) of a ground state. Then

$$\begin{aligned} A(\tilde{f}) &= \int \alpha_t\,(B(h))\,\tilde{f}(f)dt \\ &= B(\int e^{itK}\,h\,\tilde{f}(t)dt) = B(f(K)h). \end{aligned}$$

The set of all $f(K)h$ with $\mathrm{supp}\, f \subset (-\infty,0)$ is dense in $(1-E_+)(\ell_2 \oplus \ell_2)$, i.e. in the set of all h satisfying $E_+h = 0$. Hence we obtain from (3.6)

$$\varphi(B(h)^*\, B(h)) = 0$$

whenever $E_+h = 0$. This coincides with the characterizing equation of a Fock state φ_{E_+} (cf. (7.13) and (7.14)).

8. Ground States of A_+

Because Θ commutes with α_t, A_+ is setwise α_t-invariant and we can talk about ground states of A_+ (relative to α_t). If φ is a ground state of A, then (3.6) is satisfied for all $A \in A$ and hence for $A \in A_+$ in particular. Therefore a restriction of any ground state of A to A_+ must be a ground state of A_+. We shall first determine all ground states of $A_+ = A_+^{CAR}$ and then find out its possible extensions to A.

We alredy have one ground state of $A_+ = A_+^{CAR}$, which is the restriction of the unique ground state φ_{E_+} of A^{CAR} to A_+^{CAR}. In the following, we shall see that this is a unique ground state of A_+ if $(\lambda,\gamma) \neq (0,\pm 1)$, and we have some more ground states of A_+ if $(\lambda,\gamma) = (0,\pm 1)$. First we shall describe these additional ground states.

If φ is a Θ-invariant state of A^{CAR}, then $\varphi(A) = 0$ for $A \in A_-^{CAR}$. Therefore the associated representation space H_φ splits into a direct sum

$$H_\varphi = H_{\varphi+} \oplus H_{\varphi-}, \quad H_{\varphi\pm} = \overline{\pi_\varphi(A_\pm)\Omega_\varphi}. \tag{8.1}$$

(Note that $(\pi_\varphi(A_+)\Omega_\varphi, \pi_\varphi(A_-)\Omega_\varphi) = \varphi(A_+^*A_-) = 0$ for $A_\pm \in A_\pm$ due to $A_+^*A_- \in A_-$.) The subspaces $H_{\varphi\pm}$ are both $\pi_\varphi(A_+^{CAR})$-invariant. If φ is a pure state in addition, then one can show by a general argument the following facts about the restrictions of $\pi_\varphi(A_+^{CAR})$ to $H_{\varphi\pm}$, which we donte by $\pi_{\varphi\pm}$ (as a representation of A_+^{CAR}):

(A) Both $\pi_{\varphi+}$ and $\pi_{\varphi-}$ are irreducible.

(B) They are disjoint.

We can apply this to φ_{E_+} which is a Θ-invariant pure state.

Since φ_{E_+} is α_t-invariant, we have $U_\varphi(t)$ satisfying (3.3) for $\varphi = \varphi_{E_+}$. Since Θ commutes with α_t, $A_\pm$ are setwise α_t-invariant and hence (3.2) implies that $H_{\varphi\pm}$ are both $U_\varphi(t)$ invariant. Since $\pi_{\varphi\pm}$ are irreducible, $U_\varphi(t)$ satisfying (3.3) for π_φ replaced by $\pi_{\varphi\pm}$ and $\underline{A}$ restricted to A_+^{CAR} is unique up to a multiplication of a complex number $e^{it\lambda}$, $\lambda \in \mathbb{R}$. Therefore its generator H_φ is also unique up to an additive constant. This means that a vector Φ in $H_{\varphi\pm}$ yields a ground state of A_+^{CAR} if and only if Φ is an eigenvector of H_φ and the eigenvalue is the infimum of the spectrum of H_φ restricted to $H_{\varphi\pm}$, respectively. In the case of π^+, this infimum is 0 and Ω_φ is an eigenvector having this eigenvalue. We now discuss the spectrum of H_φ and relevant eigenvectors.

The Hilbert space H_φ for the Fock state $\varphi = \varphi_{E_+}$ of A^{CAR} is known to have the following structure

$$H_\varphi = \sum_{n=o}^{\infty}{}^{\oplus} H_{\varphi n}, \quad H_{\varphi n} = \mathrm{Asym}(L)^{\otimes n} \tag{8.2}$$

where L is the range of E_+ (as a Hilbert space). Asym denotes the totally antisymmetric part (under permutation) of n-fold tensor product of L with itself, and $H_{\varphi n}$ is generated by $\pi_\varphi(B(h_1) \ldots B(h_n))\Omega_\varphi$ with $E_+h_j = h_j$, which belongs to $H_{\varphi+}$ or $H_{\varphi-}$ according as n is even or odd.

Relative to the above decomposition,

$$U_\varphi(t) = \sum_{n=o}^{\infty}{}^{\oplus} (e^{iK_+t})^{\otimes n} \tag{8.3}$$

where $K_+ = KE_+$ acting on L. Since K_+ does not have eigenvalue 0 and is positive, Ω_φ (corresponding to n = 0 in the above decomposition) is the only eigenvector of H_φ belonging to 0. Furthermore,

$$\inf \operatorname{Spec}(H_\varphi | H_{\varphi n}) = n\,(\inf \operatorname{Spec} K_+)$$

where $H_\varphi | H_{\varphi n}$ is the restriction of H_φ to $H_{\varphi n}$. If $(\lambda,\gamma) \neq (0,\pm 1)$, then K_+ has an (absolutely) continuous spectrum, and hence $H_\varphi | H_{\varphi n}$, whose spectrum is obtained by n-fold "convolution" of Spec K_+, also has an (absolutely) continuous spectrum. Hence there are no vector other than (a multiple of) Ω_φ which gives rise to a ground state.

On the other hand, if $(\lambda,\gamma) = (0,\pm 1)$, then Spec $K_+ = \{4J\}$ and hence $\operatorname{Spec}(H_\varphi | H_{\varphi n}) = 4nJ$. Therefore all vectors in $H_{\varphi 1}$ (and only these vectors), which are in $H\varphi_-$, are eigenvectors of H_φ belonging to an eigenvalue 4J which is the infimum of $\operatorname{Spec}(H_\varphi | H_{\varphi-}) = \{1,3,5,\ldots\}$. Hence they produce pure ground states of A_+^{CAR}. These vectors are of the form $\pi_\varphi(B(h))\Omega_\varphi$ with $E_+h = h$ and $\|h\| = 1$. We denote the corresponding state by φ_h.

Finally we indicate how one can show that states described above exhaust all ground states of A_+^{CAR}. Suppose supp $f_j \subset (-\infty,0)$ $(j = 1,2)$. It can be shown that

$$A(\tilde{f}) \equiv \int_{\infty}^{\infty} \alpha_t(B(f_1(K)h_1)B(f_2(K)h_2))\tilde{f}(t)dt$$

coincides with A_g whenever $\tilde{f} = \tilde{g}$ on $(\text{supp } f_1) + (\text{supp } f_2) \subset (-\infty,0)$, where notation (3.4) is used. In particular, we may choose g satisfying supp $g \subset (-\infty,0)$. By (3.6), a ground state of φ of A_+^{CAR} satisfies

$$\varphi(A(\tilde{f})^*A(\tilde{f})) = \varphi(A(\tilde{g})^*A(\tilde{g})) = 0.$$

Since f is arbitrary, we may take the limit of $f \to 1$ (for example, $f(p) = e^{-\varepsilon p^2}$ with $\varepsilon \to +0$) and obtain

$$\varphi(A^*A) = 0$$

for $A = B(f,(K)h_1)B(f_2(K)h_2)$ and hence for all $A = B(h_1)B(h_2)$ with $E_+h_1 = E_+h_2 = 0$.

Let $\hat{\varphi}$ be the Θ-invariant extension of φ to A^{CAR}:

$$\hat{\varphi}(A_+ + A_-) = \varphi(A_+) , \quad A_\pm \in A_\pm^{CAR} .$$

Then $\Omega_{\hat{\varphi}}$ satisfies

$$\pi_{\hat{\varphi}}(B(h_1))\pi_{\hat{\varphi}}(B(h_2))\Omega_{\hat{\varphi}} = 0$$

whenever $E_+h_1 = E_+h_2 = 0$. This yields the following two alternatives:

(i) $\pi_{\hat{\varphi}}(B(h_2))\Omega_{\hat{\varphi}} = 0$ for all h_2 satisfying $E_+h_2 = 0$.

Then $\hat{\varphi}$ is the Fock state φ_{E_+} and φ is the restriction of φ_{E_+} to A_+^{CAR}.

(ii) $\pi_{\hat{\varphi}}(B(h_2))\Omega_{\hat{\varphi}} \neq 0$. In this case $\pi_{\hat{\varphi}}(B(h_2))\Omega_{\hat{\varphi}}$, being annihilated by all $\pi_{\hat{\varphi}}(B(h_1))$ with $E_+h_1 = 0$, produces the Fock state φ_{E_+} after proper normalization. This already shows that $\pi_{\hat{\varphi}}$ contains the Fock representation π_E. By a general argument, one can show that $\pi_{\hat{\varphi}}$ is a cyclic representation for a ground state of A^{CAR} hence must coincide with π_{φ_E}, $E = E_+$. Therefore φ mus be a vector state of the representation $\pi_{\varphi_E}^-$ of A_+^{CAR}. This exists only if $(\lambda,\gamma) = (0,\pm 1)$ and then $\varphi = \varphi_h$ as determined above

9. Ground States of A

Let φ be a ground state of A. Since Θ commutes with α_t, the condition (3.6), being satisfied by φ, will be satisfied also by $\varphi \circ \Theta$. This means that $\varphi \circ \Theta$ and hence

$$\overline{\varphi} = (\varphi + \varphi \circ \Theta)/2 \tag{9.1}$$

are ground states of $\mathcal{A}$. In addition $\overline{\varphi}$ is Θ-invariant: $\overline{\varphi}(A_+ + A_-) = \overline{\varphi}(A_+)$ if $A_\pm \in \mathcal{A}_\pm$. Thus $\overline{\varphi}$ is uniquely determined by its restriction to $\mathcal{A}_+$ which is a ground state of $\mathcal{A}_+$ and hence is already known to us. Therefore we can find all pure ground states of $\mathcal{A}$ by finding out all pure state decompositions (9.1) of the Θ-invariant extension $\overline{\varphi}$ of $\mathcal{A}_+$ to $\mathcal{A}$ and check that $\overline{\varphi}$ is in fact a ground state of $\mathcal{A}$. (Then φ in (9.1) is a ground state of $\mathcal{A}$ as already discussed in section 3.) It turns out that Θ-invariant extension of ground states of $\mathcal{A}_+$ to $\mathcal{A}$ is always a ground state of $\mathcal{A}$ in our model. If $(\lambda,\gamma) \neq (0, \pm 1)$, $\overline{\varphi}$ is unique by our uniqueness result about ground states of $\mathcal{A}_+$. Hence the known existence of ground states of a spin lattice system (as accumulation point of β-KMS states as $\beta \to +\infty$, β-KMS states themselves as accumulation points of finite interval Gibbs states as the interval approaches $\mathbb{Z}$, and the existence of accumulation points due to compactness of the set of states) implies that the unique $\overline{\varphi}$ must be a ground state.

The representation space of $\mathcal{A}$ associated with $\overline{\varphi}$ splits as in (8.1) and defines two representations $\pi_\pm$ of $\mathcal{A}_+$. If $\overline{\varphi}$ is a Θ-invariant extension of a pure ground state φ_+ of $\mathcal{A}_+$ which is necessarily the case for the above scheme, then π_+ which is the cyclic representation of $\mathcal{A}_+$ associated with φ_+ is irreducible. One can then show by quite a general arguemnt that π_- is also irreducible and $\overline{\varphi}$ has one of the following two alternative structures:

(a) $\overline{\varphi}$ is pure, in which case it is a pure ground state of $\mathcal{A}$. If $(\lambda,\gamma) \neq (0,\pm 1)$, this implies that the ground state φ of $\mathcal{A}$ is unique (and $\varphi = \varphi \circ \Theta = \overline{\varphi}$).

(b) $\overline{\varphi}$ is an average of two disjoint pure ground states of $\mathcal{A}$. If $(\lambda,\gamma) \neq (0,\pm 1)$, this implies that there are exactly two disjoint pure ground states $\varphi = \varphi_\pm$ of $\mathcal{A}$ such that $\varphi_\mp = \varphi_\pm \circ \Theta$.

In the rest of the paper, we concentrate on the case $(\lambda,\gamma) \neq (0,\pm 1)$ and explain how one can find out which of (a) and (b) occurs for each value of (λ,γ). An abstract criterion, derived by a general argument, is as follows:

(a) occurs if and only if π_+ and π_- are disjoint.

(b) occurs if and only if π_+ and π_- are equivalent.

In order to compare π_+ and π_-, we consider the state

$$\hat{\varphi}(A_+ + A_- + B_+T + B_-T) = \varphi_{E_+}(A_+) \tag{9.2}$$

of $\hat{A}$, where $A_\pm$, $B_\pm \in A_\pm^{CAR}$. (It is easily seen to be a state.) The corresponding representation space splits as

$$\hat{H} = H_+ \oplus H_- \oplus H_{+T} \oplus H_{-T} , \qquad (9.3)$$

$$H_\pm = \hat{\pi}(A^{CAR})\hat{\Omega} , \quad H_{\pm T} = \hat{\pi}(A^{CAR})\hat{\pi}(T)\hat{\Omega} . \qquad (9.4)$$

The vector state by $\hat{\Omega}$ on A^{CAR} is φ_{E_+}, while the vector state by $\hat{\pi}(T)\hat{\Omega}$ is

$$(\hat{\pi}(T)\hat{\Omega},\ \hat{\pi}(A)\hat{\pi}(T)\hat{\Omega}) = \varphi_{E_+}(TAT) = \varphi_{E_+}(\Theta_-A) = \varphi_F(A) \qquad (A \in A^{CAR})$$

where Θ_- is extended to $\hat{A}$ by $\Theta_-(T) = T$ (and hence defined on A^{CAR}) and $F = \theta_- E_+ \theta_-$, θ_- being an operator in $\ell_2 \oplus \ell_2$ defined by $(\theta_- h)_j = s(j)h_j$, $s(j)$ defined in Section 5. The cyclic representation assocated with Fock states φ_E and φ_F (with $E = E_+$) are given on $H_+ \oplus H_-$ and $H_{+T} \oplus H_{-T}$, respectively, and the restriction of each to A_+^{CAR} splits into a direct sum of disjoint irreducible represenations on $H_\pm$ and $H_{\pm T}$, respectively, by (A) and (B) of Section 8. Let us denote these irreducible representations by $\pi_{E\pm}$ and $\pi_{F\pm}$, respectively.

The cyclic representation of A associated with $\overline{\varphi}$ (a Θ-invariant extension of $\varphi_{E_+}|A_+$ to A) is the restriction of $\hat{\pi}(A)$ to $H_+ \oplus H_{-T}$ $(= \overline{\hat{\pi}(A)\hat{\Omega}})$. Hence $\pi_+ = \pi_{E+}$ and $\pi_- = \pi_{F_-}$. A result in [2] then gives the following criterion for their equivalence:

$\pi_+ \sim \pi_-$ if and only if

(α) $E - F$ is in the Hilbert-Schmidt class.

(β) $(-1)^{\dim(E \wedge (1-F))} = -1$.

The rest is to check (α) and (β) using an explicit from of $E = E_+$ given by (7.15) and (7.16) and $F = \theta_- E_+ \theta_-$. As long as $k(\theta,\lambda,\gamma) \neq 0$ for all θ, we find $\|E - F\|_{H.S.} < \infty$. This condition is satisfied in cases (1), (3) and (4) of Section 4. In case (2) (i) and (ii), $\|E - F\|_{H.S.} = \infty$ while in case (2) (iii) as well as all cases $\gamma = 0$, $|\lambda| > 1$, E_+ commutes with θ_- and hence $E - F = 0$.

If $\|E - F\|_{H.S} < \infty$, then the quantity in (β) above, called Z_2-index, is continuous (i.e. constant) in E and F in the norm topology of E

and F. Since E and hence $F = \theta_- E \theta_-$ depend on (λ,γ) continuously in any of connected components of case (1) and cases (3) and (4) together, respectively, we can determine it at one point in each connected component. For case (1), we take $\gamma = 0$, $|\lambda| > 1$ (as well as $\gamma = 0$, $|\lambda| = 1$ for case (2) (iii)) and we find out $E\wedge(1-F) = 0$ due to $E = F$. Thus in case (2) (i) and (ii), (α) is not satisfied and in cases (1) and (2) (iii), (β) is not satisfied while (α) is satisfied. Hence in cases (1) and (2), we have a unique ground state of A.

In cases (3) and (4), we evaluate the Z_2-index at $(\lambda,\gamma) = (0,\pm 1)$ and, after some explicit computation, we find that $\dim(E\wedge(1-F)) = 1$. Hence (α) and (β) is satisfied. Hence we have two pure ground states of A for the case (3).

Discussion on case (4) as well as other information about ground states such as ergodic properties and correlation functions will be given in a forthcoming full account of this work [4].

Acknowledgement

This work was completed while one of the authors (H.A.) was at the Zentrum für interdisziplinäre Forschung, Universität Bielefeld, D-4800 Bielefeld 1, FR Germany. The financial and secretarial support of ZiF is gratefully acknowledged.

References

[1] Araki, H.: On quasifree states of CAR and Bogoliubov automorphisms. Publ. RIMS, Kyoto Univ. 6, 384-442 (1970).

[2] Araki, H. and Evans, D.E.: On a C*-algebra approach to phase transition in the two dimensional Ising model. Commun. Math. Phys. 91, 489-503 (1983).

[3] Araki, H.: On the XY-model on two-sided infinite chain. Publ. RIMS Kyoto Univ. 20, 277-296 (1984).

[4] Araki, H. and Matsui, T.: Ground states of the XY-model, to be published.

[5] Bratteli, O. and Robinson, D.W.: Operator algebras and quantum statistical mechanics II. Springer-Verlag, New York-Berlin-Heidelberg, 1981.

[6] Lieb, E., Schultz, T. and Mattis, D.: Two soluble models of an antiferromagnetic chain. Annals of Phys. 16, 407-466 (1961).

Bulk Diffusion for Interacting Brownian Particles

M. Z. Guo and G. Papanicolaou*
Courant Institute, New York University

1. Introduction.

In a recent paper [1], H. Spohn analyzed the equilibrium fluctuations in space time for a system of interacting Brownian particles in the hydrodynamical limit.

The hydrodynamical limit is a rescaling of space and time that leads to a reduced or collective description of the infinite particle system, much in the spirit of passage from the Boltzmann equation to the equations of gas dynamics. Rost [2] formulated the relevant mathematical problem for interacting Brownian particles. Using the fluctuation dissipation theorem, he showed how the bulk diffusion coefficient can be identified within the framework of an equilibrium theory. This is important because the study of the infinite particle system at equilibrium is much simpler. There seems to be no mathematical analysis in a nonequilibrium situation for interacting Brownian particles.

In this note we will outline a proof of Spohn's theorem that is relatively simple, and therefore has wider applicability. Details of our proof follow [3] and will be published separately. The basic ideas are laid out in [4] for similar problems in the context of random media.

Our discussion below will be formal since we will be dealing with differential operators in infinite dimensions and other objects that must be defined carefully. The key steps make sense, however, as they are described here.

*Supported by the National Science Foundation.

2. The Process.

Let $\Phi(x)$ be a smooth nonnegative function on R^d that is even and consider the system of diffusions that satisfy

$$dx_j(t) = -\sum_{k\neq j} \nabla\Phi(x_j(t)-x_k(t))\, dt + \sqrt{2}\, dw_j(t) \,. \tag{2.1}$$

Here $[w_j(t)]$ are independent Brownian motions on R^d.

The system (2.1) of coupled diffusions has been analyzed by Lang [5] and more recently by Fritz [6]. We will deal with (2.1) formally here. We denote by $\tilde{\nu}_t$ the interacting particle process regarded as a Radon point measure with

$$(\phi,\tilde{\nu}_t) = \sum_j \phi(x_j(t)) \tag{2.2}$$

for any $\phi \in C_0^\infty(R^d)$. The formal generator of the process is given by

$$\tilde{L} = \sum_j \{\nabla_j^2 - \sum_{k\neq j} \nabla\Phi(x_j-x_k)\cdot\nabla_j\} \tag{2.3}$$

wher ∇_j is the gradient operator with respect to the coordinates of the jth particle.

The process defined by (2.1) is reversible relative to the Gibbs measure consuructed with the potential Φ. We will assume throughout that the particle density is low so that this equilibrium Gibbs measure is uniquely defined. We denote it by μ and we have, at equilibrium,

$$E^\mu\{(\phi,\tilde{\nu}_t)\} = \rho(\phi,1) \tag{2.4}$$

where ρ is the mean particle density.

It is useful to think of the Gibbs measure as one associated with the formal density.

$$a = \exp\Big\{-\frac{1}{2}\sum_{j\neq k} \Phi(x_j-x_k)\Big\} \,. \tag{2.5}$$

Then the generator has the selfadjoint form

$$\tilde{L} = \frac{1}{a}\sum_j \nabla_j\cdot(a\nabla_j) \tag{2.6}$$

and the associated bilinear form is

$$B(f,g) = -(f,\tilde{L}g) = -E^{\mu}\{f\,\tilde{L}g\} = E^{\mu}\{\sum_j \nabla_j f\cdot\nabla_j g\} \tag{2.7}$$

One can base the study of (2.1) on this bilinear form which is first defined on a suitable class of test functions [3].

If f is a function of $\tilde{\nu}$, i.e. a symmetric function of the coordinates, then we can write

$$\nabla_j f(\tilde{\nu}) = \int_R -\frac{\delta f(\tilde{\nu})}{\delta\nu(x)}\nabla\delta(x-x_j)\,dx = \left(\nabla\frac{\delta f(\tilde{\nu})}{\delta\nu(x)}\right)_{x=x_j}$$

With this operator notation the bilinear form (2.7) becomes, for f and g test functions,

$$B(f,g) = E^{\mu}\{(\nabla\frac{\delta f}{\delta\nu}\cdot\nabla\frac{\delta g}{\delta\nu},\tilde{\nu})\} \tag{2.8}$$

In this notation, for example, the weak form of the resolvent equation

$$\lambda u - \tilde{L}u = f\,,\quad \lambda > 0$$

becomes

$$\lambda E^{\mu}\{uv\} + B(u,v) = E^{\mu}\{fv\}\ . \tag{2.9}$$

The equation (2.9) is to hold for all v in a suitable class.

3. Scaling.

The scaled process is defined by

$$\nu_t^{\varepsilon} = \varepsilon^d\sum_j \delta(x-x_j^{\varepsilon}(t))\,,\quad x_j^{\varepsilon}(t) = \varepsilon x_j(t/\varepsilon^2)\,, \tag{3.1}$$

so that for any $\phi\in C_0^{\infty}$

$$(\nu_t^{\varepsilon},\phi) = \sum_j \varepsilon^d\phi(\varepsilon x_j(t/\varepsilon^2)) \tag{3.2}$$

The formal generator (2.6) in scaled variables has the form

$$L^{\varepsilon} = \sum_{j} \{ \nabla_j^2 - \frac{1}{\varepsilon} \sum_{k \neq j} \nabla\Phi (\frac{x_j - x_k}{\varepsilon}) \nabla_j \}. \tag{3.3}$$

The scaled Gibbs equilibrium measure is denoted by μ^{ε}. We may think of it as associated with the formal density

$$a^{\varepsilon} = \exp \{ -\frac{1}{2} \sum_{j \neq k} \Phi (\frac{x_j - x_k}{\varepsilon}) \} \tag{3.4}$$

If at time $t = 0$, the process is distributed with the Gibbs measure μ^{ε} then at any time $t > 0$, ν_t^{ε} will have μ^{ε} as its law. When ρ, the mean particle density, is small enough we know that μ^{ε} tends to a point probability measure concentrated on the measure $\rho\, dx$. We introduce the fluctuation process N_t^{ε} by

$$\nu_t^{\varepsilon} = \rho\, dx + \varepsilon^{d/2} N_t^{\varepsilon} . \tag{3.5}$$

Let $\tilde{\mu}^{\varepsilon}$ be the law of N_t^{ε} at equilibrium which is simply the Gibbs measure μ^{ε} translated and scaled according to (3.5). One can show (for example in [3]) that the measure $\tilde{\mu}^{\varepsilon}$ converges weakly to a Gaussian measure on $S'(R^d)$, the generalized function on R^d. Let G denote this Gaussian measure. Its covariance is given by

$$E^G\{(f,N)(g,N)\} = \rho_2 \int_{R^d} f(x)\, g(x)\, dx \tag{3.6}$$

where

$$\rho_2 = \rho + \int_{R^d} \omega(z)\, dz \tag{3.7}$$

and $\omega(z)$ is the second cluster function [7] of the Gibbs measure μ (unscaled).

The central limit theorem for N_t^{ε} at equilibrium amounts to the statement

$$E^{\tilde{\mu}^{\varepsilon}}\{e^{i(N,\phi)}\} \to E^G\{e^{i(N,\phi)}\} = e^{-\rho_2(\phi,\phi)/2} \tag{3.8}$$

4. Space-time Limit Theorem

We want to characterize the limit law of N_t^{ε} as a process and not at one time point. On the other hand we want to consider N_t^{ε} only when initially it has the equilibrium law $\tilde{\mu}^{\varepsilon}$. We consider then the resolvent equation for N_t^{ε}. If

$$u^{\varepsilon}(N) = \int_0^{\infty} e^{-\lambda t} E\{f(N_t^{\varepsilon}) \mid N_0^{\varepsilon} = N\}\, dt \tag{4.1}$$

then $u^{\varepsilon}(N)$ satisfies

$$\lambda\, E^{\tilde{\mu}^{\varepsilon}}\{u^{\varepsilon} v\} + E^{\tilde{\mu}^{\varepsilon}}\{(\nabla \frac{\delta u^{\varepsilon}}{\delta N} \cdot \nabla \frac{\delta v}{\delta N}, \nu)\} = E^{\tilde{\mu}^{\varepsilon}}\{fv\} \tag{4.2}$$

for any test function $v(N)$ in a suitable class and each $\lambda > 0$.

We want to show the following

<u>Theorem</u>. As ε tends to zero, $u^{\varepsilon}(N)$ converges to $u(N)$ weakly, i.e. $E^{\tilde{\mu}^{\varepsilon}}(u^{\varepsilon} v) \to E^{G}(uv)$ for all text functions v, and $u(N)$ satisfies the weak resolvent equation

$$\lambda E^{G}\{uv\} + \rho E^{G}\{(\nabla \frac{\delta u}{\delta N}, \nabla \frac{\delta v}{\delta N})\} = E^{G}\{fv\}, \quad \lambda > 0 . \tag{4.3}$$

From this theorem, the form of (4.3) in particular, we see that if N_t denotes the Ornstein-Uhlenbeck process on $S'(R^d)$ formally associated with the stochastic equation

$$d(N_t,\phi) = \frac{\rho}{2\rho_2}(N_t,\Delta\phi)\, dt + \sqrt{2\rho}\,(\nabla\phi, dw_t) \tag{4.4}$$

then N_t^{ε} in fact converges weakly to N_t. In (4.4) w_t is the Brownian motion in $S(R^d)$ with unit spatial covariance.

As Rost [2] noted, the fluctuation dissipation argument (explained in [2]) leads to the identification of the bulk diffusion coefficient

$$D_B = \frac{\rho}{2\rho_2} . \tag{4.5}$$

The principal interest in the theorem is this identification.

We recall that throughout, and in particular in the theorem, the density ρ is assumed small enough so the Gibbs measure μ is uniquely defined.

5. <u>Outline of the Proof</u>.

With $\lambda > 0$ fixed, we may pass to the limit in (4.2) along subsquences. We conclude that there are functions u(N) and w(N,x) such that

$$\lambda E^G\{uv\} + \rho E^G\{(w, \nabla \frac{\delta v}{\delta N})\} = E^G\{fv\} \tag{5.1}$$

At this stage we do not know that $w = \nabla \delta u/\delta N$. In fact this is the main thing to be shown. Then (5.1) is the same as (4.3) and we are finished.

To show that $w = \nabla \delta u/\delta N$, three steps are needed. First we reduce the identification to a simpler problem by using special test functions v. Then we use the ergodicity of the operator $\tilde{L}$ in (2.6) in a suitable form and finally we use the identity

$$\frac{\rho}{2\rho_2} = 1 - \frac{1}{2}\int_{R^d} z\, \nabla\Phi(z)\, \omega(z)\, dz \tag{5.2}$$

Here ω is the second cluster function as in (3.7) and the integral on the right is an isotropic tensor.

The identity (5.2) is proved directly by working with properties of the Gibbs measure μ. We have not seen this identity in the literature before, although it is implied in Rost´s paper [2]. It is not used there. We will not prove (5.2) here but we will use it.

Step one, the reduction, is accomplished by letting

$$v = (\psi, N)\, g(N) \tag{5.3}$$

with

$$\psi(x) = e\cdot x\, \phi(x) \tag{5.4}$$

Here e is a fixed vector in R^d and $\phi \in C_0^\infty(R^d)$ while g(N) is a test function on $S(R^d)$ (i.e. in $S(R^d)$). Using this v in (4.2) and in (5.1) we see that the identification amounts ot showing that

$$\lim_{\varepsilon\downarrow 0} E^{\tilde{\mu}^\varepsilon}\{(\nabla \frac{\delta u^\varepsilon}{\delta N}\cdot e\phi, \nu)g\} = \rho\, E^G\{(\nabla \frac{\delta u}{\delta N}\cdot e, \phi)g\} \tag{5.5}$$

Now we integrate by parts on both sides of (5.5). From the right we get

$$\rho E^{G}\{(\nabla \frac{\delta u}{\delta N} \cdot e, \phi) g\} = \rho \, E^{G}\{u(\nabla\phi \cdot e, \frac{\delta g}{\delta N})\}$$

$$- \frac{\rho}{2\rho_2} E^{G}\{ug(\nabla\phi \cdot eN)\} \qquad (5.6)$$

From the left we get

$$\begin{aligned} &E^{\tilde{\mu}^{\varepsilon}}\{(\nabla \frac{\delta u^{\varepsilon}}{\delta N} \cdot e\phi, \nu) g\} \qquad (5.7) \\ &= E^{\tilde{\mu}^{\varepsilon}}\{g\, \varepsilon^{d/2} \sum_{j=1}^{\infty} \phi(x_j) e \cdot \nabla_j u^{\varepsilon}\} \\ &= - E^{\tilde{\mu}^{\varepsilon}}\{u^{\varepsilon} g\, \varepsilon^{d/2} \sum_{j=1}^{\infty} e \cdot \nabla\phi(x_j)\} \\ &\quad - E^{\tilde{\mu}^{\varepsilon}}\{u^{\varepsilon} \varepsilon^{d} \sum_{j=1}^{\infty} \phi(x_j) e \cdot (\nabla \frac{\delta g}{\delta N})_{x_j}\} \\ &\quad - E^{\tilde{\mu}^{\varepsilon}}\{u^{\varepsilon} g\, \varepsilon^{d/2} \sum_{j=1}^{\infty} \phi(x_j) \sum_{k \neq j} - \frac{1}{\varepsilon} e \cdot \nabla\Phi(\frac{x_j - x_k}{\varepsilon})\} \\ &\rightarrow - E^{G}\{ug(\nabla\phi . e, N)\} + \rho\, E^{G}\{u(\nabla\phi \cdot e, \frac{\delta g}{\delta N})\} + \lim_{\varepsilon \downarrow 0} J_{\varepsilon} \end{aligned}$$

where, after symmetrization,

$$J_{\varepsilon} = \frac{1}{2} E^{\tilde{\mu}^{\varepsilon}}\{u^{\varepsilon} g\, \varepsilon^{d/2} \sum_{j,k} (\phi(x_j) - \phi(x_k)) \frac{1}{\varepsilon} \nabla\Phi\, (\frac{x_j - x_k}{\varepsilon}) \cdot e\} \qquad (5.8)$$

Expanding in a Taylor series, we see that

$$\begin{aligned} J_{\varepsilon} &\sim \frac{1}{2} E^{\tilde{\mu}^{\varepsilon}}\{u^{\varepsilon} g\, \varepsilon^{d/2} \sum_{j=1}^{\infty} \nabla\phi(x_j) \cdot \sum_{k=1}^{\infty} (\frac{x_j - x_k}{\varepsilon})\, \nabla\Phi(\frac{x_j - x_k}{\varepsilon}) \cdot e\} \\ &= \tilde{J}_{\varepsilon} \end{aligned} \qquad (5.9)$$

The main part of the proof is to show that

$$\lim_{\varepsilon \to 0} J_{\varepsilon} = \lim_{\varepsilon \downarrow 0} \tilde{J}_{\varepsilon} = H\, E^{G}\{ug(\nabla\phi \cdot e, N)\} , \qquad (5.10)$$

$$H = \int z\, \nabla\Phi(z)\, \omega(z)\, dz . \qquad (5.11)$$

If (5.10) holds, then (5.6) and the limit in (5.7) are equal, in view of the identity (5.2). This means that (5.5) holds and the theorem is proven.

The proof that $\tilde{J}_{\varepsilon} \rightarrow H\, E^{G}\{ug(\nabla\phi \cdot e, N)\}$ requires (i) the ergodicity of the process (2.1) and (ii) the fact that

$$E^{\tilde{\mu}^{\varepsilon}}\{(\nabla \frac{\delta u^{\varepsilon}}{\delta N} \cdot \nabla \frac{\delta u^{\varepsilon}}{\delta N}, \nu)\} \leq C \tag{5.12}$$

with C a constant independent of ε. This follows from (4.2). The ergodicity of the process (2.1) is used in order to construct a suitable solution to the equation

$$LU - sU = \sum_{k=1}^{\infty} (y_j - y_k) \nabla\Phi(y_j - y_k) - H \ , \ s > 0 \tag{5.13}$$

this is then used in (5.9) in much the same way as in [4] in pages 856-857.

<u>References</u>

[1] H. Spohn, Equilibrium Fluctuations for Interacting Brownian Particles, to appear.

[2] H. Rost, Hydrodynamik gekoppelter Diffusionen: Fluctuationen im Gleichgewicht, Lecture Notes Mathematics 1031, 1983.

[3] M. Z. Guo, Dissertation, New York Univ., 1984.

[4] G. Papanicolaou and S.R.S. Varadhan, Boundary value problems with rapidly oscillating coefficients, in "Random Fields" Vol. II, North-Holland, 1981, pp. 835-873.

[5] R. Lang. Unendlichdimensionale Wiener prozesse in Wechselwizkung I, II, Z. Wahr. Verw. Geb. <u>38</u> (1977) pp. 55-72; 277-299.

[6] J. Fritz, preprint, 1984.

[7] D. Ruelle, Statistical mechanics, rigorous results, New York, Benjamin, 1969.

[8] M.Z. Guo and G. Papanicolaou, Fluctuation theory and bulk diffusion for interacting Brownian particles, to appear.

SPECTRAL PROPERTIES OF RANDOM AND ALMOST PERIODIC DIFFERENTIAL AND FINITE-DIFFERENCE OPERATORS

L. A. Pastur

1. Introduction

In the last decade there has been a growing interest in the spectral properties of the Schrödinger equation and other differential and finite difference equations in $L_2(\mathbb{R}^d)$, ($\ell_2(\mathbb{R}^d)$ respectively) with random metrically transitive or almost periodic coefficients. In addition to the obvious mathematical reasons for such interest caused by the intuitive logics of the development of the spectral theory of operators, probability theory and mathematical physics, these questions have a considerable importance to a large variety of problems in theoretical physics, first of all in the physics of condensed matter. Theoretical physics is the source of many problems and methods of the spectral theory of the discussed classes of operators and has, up to now, a substantial influence on the development of this region of mathematics.

2. General metrically transitive operators (MTO)

A rather general mathematical scheme describing the class of the operators under discussion is the following (Pastur [39, 40]).

Let (Ω, F, P) be a probability space and $H(\omega)$ be a random operator, i.e. a measurable function with values in the set of linear operators on a Hilbert space H. Let $T=\{T\}$ be a group of measure preserving automorphisms of Ω and $U_T=\{U_T\}$ be a unitary irreducible representation of T in H. We call the random operator $H(\omega)$ metrically transitive (MTO) if it satisfies the relation

$$U_T H(\omega) U_T^* = H(T\omega) \pmod{P} .$$

Remarks. 1) The case of selfadjoint operators is of primary interest. 2) In the case of unbounded operators we suppose that there exists a dense linear manifold $\mathcal{D}\subset\mathcal{H}$ such that $U_T\mathcal{D}\subset\mathcal{D}$, $\forall T\subset\mathcal{T}$, $H(\omega)\mathcal{D}\subset\mathcal{D}$ (mod P) and $H(\omega)$ is symmetric or essentially selfadjoint on $\mathcal{D}$ with probability 1 (resp. symmetric or selfadjoint MTO).

As an important, paradigmatic example of MTO let us consider the one-dimensional Schrödinger operator $H = -\frac{d^2}{dx^2}+q$ in $\mathcal{H} = L_2(\mathbb{R})$ with a random potential $q(x,\omega) = Q(T_x\omega)$, where $Q(\omega)$ is a measurable function on Ω, i.e. $q(x,\omega)$ is a stationary and metrically transitive random process (one-dimensional random field) and

a) $(\Omega,\mathcal{F},P) = (S^1,\mathcal{B}_1,\frac{d\omega}{2\pi})$, where $\mathcal{B}_1$ is the Borel σ-algebra on the unit circle S^1, $\mathcal{T}=\{T_x\}_{x\in\mathbb{R}}$ is the group of rotations of S^1: $T_x\omega = \omega+x \pmod{2\pi}$, acting as $q(x,\omega)=Q(x+\omega)$, $Q(\omega)$ is a 2π-periodic function, i.e. it is a periodic potential with random, uniformly distributed origin. The shift of the origin on the x-axis in the spectral theory of the periodic Schrödinger operator proved to be a rather useful device in the modern theory of completely integrable systems (see Marchenko [34]).

b) $(\Omega,\mathcal{F},P) = (\mathrm{Tor}^{(n)},\mathcal{B}^n,(2\pi)^{-n}d\omega_1\ldots d\omega_n)$, $(T_x\omega)_i = \omega_i+\alpha_i x \pmod{2\pi}$, α_i are rationally independent numbers, and the corresponding potential is a quasi-periodic function, whose module is generated by $(\alpha_1,\ldots,\alpha_n)$.

c) $q(x)$ is an ergodic Markov process.

Similar examples can be constructed easily in the multidimensional and discrete cases (elliptic differential operators in $L_2(\mathbb{R}^d)$ and matrix operators in $\ell_2(\mathbb{Z}^d)$). In particular, the discrete analog of the Schrödinger operator in which $q(x)$, $x\in\mathbb{Z}^d$ are independent, identically distributed random quantities is called the Anderson model.

These examples represent three main groups of MTO: periodic operators, whose spectral properties are well known, almost periodic operators and random operators.

In the framework of the described abstract scheme the following facts were established:

1) If $H(\omega)$ is a symmetric MTO then both of its deficiency indexes equal 0 or ∞ with probability 1 (Figotin, Pastur [15]).

2) The spectrum σ of $H(\omega)$ as well as each of its components (point, continuous, absolutely continuous etc.) are almost surely nonrandom sets (Pastur [39], Kunz, Souillard [28], Kirsch, Martinelli [25]).

3) The event: any fixed point λ is an eigenvalue of finite multiplicity has zero probability (Pastur [39]).

4) The spectral multiplicity of every fixed interval $\Delta\subset\sigma$ is not random (Figotin, Pastur [17]).

5) The point spectrum, if it exists, is a dense set.

3. Integrated density of states (IDS)

For differential and finite difference MTO the integrated density of states can be defined. This is the simplest nontrivial spectral characteristics of this class of MTO and was extensively studied because of its relevance both from mathematical and physical points of view. Namely, let $\{\Lambda\}$ be a family of bounded domains in $\mathbb{R}^d$ $(\mathbb{Z}^d)$ which sufficiently regularly exhausts $\mathbb{R}^d$ $(\mathbb{Z}^d)$ and let H_Λ be the operator defined by the same operator inside Λ and some selfadjoint boundary conditions on $\partial\Lambda$. Denote by $N_\Lambda(\lambda)$ the number of eigenvalues of H_Λ, which are less than λ, divided by the measure $|\Lambda|$ of Λ. Then under some mild conditions on the coefficients of the matrix MTO and elliptic MTO of arbitrary order, there exists a nonrandom, nondecreasing function $N(\lambda)$ such that with probability 1 at every continuity point of $N(\lambda)$

1) $\lim\limits_{\Lambda\to\mathbb{R}^d} N_\Lambda(\lambda) = N(\lambda)$

(Benderski, Pastur [6], Figotin, Pastur [15], Shubin [45]).

2) $N(\lambda) = M\{E(0,0,\lambda)\}$

where $E(x,y,\lambda)$ is the kernel of the resolution of identity E_λ of the selfadjoint MTO $H(\omega)$, and $M\{\ldots\}$ denotes the expectation with respect to the probability measure P (Pastur [38], Figotin, Pastur [15]).

3) The spectrum σ of $H(\omega)$ is the set of points of nonconstancy of $N(\lambda)$ (Pastur [39]).

The last property illustrates the role of IDS in the spectral analysis of MTO. Namely, it answers one of the main questions: what is the spectrum of MTO as a set on the real axis.

A large group of results concerns the asymptotic behaviour of $N(\lambda)$ near the bottom of the spectrum. Here are some of them.

1) For the Schrödinger operator $H = -\Delta + q(x)$ and

a) $y(x) = \sum n(x-x_j)$, where $n(x)$ is a sufficiently smooth integrable function and $\{x_i\}$ are the points of a Poisson process in $\mathbb{R}^d$ with density ν.

$$\ln N(\lambda) = (1+o(1)) \begin{cases} -\dfrac{\lambda}{n(o)} \ln \dfrac{\lambda}{n(o)}, & n \le 0, \quad \lambda \to \infty \\[2ex] -C_1(d)\nu\lambda^{-d/2}, & n \ge 0, \quad n = o(|x|^{-d-2}), \quad \lambda \to 0 \\[2ex] -C_2(d,\alpha)\lambda^{\alpha/\alpha-d}, & n \ge 0, \quad n = \dfrac{n_0}{|x|^{\alpha}}, \quad |x| \to \infty, \quad \lambda \to 0 \end{cases}$$

(Pastur [39-41]). $C_1(d)$, $C_2(d,\alpha)$ are known constants.

b) The "Poisson" potential from the previous case can be represented in the form

$$q(x) = \int_{\mathbb{R}^d} n(x-y)\, m_p(dy)$$

where $m(dy)$ is the random, shift invariant measure, such that $m(A_1)$ and $m(A_2)$ are independent random variables of $A_1 \cap A_2 = \emptyset$ and

$$\mathrm{Prob}\{m(A)=m\} = e^{-\nu|A|} \frac{(\nu|A|)}{m!}$$

If instead of this Poisson distribution for $m(A)$ we take the so called Γ-distribution

$$\text{Prob}\{m(A) \in (m, m+dm)\} = \nu^{|A|}\Gamma^{-1}(|A|)m^{|A|-1}e^{-\nu m}$$

then if $n \geq 0$ and $n(x) = o(|x|^{-d-2})$

$$\ln N(\lambda) = -C_3(d)\lambda^{-d/2}|\ln \lambda|(1+o(1)), \quad \lambda \to 0$$

and if $m(dy) = g^2(y)dy$, where $g(x)$ is the Gaussian process in $\mathbb{R}^d$ with correlation function $\exp\{-\sum_1^d |x_i|\}$, then

$$\ln N(\lambda) = -C_4(d)\lambda^{-\frac{d}{2}}\lambda^{-1}|\ln \lambda|^{2(d-1)}(1+o(1)), \quad \lambda \to 0$$

(see Figotin, [13], [14] for these and more general cases, which were studied using a generalized form of known results of Donsker and Varadhan).

2) A more general but weaker result was obtained by Kirsch and Martinelli [23] in the continuous case (Schrödinger equation) and by Simon [46] in the corresponding discrete case. These authors assumed that the nonnegative potential has sufficiently good mixing properties and the probability to take small values is not too small, and proved that

$$\lim_{\lambda \to 0} \frac{\ln|\ln N(\lambda)|}{\ln \lambda} = -\frac{d}{2}.$$

The results for 1b) show that this asymptotic formula cannot be sharpened in general. The number $-\frac{d}{2}$ in the r.h.s. is the so called Lifschitz exponent.

3) For one-dimensional Schrödinger operators with potential

$$q(x) = k_0 \sum \delta(x-x_i),$$

where $k_0 > 0$, $x_{j+1} - x_j \equiv y_j \geq 0$, $y_j = \phi(j\beta)$, β is irrational, $\phi(z)$ is monotonic, unbounded at $z=1$ and has period 1; we have

$$N(\lambda) = C_5(\nu,\alpha)\lambda^{\alpha}(1+o(1)), \quad \lambda \to 0$$

if $\phi(z) \sim (1-z)^{-1/\alpha}$, $z \to 1$, $\alpha > 0$, $\nu^{-1} = \int_0^1 \phi dz$ (Gredeskul, Pastur [22]).

All these asymptotics are called now Lifschitz tails

and correspond to the so called fluctuation region of the spectrum, where it exists due to the large deviation part of the potential. These fluctuations have the form of rather deep and wide wells with large distance between them. Due to the random nature of the potential, these wells have different forms and therefore the resonance tunneling between them is extremely weak. The asymptotics of $N(\lambda)$ is determined by the largest well for which the given small λ is the lowest eigenvalue. This fluctuation ideology proposed by I. Lifschitz [30] gives the possibility to obtain a large variety of asymptotics for $N(\lambda)$ and related quantities and is widely used by physicists (see Lifschitz et al. [31] and references therein). It also shows fairly well the qualitative structure of the eigenfunctions, which should be localized in essence in the neighbourhood of the corresponding well and suggests the point nature of the spectrum in the fluctuation region.

4) Another type of spectral boundaries is the stable one (Lifschitz et al. [31]). Here the spectrum exists due to all typical parts of the random potential. In the vicinity of such a boundary, $N(\lambda)$ behaves as the same function of some equation with constant coefficients (which corresponds to the so called effective medium). Here are some examples.

a) Schrödinger equation. Stable boundary is the point $\lambda=\infty$ and its vicinity,

$$N(\lambda) = N_0(\lambda)(1+o(1)) ,$$

where $N_0(\lambda) = C(d)\lambda^{d/2}$ is the IDS for $H_0=-\Delta$.

b) $H = -\frac{1}{m(x)}\frac{d}{dx}(k\frac{d}{dx}\cdot)$, where $m(x)$ and $k(x)$ are strictly positive metrically transitive processes. Here the stable boundary is the point $\lambda=0$. In the simplest case $m(x)=m$, $k(x)=k$, m and k are constants; the corresponding IDS is $N_0(m,k,\lambda) = \frac{1}{\pi}(\frac{m\lambda}{k})^{1/2}$. In the general case it turns out that

$$N(\lambda) = N_0(\mu,\text{æ},\lambda)(1+o(1)) , \quad \lambda\to 0$$

where $\mu = M\{m(x)\}$, $\varkappa = \mu^{-1}\{k^{-1}(x)\}$ (Lifschitz et al. [31]).

4. The nature of the spectrum of random operators

As it was mentioned in the previous section, $N(\lambda)$ shows where does the spectrum lay. But it does not tell anything about the nature of the spectrum, i.e. about the structure of eigenfunctions. The answer to this important question of spectral analysis as well as other interesting features of MTO depend on a more detailed structure of the coefficients, such as whether they are almost periodic, or have good mixing properties, etc.

The main body of known results concerns the one-dimensional case. In this case the Liapunov index plays an important role. It is defined as

$$\gamma(\lambda) = \lim_{|x|\to\infty} |2x|^{-1} \ln(u^2(x)+u'^2(x))$$

where $u(x)$ is the Cauchy solution of the Schrödinger equation, i.e. $-u''+qu = \lambda u$, $u(o) = \sin\alpha$, $u'(o) = \cos\alpha$, $\alpha\in[0,\pi)$. The following facts are true:

1) The limit in the r.h.s. exists with probability 1 for every pair (λ,α) and is nonrandom in the general case of a metrically transitive $q(x)$ (Oseledets [37]).

2)

$$\gamma(\lambda) = \gamma_0(\lambda)+\int \ln|\lambda-\mu|\, d(N(\mu)-N_0(\mu))$$

where $\gamma_0(\lambda) = {}_{+}\sqrt{-\lambda}$, $N_0(\lambda) = \pi^{-1}{}_{+}\sqrt{\lambda}$ are the Liapunov exponent and the IDS for the Schrödinger equation with $q=0$ (Thouless [47], Avron, Simon [2], Figotin, Pastur [18], Craig, Simon [10]).

3) $\gamma(\lambda) = 0$ for almost every λ from a Lebesgue measurable set $\Delta\subset\sigma$ if and only if $\Delta\cap\sigma_{ae} \neq 0$, where σ_{ae} denotes the absolutely continuous part of the spectrum (if part: Casher, Lebowitz [7], Ishii [24], Pastur [39]; only if part: Kotani [27]).

4) For random MTO (i.e. if $q(x)$ is a Markov process) $\gamma(\lambda)>0$, $\forall\lambda\in\sigma$ (Ishii [24], Pastur [39], Molchanov [35]). This property is opposite to the property of periodic Schrö-

dinger operators, for which $\sigma = \{\lambda : \gamma(\lambda) = 0\}$.

From properties 3) and 4) it follows that the spectrum of the random Schrödinger operator (and the random Jacobi matrix as well) has no absolutely continuous part. But in the random case much more is known. Namely, the spectrum of the one-dimensional Schrödinger equation with a potential of the form $q(x) = Q(\xi_x)$, where ξ_x is an ergodic Markov process on a smooth Riemann compact manifold and $Q(\xi)$ is a smooth and nondegenerate function, is pure point (Goldsheidt et al. [21], Carmona [8]) and all eigenfunctions decrease exponentially (Molchanov [35], Carmona [8]). The same assertions hold in essence for any one-dimensional second order operator (see Kunz, Souillard [28], Delyon et al. [11]) and for some one-dimensional operators of higher order (Goldsheidt [20], Lacroix [32, 33]). The main conditions are the good mixing property of coefficients (e.g. Markov in continuous case, i.i.d. in discrete one) and smoothness of their distributions (the last condition is needed to exclude the so called quantum resonance effects (resonance tunneling).

We see that the spectral theory of one-dimensional random MTO is at the present time in a state of completeness not too far from the same theory of periodic operators. These two cases represent the "end" (extremal) points, the pure cases of a large variety of possible spectral situations which is possible for MTO and can be described symbolically as the points of segment $[0,1]$:

	mixing property of pot.	
periodic	$\bullet\!\longrightarrow\!\bullet$	random
	$\longleftarrow$ resonance effects	
pure abs. cont. spectrum of multipl. 2		pure point spectrum of multipl. 1

Concluding this short and fragmentary review of the spectral theory of MTO, I want to make some remarks.

1. It is widely believed and sufficiently well argumented by physicists that at the vicinity of the fluctuation boundary (deep fluctuation region) the spectrum should be pure point and the eigenfunctions are exponentially lo-

calized in any dimension. In contrast to this, in the vicinity of the stable boundary, in particular in the high-energy region of the Schrödinger equation, the spectrum of random operators should be continuous or even absolutely continuous (at least for $d \geq 3$). At present there exists only a few rigorous results, supporting these believes (more exactly facts of theoretical physics). These are the results on the absence of diffusion in the vicinity of the fluctuation region in the Anderson model (Fröhlich, Spencer [19]), the announcement of Goldsheidt and Fröhlich, Spencer on the existence of point spectrum and the results by Kunz, Souillard (Kunz, Souillard [29]) on the Anderson model on the Cayley tree (Bethe lattice).

2. The two-dimensional case is of considerable interest and rather intriguing. Here, according to the just as beautiful as nonrigorous arguments, the spectrum may be singular in the high energy region with a power law fall off of the generalized eigenfunctions.

3. Even in the one-dimensional case, one has to answer (if one has a theoretical physicist's taste) not only the traditional questions of the spectral theory, but also to analyse a variety of quantities constructed from the eigenelements of random Schrödinger or Jacobi operators. These quantities have a sufficiently complicated structure and therefore such calculations, as a rule, are possible only as asymptotic or approximate ones. An important example of such quantities is the zero-temperature conductivity in the homogeneous electric field of frequency ν :

$$\sigma_\lambda(\nu) = M\left\{\int \frac{\partial}{\partial y}\, e(x,y,\lambda+\nu)\frac{\partial}{\partial y}\, e(y,x,\lambda)\Big|_{x=0}\right\}$$

where $e(x,y,\lambda)$ is the kernel of the operator $\frac{dE(\lambda)}{d\lambda}$ and $E(\lambda)$ is the resolution of identity of the Schrödinger operator. Physicists have worked out some approximation methods for handling this and similar quantities, at last for some range of the parameters λ, ν and for some characteristics of random potential, which is called the semiclassical region (see our book [31] and the references therein). But these methods require substantial foundations. Besides, the similar methods are badly needed for treating many other relations between the parameters.

4. It is of great interest to analyze the picture of the spectrum in the case, when in addition to the random potential, an electric or (and) a magnetic field are present. The latter case demonstrates a very interesting interplay of randomness and almost periodicity in the spectral and related questions.

5. Almost-periodic operators

In recent years the "hot point" of the MTO spectral theory has been shifted to the almost periodic case. Such operators model the so called incommensurable systems and some other physical situations. Besides they are filling partly in the gap between periodic operators and random ones, demonstrating a wide variety of types of spectral behaviour. As a consequence, the study of this class of MTO turns out to be very useful for the spectral theory of MTO as a whole.

To my opinion the main discovery here is the essentially non-algebraic nature of the spectral picture in contrast to the periodic case, where a mere commutativity of the corresponding operator with the translation operator gives readily the main features of the spectral picture. This discovery was made and can be demonstrated upon the rather simple case of the so called almost-Mathieu equation

$$\psi(x+1)+\psi(x-1)+2g\cos(2\pi\alpha x+\omega)\psi(x) = \lambda\psi(x), \quad x \in \mathbf{Z}$$

where $g>0$ and α is an irrational number. A very important and surprising observation was made by Aubry and Andre [1]. Namely, they argued that for $g>1$ the Liapunov exponent (in the discrete case it is defined as $\lim_{|x|\to\infty} |2x|^{-1}\ln(u^2(x)+u^2(x+1))$, where $u(x)$ is the solution with $u(1)=\sin\alpha$, $u(0)=\cos\alpha$) of this equation should be strictly positive, moreover that the following estimate holds

$$\gamma(\lambda) \geq \ln g .$$

As a consequence, for $g>1$, σ_{ae} is empty. The arguments of Aubry and Andre were made rigorous by Avron, Simon [2] and Figotin, Pastur [18]. Hermann [23] gave a general proof, based on ergodic theory of smooth dynamical systems, according to which the estimate $\gamma \geq \ln|a_n|$ holds for potential $q(x)=p(2\pi x\alpha+\omega)$, where $p(\phi) = \sum_{|k|\leq m} a_k e^{ik\phi}$ $a_k^* = a_{-k}$. From the other side, J. Bellisard et al. [4], generalizing the results by Dinaburg, Sinai [12] and Belokolos [5] showed that if α is not well approximated by rationals then for $g>>1$ the spectrum of almost-Mathieu operators has a point component whose closure has positive Lebesgue measure, and for $g<<1$ the spectrum has a massive absolutely continuous component. At last, Avron, Simon [2], based on Gordon's result proved that if α is a Liouville number (p_n, q_n are integers and $\left|\alpha-\frac{p_n}{q_n}\right| \leq C n^{-q_n}$, $n\to\infty$) and $g>1$, then the spectrum is singular continuous. We see that the almost-Mathieu operator demonstrates practically all possible types of spectral behaviour depending only on the arithmetic properties of the parameter α (quasiperiod) but also on the magnitude of the potential, i.e. parameter g.

Now I report on the results of the study of two more classes of almost-periodic operators. These results develop and add new features to the outlined picture.

A) Limit periodic operators

Recall that an almost-periodic function $q(x)$, $x \in \mathbb{R}$ is called limit periodic if there exists a sequence of periodic functions $q_n(x)$, $q_n(x+a_n) = q_n(x)$ with infinitely growing periods a_n such that

$$\sup_{x\in\mathbb{R}} |q(x)-q_n(x)| = \|q-q_n\|_C \to 0 \quad , \quad a_n \to \infty \; .$$

The class of the Schrödinger operators with such potentials was studied first by Moser [36], Avron, Simon [2], Chulaevsky [9]. These authors showed that

i) for a dense G_δ set of limit-periodic potentials (they form a complete metric space) the spectrum is a (nowhere dense) Cantor set of positive Lebesgue measure;

ii) for a dense set of potentials the spectrum is a Cantor set and pure absolutely continuous.

The proofs of these results use the natural idea of the approximations of limit-periodic operators by periodic ones, i.e. "closing" the spectral theory of periodic Schrödinger operators. But the above mentioned authors used the classical part of this theory. We (Pastur, Tkachenko [42]) used its more modern and refined form, especially the theory of Marchenko, Ostrovsky (see Marchenko [34]) containing in particular the complete solution of the inverse spectral problem.

According to this theory, every square integrable periodic potential determines the Nevanlina function $\theta(\mu)$, $\mu^2=\lambda$ (this is the quasimomentum as a function of the square root of energy) and the sequence of complex numbers $\{\theta_k\}_0^\infty$, $\mathrm{Im}\,\theta_0 = 0$, $|\theta_0|^2+\sum_{k\geq 1} k^2|\theta_k|^2 < \infty$, and in their turn, $\theta(\mu)$ and $\{\theta_k\}$ determine the potential uniquely. One can say that they form a complete set of the spectral data. In particular, the function $\theta(\mu)$ gives the conformal mapping of the upper half-plane $\mathbb{C}^+ = \{\mu: \mathrm{Im}\,\mu \geq 0\}$ onto the domain $\Pi^+ = \mathbb{C}^+ \setminus \bigcup_{k\neq 0} \{k\pi, k\pi+i|\theta_{|k|}|\}$ (comb), and for every such function there exists a periodic operator whose spectrum is the set $\{\lambda: \mathrm{Im}\,\theta(\sqrt{\lambda-\theta_0})=0\}$.

Consider now the space B^2 of the Besikovitch limit-periodic functions which is generated by the norm

$$\|q\|_{B^2} \equiv \lim_{T\to\infty} \left(T^{-1}\int_0^T |q(x)|^2dx\right)^{1/2} .$$

Denote by

a) $A = \{a_n\}_1^\infty$ a sequence of positive numbers such that $\frac{a_{n+1}}{a_n} \in \mathbb{Z}_+$, $a_n\to\infty$;

b) $Q_\infty(A)$ the class of Besikovitch limit-periodic potentials $q(x)$ such that for some sequence of a_n-periodic functions $q_n(x)$

$$\lim_{n\to\infty} \|q-q_n\|_B\, e^{Ca_{n+1}} = 0 , \quad \forall C > 0$$

(we call such potentials superexponentially approximated);

c) $\mathcal{R}_n$ the set of real numbers of the form $r = \frac{\pi k}{a_n}$ $k \in \mathbb{Z}$ and by $\mathcal{R}(A) = \bigcup_{n \geq 1} \mathcal{R}_n$;

d) $K_\infty(A)$ the set of complex-valued sequences $\{\kappa_r\}_{r \in \mathcal{R}}$ (functions on $\mathcal{R}$) such that

$$\lim_{n\to\infty} \left\{ \sum_{r \in \mathcal{R}(A) - \mathcal{R}_n} r^2 |\kappa_r|^2 \right\}^{1/2} e^{Ca_{n+1}} = 0 , \quad \forall\, C>0 .$$

Then the following facts are true.

1) Every potential $q \in Q_\infty(A)$ determines uniquely an element of $K_\infty(A)$ and vice versa, i.e. the set $K_\infty(A)$ consists of the elements κ which are the complete sets of spectral data for the potentials from $Q_\infty(A)$.

2) A closed set $\sigma \in \mathbb{R}$ is the spectrum of some Schrödinger operator with potential $q \in Q_\infty(A)$ if and only if $\sigma = \{\lambda: \lambda \geq \lambda_0 , \operatorname{Im} \kappa(\sqrt{\lambda-\lambda_0})=0\}$, where the function $\kappa(\mu)$ is a conformal mapping of $\mathbb{C}^+$ onto $\Pi^+_\infty = \mathbb{C}_+ \setminus \bigcup_{r \in \mathcal{R}(A)} [r, r+i|\kappa_{|r|}|]$ (dense comb) specified by the conditions: $\kappa(0) = 0$, $\lim_{y\to\infty} (iy)^{-1} \kappa(iy) = \pi$.

This statement gives the possibility to construct limit periodic operators with $q \in Q_\infty(A)$ which have a prescribed spectrum, i.e. it solves the inverse spectral problem for this class of potentials.

3) $N(\lambda) = \pi^{-1} \operatorname{Re} \kappa(\sqrt{\lambda-\lambda_0})$, $\gamma(\lambda) = \operatorname{Im} \kappa(\sqrt{\lambda-\lambda_0})$.

4) The spectrum of the Schrödinger operator with $q \in Q_\infty(A)$ is absolutely continuous and has multiplicity 2 . For almost every $\lambda \in \sigma$ there exists two linearly independent solutions $\psi^\pm(\lambda,x) = e^{\pm i\kappa x} u^\pm(\mu,x)$, where $u^\pm(\mu,\cdot) \in \mathcal{B}^2$ and for some sequence of a_n -periodic functions $u_n^\pm(\mu,x)$ and every Borel set $\Delta \in \mathbb{R}^+$ and $C>0$

$$\lim_{n\to\infty} \int_\Delta \| u^\pm(\mu,\cdot) - u_n^\pm(\mu,\cdot) \|^p_{\mathcal{B}^2} \, d\mu \cdot e^{Ca_{n+1}} = 0 ,$$

$\forall p \in (1,2)$.

We see that this class of almost-periodic operators is maximally close to the periodic one. The only novel feature is the possibility to have nowhere dense spectrum which is always absolutely continuous, because this Cantor set has always a positive Lebesgue measure. This picture is a "typical" one in the sense that the corresponding potentials are dense. One should only use the dense comb Π_∞^+ for it, i.e. a function $\kappa \in K_\infty(A)$ which has a dense support in $\mathbb{R}$.

B) Quasi-periodic operators

This class of operators is, on the contrary to the previous one, rather close to the random MTO. Recall that an almost-periodic function $f(x)$ is called quasi-periodic if it has a form $f(x) = F(2\pi\alpha_1 x, \ldots, 2\pi\alpha_n x)$ where $F(x_1, \ldots, x_n)$ is a 2π -periodic function for every variable $x_i \in \mathbb{R}$, $i = 1, \ldots, n$ and $\alpha_1, \ldots, \alpha_n$ are rationally independent numbers. In the discrete case the simplest quasi-periodic function has a form $f(x) = F(2\pi\alpha x)$, $x \in \mathbb{Z}$ where $F(x)$ is a 2π -periodic function, $F(x) = F(x+1)$ (e.g. $F(x) = g \cos x$ as in the almost-Mathieu equation).

Let us consider now the matrix operators H_d which act in $\ell_2(\mathbb{Z}^d)$ according to the rule:

$$(H_d\psi)(x) = \sum_{y \in \mathbb{Z}^d} w_{x-y}\psi(x) + g \operatorname{tg} \alpha_x \cdot \psi(x) \quad , \quad x \in \mathbb{Z}^d$$

where w_x is a sequence of complex numbers satisfying the conditions

$$w_\kappa^* = w_{-\kappa} \quad , \quad |w_\kappa| \leq Ce^{-\rho|\kappa|} \quad , \quad C, \rho > 0$$

and

$$\alpha_x = \pi(\alpha, x) + \omega \quad , \quad \alpha \in \mathbb{R}^d \, , \quad \omega \in [0, \pi) \quad , \quad (\alpha, x) = \sum_1^d \alpha_i x_i \ .$$

This class of operators, whose elements are obviously MTO for any $\alpha \in \mathbb{R}^d$ and are almost-periodic for rationally independent "frequencies" $\alpha_1, \ldots, \alpha_d$, is the natural generalization of the remarkable class of one-dimensional operators proposed by Prange et al. [44]. These authors proved that if α is approximated by rationals sufficiently badly, then H_2 has a dense set of eigenvalues on $\mathbb{R}$ and the corresponding eigenfunctions are exponentially localized. However their method was somewhat indirect from the point of view of the spectral theory. We (Figotin, Pastur [16]) found, using rather simple and essentially algebraic arguments, a convenient representation for the resolvent of the operator H_d and proved the following properties.

1) The IDS of H_d i.e. $N(\lambda)$, is an absolutely continuous and even real analytic function and its density $\rho(\lambda)$ has the form

$$\rho(\lambda) = (2\pi)^{-d} \int_{Tor^d} \frac{g}{g^2+(W(k)-\lambda)^2}\, dk$$

$$W(k) = \sum_{x\in\mathbb{Z}^d} e^{2\pi i(k,x)} w_x \ , \quad k \in Tor^d \ .$$

It is rather surprising that this $N(\lambda)$ coincides with the same function for the random operator of type H_d such that α_x, $x\in\mathbb{Z}^d$ are independent random variables with a common uniform distribution on $[0,\pi)$. Thus we have here two examples of rather different operators (almost periodic and random) which have the same IDS.

2) If the frequencies $\alpha_1,\dots,\alpha_d$ satisfy the condition

$$|(\alpha,x)-m| \ge C|x|^{-\beta}, \quad C,\beta > 0 \ , \quad x \ne 0$$

then the spectrum of H_d is the set of solutions of the equation

$$N(\lambda) = \frac{1}{2} + \frac{\omega}{\pi} + (\alpha,r) \ , \quad \forall\, r\in\mathbb{Z}^d$$

and as a consequence of continuity and strict monotonicity of $N(\lambda)$, the spectrum is dense in $\mathbb{R}$.

3) To every eigenvalue λ_r, $r\in\mathbb{Z}^d$ there is a unique eigenfunction $\psi_r(x)$, $r,x\in\mathbb{Z}^d$. It has the form

$$\psi_r(x) = \chi(x, \lambda_r) \ ,$$

$$\chi(x,\lambda) = (2\pi)^{-d} \int_{(Tor)^d} \frac{\exp[-2\pi i(k,x)-it(k,\lambda)]}{W(k)-\lambda-ig}\, dk$$

where $t(z,k)$ is the unique and (z,k) holomorphic solution of the equation

$$f(z,k) = f_0(z)+t(z,k)-t(z,k+\alpha) \ , \ z \in \mathbf{C} \ ,$$

$$f(k) = (2\pi)^{-d} \int_{(Tor)^d} f(z,k)dk, \quad f(z,k) = -\frac{W(k)-z+ig}{W(k)-z-ig} \ .$$

4) These eigenfunctions form a complete set in $\ell_2(\mathbb{Z}^d)$. Thus the spectrum of H_d is simple pure point and the corresponding eigenfunctions are exponentially decreasing as $|x|\to\infty$.

5) If $d=1$ then for any irrational α the spectrum of H_1 has no absolutely continuous component.

6) If $d>1$ and α is a Liouville number then the spectrum of H_1 is singular continuous.

7) Statement 2) remains valid for the operators

$$\tilde{H}_d = H_d + \varepsilon V$$

where

$$(V\psi)(x) = V(\exp(2\pi i(\alpha,x))+2i\omega)\psi(x) \ , \quad \varepsilon<<1$$

and $V(z)$, $z \in \mathbf{C}$ is an analytic function in the vicinity of the unit circle $|z|=1$, mapping it into the real line.

The listed results were independently and approximately in the same time discovered by Simon [46].

Properties 1)-4) seem rather natural in the one-dimensional case $(d=1)$ because if α is not too well approximated by rationals, then the potential $g\,\mathrm{tg}\,(\pi x\alpha+\omega)$, $x \in \mathbb{Z}$ takes on arbitrary large values on a sequence of points going to infinity. It is, however, interesting that these peaks are also sufficiently thickly and irregularly placed even in the multidimensional case, so that for any energy a particle still becomes "entangled" between them and cannot escape to infinity. Due to this property of the potential $g\,\mathrm{tg}\,(\pi(\alpha,x)+\omega)$, they are in essence supressed and thus the spectrum in this case looks very similar to the deep fluctuation spectrum for random MTO. This similarity is supported by the analogous asymptotic behaviour of conductivity in the low-frequency regime. Namely, according to

Figotin, Pastur [17] and Prange et al. [44]

$$\sigma(\nu) \leq C_1 \exp\left\{-\left(\frac{\nu_0}{\nu}\right)^{1/\beta}\right\}, \quad \nu \to 0 .$$

As it is well known (see for example Lifschitz et al. [31]), the low frequency conductivity in the random case has a similar behaviour:

$$\sigma(\nu) \sim \left(\frac{C_2}{\nu}\right)^2 \exp\left\{-\frac{C_2}{\nu}\right\}, \quad \nu \to 0 .$$

References

[1] Aubry S., Andre G. Ann. Isr. Phys. Soc. 3, 139 (1980).

[2] Avron J., Simon B. Comm. Math. Phys. 82, 101 (1982).

[3] Avron J., Simon B. Duke Math. J. 50, 369 (1983).

[4] Bellisard J., Lima R., Testard D. Comm. Math. Phys. 88, 207 (1983).

[5] Belokolos E. Theor. Math. Phys. 25, 344 (1975).

[6] Bendersky M., Pastur L. Mat. Sborn. (1971).

[7] Casher A., Lebowitz J. J. Math. Phys. 12, 1701 (1970).

[8] Carmona R. Duke Math. J. 49, 191 (1982).

[9] Chulaevsky V. Usp. Mat. Nauk 36, 143 (1981).

[10] Craig W., Simon B. Duke Math. J. 50, 551 (1983).

[11] Delyon F., Kunz H., Souillard B. J. Phys. A16, 25 (1983).

[12] Dinaburg E., Sinai Ja. Func. Anal. Appl. 9, 279 (1975).

[13] Figotin A. Funct. Theory Appl. (Kharkov) 30, 132 (1980).

[14] Figotin A. Doklady Ukr. Acad. Sci. N6, 27 (1981).

[15] Figotin A., Pastur L. In Dif. eq. and Funct. Anal., Kiev, Naukova Dumka, 117 (1978).

[16] Figotin A., Pastur L. JETP Letters 37, 575 (1983), and to be published in Comm. Math. Phys. (1984).

[17] Figotin A., Pastur L. (unpublished) (1983).

[18] Figotin A., Pastur L. J. Math. Phys. (to be published) (1984).

[19] Fröhlich J., Spencer T. Comm. Math. Phys. 88, 151 (1983).

[20] Goldsheidt Ja. Doklady Acad. Sci. USSR (1979).

[21] Goldsheidt Ja., Molchanov S., Pastur L. Funct. Anal. Appl. 11, 1 (1977).

[22] Gredeskul S., Pastur L. Theor. Math. Phys. (to be published) (1984).

[23] Herman M. Preprint École Polytichnique (1982).

[24] Ishii K. Progress Theor. Phys. Suppl. 53, 77 (1973).

[25] Kirsch W., Martinelli F. Comm. Math. Phys. 85, 329 (1982).

[26] Kirsch W., Martinelli F. Comm. Math. Phys. (to be published) (1983).

[27] Kotani S. Proc. Kyoto Conf. on Stoch. Proc. (1982).

[28] Kunz H., Souillard B. Comm. Math. Phys. 78, 201 (1980).

[29] Kunz H., Souillard B. J. Phys. (Paris) Letters 44, L411 (1983).

[30] Lifschitz I. Adv. Phys. 13, 485 (1964).

[31] Lifschitz I., Gredeskul S., Pastur L. Introduction to the Theory of Disordered Systems. Moscow, Nauka (1982) and to be published by Wiley and Sons in 1985.

[32] Lacroix J. Ann. Inst. E. Costan N 7 (1983).

[33] Lacroix J. Ann. Inst. H. Poincaré A38, N 4 (1983).

[34] Marchenko V. Sturm-Liouville operators and their application. Kiev, Naukova Dumka (1977), and to be published by Riedel in 1985.

[35] Molchanov S. Math. USSR. Izv. 12, 69 (1978).

[36] Moser J. Comm. Math. Helv. 56, 198 (1981).

[37] Oseledec V. Trans. Mosc. Math. Soc. 19, 197 (1968).

[38] Pastur L. Math. USSR Uspekhi 28, 3 (1973).

[39] Pastur L. Preprint Low Temp. Inst. Kharkov (1974).

[40] Pastur L. Comm. Math. Phys. 75, 179 (1978).

[41] Pastur L. Theor. Math. Phys. (1978).

[42] Pastur L., Tkachenko V. Doklady Acad. Sci. USSR (to be published) (1984).

[43] Prange R., Grempel D., Fishman S. Phys. Rev. Lett. 49, 833 (1982).

[44] Prange R., Grempel D., Fishman S. Preprint of Maryland University (1984).

[45] Shubin M. Trudy Sem. Petrovskii 3, 243 (1978).

[46] Simon B. Caltech. Preprint and to be published in J. Stat. Phys. (1984).

[47] Thouless D. J. Phys. C5, 511 (1972).

EQUILIBRIUM FLUCTUATIONS FOR SOME STOCHASTIC PARTICLE SYSTEMS

Herbert Spohn

1. Introduction

Our knowledge about the microscopic structure of macroscopic systems in thermal equilibrium is encoded as equilibrium time correlations. These objects are measured in scattering experiments : a purely angular resolution of the scattered beam translates into static correlations and its frequency (energy) resolution into time correlations. Despite their central physical importance our mathematical understanding of equilibrium time correlations is very modest, quite in contrast to their static counterpart on which we have a wealth of qualitative information.

If one insists on Hamiltonian dynamics, the problem of understanding the structure of equilibrium time correlations is indeed a difficult one. This should come as no surprise, since even in statics very few results pertain to physically realistic Hamiltonians. Our beloved Ising model is certainly a crude approximation to the "real life" ferromagnet $Tb(OH)_3$, say, although it catches a great deal of essential physics. Following the philosophy of approximation in static equilibrium models leads to particle or spin models with stochastic dynamics. This is a well accepted procedure within statistical physics and has been applied with great success to critical dynamics, to the dynamics of phase segregation and to many other problems. I will follow this practice here and consider only models with a stochastic time evolution.

I started to become interested in equilibrium time correlations for stochastic particle models about three years ago. To my surprise the only available result at that time was the one of Holley and Strook [1,2] who prove an exponential decay of correlations in space-time for the stochastic Ising model at high temperatures. They use an in essence perturbative argument exploiting that the stochastic Ising model has <u>no</u> conservation law. In this case the model resembles an Ising model in d+1 dimensions and exponential decay is the well-known high temperature behavior. To my knowledge the only extension is the one by Wick [3] to the stochastic Heisenberg model. Further generalizations should be feasible.

The natural question to be asked next concerns models with one or several conservation laws. This is the topic of the present article. To summarize very briefly : we have now a technique which enables us to prove the asymptotic behavior of correlations in space-time for sufficiently small interaction. The technique is in no sense perfected and there are many open problems some of which will be listed at the end.

The work to be described here developed in close cooperation with T. Brox, A. DeMasi, A. Galves, C. Kipnis, E. Presutti, H. Rost and D. Wick. Without their contributions none of the results would have materialized. I would like to draw the attention of the reader to [4], which is of a more programmatic nature and where some aspects are mentioned which I will not repeat here, and to the very extensive and most recommendable expository article [5] which however predates the results to be described here.

2. Some Models

I describe three classes of models of physical interest. For finite volume the existence of the dynamics is standard Markov process theory, for infinite volume cf. remarks at the end of this section.

A) Stochastic lattice gases (exclusion processes with speed change, spin exchange dynamics [6,7]).

In a stochastic lattice gas particles jump on a bounded set $\Lambda \subset Z^d$, the d-dimensional hypercubic lattice. There is at most one particle per lattice site. The space of configurations is then $\{0,1\}^\Lambda$. A configuration is denoted by η and η_x is its value at site x. $\eta_x = 0(1)$ corresponds to site x being empty (occupied). The dynamics is specified through the jump rates $c(x,y,\eta) \geqq 0$ which give the rate of exchange of the occupations at sites x and y when the configuration is η. Clearly $c(x,y,\eta) = c(y,x,\eta)$. c is assumed to be short ranged and translation invariant. To avoid degeneracies we also assume $c(x,y,\eta)>0$ for $|x-y| = 1$ and $\eta_x \neq \eta_y$. The generator of the jump process is

$$L\psi(\eta) = \frac{1}{2} \sum_{x,y \in\Lambda} c(x,y,\eta)\{\psi(\eta^{xy}) - \psi(\eta)\} \quad , \tag{2.1}$$

where η^{xy} denotes the configuration η with occupations at sites x and y exchanged. The jump process for L is denoted by η_t.

Let

$$H(\eta) = \sum_{A\subset\Lambda, |A|\geq 2} \Phi(A) \prod_{x\in A} \eta_x \tag{2.2}$$

be the energy constructed from the short range and translation invariant potential Φ. We impose then the condition of detailed balance as

$$c(x,y,\eta) = c(x,y,\eta^{xy}) \exp\{ - (H(\eta^{xy}) - H(\eta)) \} \quad . \tag{2.3}$$

Then η_t with initial measure

$$\frac{1}{Z} \exp \{ - H(\eta) + \mu \sum_x \eta_x \} \tag{2.4}$$

is reversible. In particular, the Gibbs measures (2.4) are time invariant. The conservation of the total number of particles is reflected by the fact that η_t has a one parameter family of stationary measures.

One may remove the restriction of single site occupancy. This case has not been studied systematically. The only model investigated in considerable detail is the zero range process [8]. There a particle at a given site x jumps with equal probability to neighboring sites with a rate, $c(\eta_x)$, depending only on the number η_x of particles at x. Again the process is reversible and admits a one parameter family of stationary measures which happen to be product measures.

B) Interacting Brownian particles.

This is a reasonable model for dilute suspensions, e.g. small polystyrene balls in water [9,10]. One has N particles in a bounded volume $\Lambda\subset R^d$. Their time evolution is described by the diffusion process

$$dx_j(t) = -\frac{1}{2} \sum_{i\neq j=1}^{N} \operatorname{grad} V(x_j(t)-x_i(t)) + d\omega_j(t) \quad , \tag{2.5}$$

$j=1,\ldots,N$, plus reflecting boundary conditions. Here $x_j(t)$ denotes the position of the j-th particle and $\{\omega_j(t)\}$ are independent Brownian motions. V is a short range, even potential. Reversibility is ensured through the drift being given as the gradient of a potential. The stationary measure is the Gibbs measure

$$\frac{1}{Z}\exp\{-\frac{1}{2}\sum_{i\neq j=1}^{N} V(x_i-x_j)\}\frac{1}{N!}\,dx_1\dots dx_N \quad . \tag{2.6}$$

If we think of the particle number as free to vary (grand canonical prescription), then the diffusion process has a one parameter family of stationary measures.

There are other versions of this model which however have not been studied in any detail. In (2.5) the diffusion matrix may be process dependent. Under some further conditions this case should be handable by our method. Sometimes one would like to give a finer dynamical description by including also velocities in (2.5). This leads to the interacting Ornstein-Uhlenbeck processes,

$$\begin{aligned} dx_j(t) &= v_j(t)dt \\ dv_j(t) &= \sum_{i\neq j=1}^{N} F(x_j(t)-x_i(t))dt - \gamma v_j(t)dt + d\omega_j(t) \;, \end{aligned} \tag{2.7}$$

$j=1,\dots,N$, plus reflecting boundary conditions. Now, time reversal includes velocity reversal and, as it stands, our method fails.

C) Time dependent Ginzburg-Landau models (Cahn-Hilliard theory [11], model B of critical dynamics [12]).

These are "field theory" type models of very wide use. We describe a lattice version which for our purposes is good enough since only the large distance behavior of correlations is of interest. To each lattice site $x \in \Lambda \subset Z^d$ we associate a real random variable ϕ_x(spin). A spin configuration is denoted by ϕ. To each nearest neighbor bond (x,y) we associate a random current from x to y, $dj_r(x,y,t)$. $\{j_r(x,x+e,t)\,|\,|e|=1, e$ positively oriented$\}$ are independent Brownian motions and we define $j_r(x+e,x,t) = -j_r(x,x+e,t)$. Then the dynamics is specified by the system of stochastic differential equations

$$d\phi_{t,x} + \sum_{x,y\in\Lambda,|x-y|=1} dj(x,y,t) = 0 \tag{2.8}$$

$$dj(x,y,t) = -\left(\frac{\partial H}{\partial\phi_y} - \frac{\partial H}{\partial\phi_x}\right)(\phi_t)dt + dj_r(x,y,t) \text{ for } |x-y|=1 \quad , \tag{2.9}$$

where the (free) energy H is given by

$$H(\phi) = \frac{1}{2}\sum_{x,y\in\Lambda,|x-y|=1}(\phi_x-\phi_y)^2 + \sum_{x\in\Lambda} V(\phi_x) \tag{2.10}$$

with $V(q) \geqq aq^2+b$, $a > 0$. (2.8) and (2.9) are written in a form to make their meaning transparent. (2.8) is the local conservation law. In (2.9) the current from x to y is the sum of a random, white noise part and a systematic part which is minus the difference in chemical potentials at x and y - the quantity the system tries to equilibrate. One readily checks that the process ϕ_t is reversible and admits the one parameter family of stationary measures

$$\frac{1}{Z} \exp\{-H(\phi) + h \sum_{x \in \Lambda} \phi_x\} \prod_{x \in \Lambda} d\phi_x \quad . \tag{2.11}$$

The model may be extended to several components, but this is not the place to go into details.

Let me summarize briefly the essential features of the stochastic dynamical models discussed so far. The interaction is short ranged and translation invariant. The basic field is locally conserved and this is the only conservation law. The dynamics is reversible and admits a one parameter family of stationary measures. These are the canonical, with respect to the particle number or the magnetization, Gibbs measures. The potential corresponding to the Gibbs measures is short ranged and translation invariant.

As a first step, we have to establish the existence of the infinite volume dynamics. Since we focus on equilibrium time correlations, the existence of the equilibrium dynamics suffices, i.e. the existence almost surely with respect to the initial Gibbs measure. For stochastic lattice gases the complete theory is developed in [7]. The equilibrium dynamics for interacting Brownian particles was constructed by R. Lang [13,14]. For Ginzburg-Landau models there are only partial results [15] which are extendable however.

3. The problem and its Expected Solution

The essential physics is contained in the equilibrium two-point function in space-time

$$\langle \eta_{t,x}\, \eta_{s,y} \rangle - \langle \eta_x \rangle \langle \eta_y \rangle \tag{3.1}$$

(and similarly for the other models), in particular its behavior at large distances, $|x-y|$ and $|t-s|$ large, which should be independent of the fine details of the interaction. At such generality we have no control over (3.1) and we therefore refrain ourselves to the more modest goal to obtain the asymptotic behavior of the two point function for small interaction. We fix once and for all the density (chemical potential, magnetic field) of the initial, time invariant Gibbs measure and set $\langle\eta_x\rangle = \rho$ (similarly for the other models). By standard high temperature expansion and by Dobrushin uniqueness, the Gibbs measure is then unique, translation invariant and has good cluster properties. $\langle\cdot\rangle_\rho$ always refers to expectations of the process with this initial measure.

We rewrite (3.1) as $\langle\eta_{t,x}(\eta_{o,o} - \rho)\rangle_\rho$ which we read as the average density at x at time t for the initial signed measure $(\eta_o - \rho)\mu_\rho$, the Gibbs measure somewhat modified at the origin. Because of the conservation law we expect that this local perturbation spreads out according to the diffusion equation. Therefore for large separation in space-time

$$\langle\eta_{t,x}\,\eta_{o,o}\rangle_\rho - \rho^2 \simeq \chi(\rho)\, e^{D(\rho)\Delta|t|/2}\,(x,0) \quad . \tag{3.2}$$

Here Δ is the (lattice) Laplacian and $\exp(t\Delta)\,(x,0)$ the corresponding transition probability. $\chi(\rho) = \Sigma_x(\langle\eta_x\eta_o\rangle_\rho - \rho^2)$ is the static compressibility which is seen to be the correct normalization upon summing both sides of (3.2) over x. The speed at which the local disturbance spreads out is determined by the <u>bulk diffusion coefficient</u> $D(\rho)$, where, for simplicity, we assumed isotropy. $D(\rho)$ may be identified by considering $\Sigma_x\, x^2(\langle\eta_{t,x}\,\eta_{o,o}\rangle_\rho - \rho^2)$. As a general result one finds the Green-Kubo formula

$$D(\rho)_{\alpha\beta} = (2\chi(\rho))^{-1} \int dt\, \langle J_\alpha(t)\, J_\beta(0)\rangle_\rho \quad , \tag{3.3}$$

$\alpha,\beta=1,\dots,d$, where $\langle J_\alpha(t)J_\beta(0)\rangle_\rho$ is the total current-current correlation, cf.[4,5] for details.

We notice that in the Ginzburg-Landau model for the total current the systematic part drops out and $J_\alpha(t) = |\Lambda|^{-1/2} \sum_{x\in\Lambda} dj_r(x,x+e_\alpha,t)$ with e_α the unit vector along the positive α-axis. Therefore $D_{GL}(m) = 1/2\chi(m)$ with $\chi(m) = \Sigma_x(\langle\phi_x\phi_o\rangle_m - m^2)$ and $m = \langle\phi_o\rangle_m$. Similarly one finds that for interacting Brownian particles $D_B(\rho) = \rho/2\chi(\rho)$. For stochastic lattice gases the situation is more complicated. The analogue of (2.9) is <u>not</u> valid in general. In (2.9) the average current given the configuration ϕ is the gradient of something else and this is why the analogue of (2.9) has been baptized "gradient condition". In one dimension stochastic lattice gases satisfying the gradient condition have been "enumerated" [16]. The bulk diffusion coefficient is now a somewhat more complicated object but may be computed without solving for the dynamics. As in the other examples the current-current correlation $\langle J_\alpha(t)J_\beta(0)\rangle_\rho = c_{\alpha\beta}\delta(t)$.

I emphasize that our method works only if the gradient condition is satisfied. This holds for interacting Brownian particles, Ginzburg-Landau models and for the zero range process, but only for a particular class of stochastic lattice gases. For non-gradient lattice gases most parts of the proof actually go through, but there is one crucial step missing [17] .

4. Fluctuation Field, Scaling, Infinite Dimensional Ornstein-Uhlenbeck Process

In the tradition of mathematical physics the first inclination might be to find upper and lower bounds which pin down the behavior of the two-point function to (3.2). This we have not achieved and it is not clear whether it constitutes a feasible program. Instead we formulate (3.2) as <u>scaling limit</u> which is a concept familiar from many dynamical problems. Because of the scale invariance of (3.2) we consider then spatial distances of the order ε^{-1} and time distances of the order ε^{-2}, $\varepsilon\to 0$, and want to show that under this scaling the two-point function tends to the fundamental solution of the diffusion equation with the appropriate $D(\rho)$ and $\chi(\rho)$. I emphasize that this hydrodynamic limit differs from the often studied mean field type and weak coupling limits. In the former only time and space are scaled whereas in the latter also the interaction. In contrast to the Grad limit the density remains finite as $\varepsilon \to 0$.

On a technical, but also conceptual, level it is convenient to introduce the <u>density fluctuation field</u> in its scaled form as

$$\xi^\varepsilon(f,t) = \varepsilon^{d/2} \sum_{x} f(\varepsilon x)(\eta_{\varepsilon^{-2}t,x} - \rho) \qquad (A)$$

$$\varepsilon^{d/2} \{\sum_j f(\varepsilon\, x_j(\varepsilon^{-2}t) - \rho\int dq\, f(\varepsilon q)\} \qquad (B)(4.1)$$

$$\varepsilon^{d/2} \sum_x f(\varepsilon x)(\phi_{\varepsilon^{-2}t,x} - m) \quad . \qquad (C)$$

$f\in S'(R^d)$ is a rapidly decreasing test function. The goal is then to show that

$$\lim_{\varepsilon\to 0} < \xi^{\varepsilon}(f,t)\xi^{\varepsilon}(g,0)>_{\rho} = \chi(\rho)\int dqf(q)(e^{D(\rho)\Delta|t|/2}g)(q) \tag{4.2}$$

(spatial isotropy assumed and ρ replaced by m for C). $\xi^{\varepsilon}(f,t)$ is a generalized random field over $R^d\times R$ which is stationary in space-time. At fixed time $\xi^{\varepsilon}(f,t)$ tends weakly to Gaussian white noise with strength $\chi(\rho)$. Since time instants become widely separated as $\varepsilon \to 0$, it is reasonable to expect that also $\xi^{\varepsilon}(f,t)$ jointly has a Gaussian limit. Let $\xi(f,t)$ be the Gaussian field with covariance (4.2). Then we would like to show that weakly

$$\lim_{\varepsilon\to 0} \xi^{\varepsilon}(f,t) = \xi(f,t) \quad . \tag{4.3}$$

Since the covariance comes from a semigroup, $\xi(f,t)$ is the infinite dimensional Ornstein-Uhlenbeck process governed by the partial stochastic differential equation

$$d\xi(q,t) = \frac{1}{2}D(\rho)\Delta\xi(q,t)dt + \mathrm{div}(dj(q,t)) \tag{4.4}$$

with initial Gaussian white noise of strength $\chi(\rho)$. Here $dj(q,t)$ is Gaussian white noise in space-time with independent components.

To the models considered one can apply the general theory of Holley and Stroock [18]. It turns out that, using their results, the same estimates which prove (4.2) also give the stronger result (4.3). (We use here that the equal time, static equilibrium fluctuations have been studied extensively. Our hypothesis on the interaction will be such that their convergence to Gaussian white noise is ensured.)

5. Some Results

We quote three theorems only to give the reader an impression of how much can be done at present.

Theorem 1 (T. Brox and H. Rost [19]). Let $\xi^{\varepsilon}(f,t)$ be the density fluctuation field for the zero range process, cf. end of Section 2A) for the definition. Let the jump rate $c:N\to R$ satisfy $c(0)=0<c(1)$ and c is increasing but at most linearly. Then with the appropriate χ and D (4.2) holds and (4.3) in the sense of weak convergence of paths measures on $D(R,S'(R^d))$.

Theorem 2 (A. DeMasi, E. Presutti, H. Spohn and D. Wick [17]). Let $\xi^{\varepsilon}(f,t)$ be the density fluctuation field for exclusion processes with speed change. Let the jump rates be such that

$$L\eta_o = \sum_{m=1}^{M} \sum_x p_m(x)\tau_x h_m(\eta) \quad , \tag{5.1}$$

where τ_x is the shift by x, $p_m(\cdot)$ are functions on Z^d with $\sum_x p_m(x) = 0$, $\sum_x xp_m(x) = 0$, and h_m are local functions on $\{0,1\}^{Z^d}$. Let be either d=1 or the potential Φ be such that $e^{\lambda}e^{\|\Phi\|}(\exp(e^{\|\Phi\|}-1)-1) \leq 0.1$ with $\lambda = \Phi(\{0\})$ and $\|\Phi\| = \sum_{A\ni 0,|A|\geq 2} |\Phi(A)|$. Then with the appropriate χ and D (4.2) holds and (4.3) in the sense of weak convergence of paths measures on $D(R,S'(R^d))$.

Theorem 3 (H. Spohn [20]). Let $\xi^{\varepsilon}(f,t)$ be the density fluctuation field for interacting Brownian particles. Let V be finite range, $V(q)=V(-q)$, $V\in C^3$, $V\geq 0$ and $V(0)>0$ (or V superstable). Let the initial measure be the Gibbs measure for the potential V with fugacity $0<z<0.28e/dq(1-e^{-V(q)})$. Then with the appropriate choice of χ and D (4.2) holds and (4.3) in the sense of weak convergence on $C(R,S'(R^d))$.

6. The Technique

For the zero range process the proof simplifies because the Gibbs measures are product measures. On a technical level stochastic lattice gases are most easily handled but the gradient condition is somewhat unintuitive. For interacting Brownian particles the line of reasoning is clouded by many technicalities. I therefore decided to illustrate the method for the case of the Ginzburg-Landau model having the extra bonus of convincing the reader that this class of models should be added to the list of Section 5. My remarks should be taken with a grain of salt because not every step has been worked out in detail. To avoid technicalities let me assume a compact single site space, say $V(q) \to \infty$ as $|q| \to 1$, and $V \geq a$, $V \in C^3(]-1,1[)$. To save writing I assume $V(q) = V(-q)$, $\langle\phi_x\rangle = 0$ (i.e. zero magnetic field) and one dimension.

ϕ_t is the stochastic process governed by (2.8), (2.9) with initial measure the Gibbs measure (2.11) (h=0) in the infinite volume limit. Expectations with respect to ϕ_t are denoted by E and $\langle\ \rangle$ refers to expectations with respect to the Gibbs measure. $t \to \xi^{\varepsilon}(\cdot,t)$ (defined in (4.1) with m=0) is considered as an $S'(R)$-valued stochastic process. It has continuous sample paths and the path space is $C(R,S'(R))$. Its path measure is denoted by P^{ε}. First one has to establish tightness of the family $\{P^{\varepsilon}|0<\varepsilon\leq 1\}$ which will not be done here. A limit point of this family is denoted by P. The idea is to show that any P has to be the solution of a martingale problem. Independently one knows that the martingale problem has as unique solution the Ornstein-Uhlenbeck process with covariance (4.2) with the appropriate coefficients χ and D. This establishes then the desired result.

The martingales to be considered are constructed by the procedure standard for Markov processes. We have the two P^{ε}-martingales

$$
\begin{aligned}
M_1^{\varepsilon}(f,t) &= \xi^{\varepsilon}(f,t) - \int_0^t ds\ \varepsilon^{-2}\varepsilon^{1/2} \sum_x \Delta_1 f(\varepsilon x)\left(\frac{\partial H}{\partial \phi_x}\right)(\phi_{\varepsilon^{-2}s}) \\
M_2^{\varepsilon}(f,t) &= M_1^{\varepsilon}(f,t)^2 - 2t\varepsilon^{-2}\varepsilon\sum_x f(\varepsilon x)\ \Delta_1 f(\varepsilon x),
\end{aligned}
\tag{6.1}
$$

where $(\Delta_1 f)(x) = f(x+1) - 2f(x) + f(x-1)$ is the lattice Laplacian and $\frac{\partial H}{\partial \phi_x} = -\phi_{x+1} + 2\phi_x - \phi_{x-1} + V'(\phi_x)$. Note that $\frac{\partial H}{\partial \phi_x}$ is a local function on $[-1,1]^Z$ and may be written as $\tau_x(\partial H/\partial \phi_o)$ with τ_x the shift by x. We want to establish that with an appropriate constant D

$$
\begin{aligned}
M_1(f,t) &= \xi(f,t) - \int_o^t ds\xi(2Df'',s) \\
M_\alpha(f,t) &= M_1(f,t)^2 - 2t\int dq f(q)f''(q)
\end{aligned}
\tag{6.2}
$$

are P-martingales. Since the Gibbs measure < • > is a stationary Markov chain with a compact state space and a smooth transition probability, $\xi^{\varepsilon}(f,t=0)$ tends to Gaussian white noise with strength $\chi = \sum_x\langle\phi_x\phi_o\rangle$ as $\varepsilon \to 0$. With this initial condition the martingale problem (6.2) has a unique solution which is the infinite dimensional Ornstein-Uhlenbeck process of Section 4 [18].

Clearly for $M_2^{\varepsilon}(f,t)$ there is nothing to do. So we are left with the problem

$$
\lim_{\varepsilon \to 0} E\left(\left|\int_o^t ds\ \varepsilon^{1/2} \sum_x f''(\varepsilon x)\left\{\frac{\partial H}{\partial \phi_x}(\phi_{\varepsilon^{-2}s}) - 2D\phi_{\varepsilon^{-2}s,x}\right\}\right|\right) = 0\ . \tag{6.3}
$$

Life would be easy if the bound, having used stationary, $t\langle|\varepsilon^{1/2}\sum_x f''(\varepsilon x)\{(\partial H/\partial\phi_x) - 2D\phi_x\}|\rangle$ would tend to zero as $\varepsilon \to 0$. However $\varepsilon^{1/2}\sum_x f''(\varepsilon x)(\partial H/\partial\phi_x)$ and $\varepsilon^{1/2}\sum_x f''(\varepsilon x)\phi_x$ tend jointly to a non-degenerate Gaussian. Therefore some time-averaging has to be kept. To do this we break [0,t] into intervals of length $\varepsilon^2 T$ with T arbitrarily large. Then after some rearrangements one finds that (6.3) is implied by

$$\lim_{t\to\infty} \sum_x E\Big(\Big(\frac{\partial H}{\partial \phi_x}(\phi_t) - 2D\phi_{t,x}\Big)\Big(\frac{\partial H}{\partial \phi_o}(\phi_{t=o}) - 2D\phi_{o,o}\Big)\Big) = 0 . \tag{6.4}$$

(6.4) is the hard problem.

Let T_t be the Markov semigroup of the stochastic evolution, i.e. $(T_t\psi)(\phi) = E(\psi(\phi_t)|\phi_{t=o} = \phi)$. T_t is a continuous, self-adjoint contraction semigroup on $L^2([-1,1]^Z,<\cdot>) \equiv L^2$. Let D_o denote the set of local functions, i.e. depending only on finitely many ϕ_x's , which are C^2 with bounded derivatives. For $\psi_1, \psi_2 \in D_o$ with $<\psi_1> = 0 = <\psi_2>$ we define the scalar product

$$\begin{aligned} <\psi_1|\psi_2> &= \sum_x <\psi_1 \tau_x \psi_2> \\ &\lim_{\Lambda\to Z} <(|\Lambda|^{-1/2}\sum_{x\in\Lambda}\tau_x\psi_1)(|\Lambda|^{-1/2}\sum_{x\in\Lambda}\tau_x\psi_2)> . \end{aligned} \tag{6.5}$$

Let H be the closure of D_o with $<\cdot|\cdot>$. Note that H has a large null space. On an abstract level we may decompose $L^2 = 1 \oplus \oint dk H(k)$ as a direct integral according to the spectral representation of the unitary induced by the shift τ_1 . Then $H = H(0)$. Since T_t commutes with translations, it respects this decomposition. In particular, T_t is a continuous, self-adjoint contraction semigroup on H . With this definition we include (6.4) by

$$\lim_{t\to\infty} <\psi_1|T_t\psi_2> = <\psi_1|\chi^{-1/2}\phi_o> <\chi^{-1/2}\phi_o|\psi_2> , \tag{6.6}$$

where $\chi^{-1/2}$ ensures that $|\chi^{-1/2}\phi_o>$ has norm one in H . Given (6.6), in (6.4) we have to set $2D = <\partial H/\partial\phi_o>/\chi = 1/\chi$ as claimed.

By the spectral theorem (6.6) we have to establish then that the T_t-invariant subspace in H is one-dimensional. Note that because we work in H and not in L^2 (6.6) differs from the well known mixing.

How do we solve this dynamical problem ? By a miracle it can be reduced to a purely static, equilibrium problem.

We define the partial dynamics in the interval [m,n] by fixing the spins outside [m,n] and by setting $dj(m-1,m,t) = 0 = dj(n,n+1,t)$ in (2.9) The dynamics of the spins in [m,n] depends on ϕ_{m-1} and ϕ_{n+1}. We denote by $T_t^{[m,n]}$ the corresponding semigroup and by $L_{[m,n]}$ its generator. Note that $< \cdot >$ is invariant under every partial dynamics and that $T_t^{[m,n]}$ is a self-adjoint contraction semigroup in L^2 . Now for $\psi_1, \psi_2 \in D_o$ with $<\psi_1> = 0 = <\psi_2>$

$$\langle\psi_1|L_{[-n,n]}\psi_2\rangle^2 = \Big(\sum_{y=-n}^{n-1}\sum_x \langle(\tau_x\psi_1)(L_{[y,y+1]}\psi_2)\rangle\Big)^2$$

$$\leq 2n\sum_{y=-n}^{n-1}\langle(L^{1/2}_{[y,y+1]}\sum_x\tau_x\psi_1)\,(L^{1/2}_{[y,y+1]}\psi_2)\rangle^2$$

$$\leq 2n\sum_{y=-n}^{n-1}\sum_{x_1,x_2}\langle(\tau_{x_1}\psi_1)\,(L_{[y,y+1]}\tau_{x_2}\psi_1)\rangle\langle\psi_2 L_{[y,y+1]}\psi_2\rangle$$

$$= 2n\,\langle\psi_1|L\psi_1\rangle\langle\psi_2\,L_{[-n,n]}\psi_2\rangle \quad . \tag{6.7}$$

Because of the exponential decay of the Markov chain $\langle\cdot\rangle$ the sums are well defined.

We extend the inequality (6.7) to ψ_1's of the form $T_t\psi_1$ (to be able to do this D_o has to be a domain of essential self-adjointness of L in H. Under our hypothesis this can be proved) and let $t \to \infty$. Then by the spectral theorem the right hand side vanishes and on the left hand side $\lim_{t\to\infty} T_t\psi_1 = \psi_o \in PH$, the T_t-invariant subspace of H. Therefore for all $\psi_o \in PH$, $\psi_2 \in D_o$ and n

$$\langle\psi_o|L_{[-n,n]}\psi_2\rangle = 0 \quad . \tag{6.8}$$

By continuity then also $\langle\psi_o|T_t^{[-n,n]}\psi_2\rangle = \langle\psi_o|\psi_2\rangle$ and by again letting $t \to \infty$

$$\langle\psi_o|P_{[-n,n]}\psi_2\rangle = \langle\psi_o|\psi_2\rangle \quad . \tag{6.9}$$

$P_{[-n,n]}$ projects onto the $T_t^{[-n,n]}$invariant subspace in L^2 and therefore

$$P_{[-n,n]}\psi = \langle\psi\|\sum_{x=-n}^{n}\phi_x\,,\ \phi_{[-n,n]^c}\rangle \tag{6.10}$$

where $\langle\psi\|\cdot\rangle$ denotes conditional expectation and $\phi_{[-n,n]^c}$ the configuration in the complement of the interval [-n,n]. Therefore (6.6) holds provided we can show that for all $\psi \in D_o$

$$\lim_{n\to\infty} P_{[-n,n]}\,\psi = (1/\chi)|\phi_o\rangle\langle\phi_o|\psi\rangle \tag{6.11}$$

in H. (6.11) constitutes the reduction to a static problem which we aimed for.

Now we are on safe grounds. Since $\langle\cdot\rangle$ is a Markov chain particular methods are applicable. The reader should keep in mind however that this is so only for the sake of presentation. In general $\langle\cdot\rangle$ will be some

Gibbs measure with a finite range and sufficiently small potential (in particular $d \geq 1$). The method to be described refers to the general case. In fact one of the main tools will be the local central limit theorem for the magnetization which was one of the topics of the previous Hungarian meeting.

Let me reintroduce the magnetic field h as in (2.11). The corresponding Gibbs measure is denoted by $\langle \cdot \rangle_h$. Then $\langle \psi_o | \psi \rangle = \frac{\partial}{\partial h} \langle \psi \rangle_h |_{h=0} = \langle \psi \rangle'_o$. The magnetization per spin in $[-n,n]$ is abbreviated as $m_n = (1/2n+1) \sum_{x=-n}^{n} \phi_x$, $m_n = \phi_o$ in H. Because of the exponential decay of correlations the convergence of local functions in H may be reexpressed, at the expense of a volume factor, as convergence in L^2 which is easier to handle. Then (6.11) is implied by

$$\lim_{n \to \infty} (2n+1) \langle \left[\langle \psi || m_n, \phi_{-n-1}, \phi_{n+1} \rangle - (1/\chi) \langle \psi \rangle'_o m_n \right]^2 \rangle = 0 . \tag{6.12}$$

Since $\psi \in D_o$ for n large enough the dependence on the configuration outside $[-n,n]$ is only through ϕ_{-n-1} and ϕ_{n+1}. $\langle \psi || m_n, \phi_{-n-1}, \phi_{n+1} \rangle \to \langle \psi \rangle = 0$ as $n \to \infty$. Its dependence on m_n should be then the second term of (6.12) with an error $o(1/\sqrt{n})$ in L^2.

Clearly, one needs a good control on the dependence of $\langle \psi || m_n, \phi_{-n-1}, \phi_{n+1} \rangle$ on m_n. The idea is to smoothen this dependence by transfering it to the magnetic field,

$$\begin{aligned} \langle \psi || m_n, \phi_{-n-1}, \phi_{n+1} \rangle &\simeq \langle \psi || \phi_{-n-1}, \phi_{n+1} \rangle_{h(m_n, \phi_{-n-1}, \phi_{n+1})} \\ &\simeq \langle \psi \rangle_{h(m_n, \phi_{-n-1}, \phi_{n+1})} \quad , \end{aligned} \tag{6.13}$$

where $h(m, \phi_{-n-1}, \phi_{n+1})$ is implicitely defined by
$\langle m_n || \phi_{-n-1}, \phi_{n+1} \rangle_{h(m, \phi_{-n-1}, \phi_{n+1})} = m$.

For untypical values of m_n, $|m_n| > a > 0$, by the theory of large deviations (6.13) is exponentially small in n . The second equality in (6.13) reflects that grand canonical expectations depend exponentially little on the boundary conditions. The first equality, for typical values of m_n, is more delicate and requires a control over the error term in the local central limit theorem for the magnetization m_n . Given the right hand side of (6.13) we expand at $m_n = 0$. The first term in the expansion reproduces $(1/\chi) \langle \psi_o \rangle' m_n$ and the remainder has an L^2-norm of the order n^{-1}.

7. Open Problems

In this field it is very easy to state well-defined and physically interesting problems which are beyond any mathematical reach. As a realistic and hopefully reachable goal I propose to prove the large space-time behavior of equilibrium time correlations for stochastic evolution models on pars with the well established expansions in equilibrium statistical mechanics. We have made a first step in this direction but much remains to be done.

(i) How does one treat models where the gradient condition is violated ?

(ii) In equilibrium statistical mechanics at low temperatures one may expand around one of the ground states. How can this analysis be extended to time correlations ?

(iii) The models discussed here are purely dissipative, in probabilistic terms the dynamics is reversible as expressed by the symmetry of the transition probability. In many physical models the drift contains in addition so-called Hamiltonian or reversible (sorry for the terminology) parts. Examples come from fluids but also from Heisenberg spins which precess in addition to the dissipative drift $-\delta H/\delta\phi$. A related example are the interacting Ornstein-Uhlenbeck processes. Common feature of these models is that the transition probability is no longer symmetric. How does one prove the asymptotic behavior of time correlations in these models ?

References

[1] R.A. Holley and D.W. Stroock, Comm.Math.Phys. 48, 249 (1976)

[2] R.A. Holley and D.W. Stroock, Z. Wahrscheinlichkeitstheorie verw. Gebiete 35, 87 (1976)

[3] D. Wick, Comm.Math.Phys. 81, 361 (1981)

[4] H. Spohn, Large Scale Behavior of Equilibrium Time Correlation Functions for Some Stochastic Ising Models. In: Stochastic Processes in Quantum Theory and Statistical Physics, ed. S. Albeverio, Ph. Combe and M. Sirugue-Collin. Lecture Notes in Physics 173, p. 304. Springer, Berlin 1982

[5] A. De Masi, N. Ianiro, A. Pellegrinotti and A. Presutti, A Survey of the Hydrodynamical Behavior of Many-Particle Systems. In: Nonequilibrium Phenomena II, eds. J.L. Lebowitz and E.W. Montroll. North-Holland, Amsterdam 1984

[6] K. Kawasaki, Kinetics of Ising Models. In: Phase Transitions and Critical Phenomena, eds. C. Domb and M. Green, vol. 2. Academic Press, New York 1972

[7] T.M. Liggett, The Stochastic Evolution of Infinite Systems of Interacting Particles. Lecture Notes in Mathematics 598. Springer, Berlin 1978

[8] T.M. Liggett, Ann.Prob. 1, 240 (1973)

[9] P.N. Pusey and R.J.A. Tough, J.Phys. A15, 1291 (1982)

[10] P.N. Pusey and R.J.A. Tough in: Dynamic Light Scattering and Velocimetry : Applications of Photon Correlation Spectroscopy, eds. R. Pecora. Plenum Press, New York 1981

[11] J.W. Cahn and J.E. Hilliard, J.Chem.Phys. 28, 258 (1958)

[12] P.C. Hohenberg and B.I. Halperin, Rev.Mod.Phys. 49, 435 (1977)

[13] R. Lang, Z. Wahrscheinlichkeitstheorie verw. Gebiete 38, 55 (1977)

[14] T. Shiga, Z. Wahrscienlichkeitstheorie verw. Gebiete 47, 299 (1979)

[15] W.H. Faris, J.Funct.Anal. 32, 342 (1979)

[16] S. Katz, J.L. Lebowitz, H. Spohn, J.Stat.Phys. 34, 497 (1984)

[17] A. De Masi, E. Presutti, H. Spohn, D. Wick, Asymptotic Equivalence of Fluctuation Fields for Reversible Exclusion Processes with Speed Change, preprint

[18] R.A. Holley and D.W. Stroock, RIMS Kyoto Publications A14, 741 (1978)

[19] T. Brox and H. Rost, Ann.Prob. to appear

[20] H. Spohn, Equilibrium Fluctuations for Interacting Brownian Particles, preprint

*
Theoretische Physik, Universität München, Theresienstrasse 37, 8 München 2, FRG

LINEAR AND RELATED MODELS OF TIME EVOLUTION IN QUANTUM STATISTICAL MECHANICS

A. G. Shuhov and Yu. M. Suhov

1. Introduction

The aim of this paper is to give a review of some recent results concerning the problem of constructing and studying time evolution for quantum systems with infinitely many degrees of freedom. The first rigorous results in this direction are due to O. E. Lanford and D. W. Robinson [17--20], see also [13], Ch. 5.3. An approach to this problem has been proposed by O. Bratteli and D. W. Robinson [12], see also [13], Ch. 6.3. Among recent papers we refer to [1-3], [10-11], [16], [24].

Now the problem of construction of the time evolution of a state of a quantum system is solved in a satisfactory way for the classes of lattice spin and lattice fermion systems where the evolution comes from a group of *-automorphisms of an appropriate C^*-algebra (time dynamics). The main feature of these systems is that they are "locally finite-dimensional". Other classes of realistic quantum systems such as quantum particle systems in euclidean space R^{ν} (both statistics) or lattice boson system are "locally infinite-dimensional"; for these classes the problem of constructing the time dynamics is solved only for a degenerated class of models ("linear" fermion models in R^{ν}) and perhaps has no positive solution in general.

In our context, the problem of studying the time evolution is reduced to the question: what is the limit behavior of the time-evolved state Q_t, $t \in R^1$, when $t \to \pm\infty$? This question is of very great interest for understanding the approach to equilibrium and thereby for the mathematical foundation of the Boltzmann-Gibbs postulate in Statistical Mechanics. Now the problem of convergence to equilibrium seems to be very hard and is solved only for a number of simple models: linear models (both statistics), one dimensional X-Y model and one dimensional hard rod model. This list is probably not a final one: in Section 5 we shall briefly discuss some conjectures which seem to be realistic.

2. Preliminaries

Let V be a complex separable Hilbert space, $H^{\pm}=H^{\pm}_{V}$ be the Fock space constructed over V

$$H^{\pm} = \oplus_{n=0}^{\infty} H_n^{\pm} , \quad H_n^{\pm} = (V^{\otimes n})_{\pm} , \quad n \in Z_+^1 \tag{1}$$

($\pm$ indicate the statistics and are omitted whenever there will be no confusion). As usually, $a^+(f)$, $a(g)$, $f,g \in V$, denote the creation and annihilation operators in H which obey the CCR (+) and CAR (-) . We shall have in mind two concrete realisations: $V=\ell_2(Z^\nu)$ (lattice systems) and $V = L_2(R^\nu)$ (continuous systems). By V^o we denote $\ell_2^o(Z^\nu)$ and $L_2^o(R^\nu)$, respectively (sequences and functions with compact support). In the usual way, one introduces CCR and CAR C^*-algebras $\mathfrak{A}^{\pm}$ over V^o (in the fermion case V^o may be replaced by V). The action of the *-automorphisms group $\{S_x\}$ of space translations on $\mathfrak{A}$ is defined by

$$S_x a^{\cdot}(h) = a^{\cdot}(s_x h) , \quad h \in V^o , \quad x \in Z^\nu \text{ or } R^\nu , \tag{2}$$

where $s_x h(\cdot) = h(\cdot - x)$

As usually, by a state of a C^*-algebra we mean a linear non-negative normalized functional. In a standard way one introduces translation invariant (t.i.), ergodic, even, gauge invariant (g.i.), locally normal (ℓ.n.) and quasi-free states (see, e.g. [13]). In the boson case one uses the notions of a C^∞-state and analytic state (see [13]). For instance, for a C^∞-state Q the values

$$Q(\Pi_{j=1}^{m} a^+(f_j) \Pi_{k=1}^{n} a(g_k)), \ f_j, g_k \in V^o, \ j=1,\dots,m; \ k=1,\dots,n; \ m,n \in Z_+^1 \tag{3}$$

are defined and for an analytic state Q these values uniquely determine the state Q itself (in the fermion case this is given for all states by definition). The values (3) define the so-called (m,n) -moment functionals on $(\mathcal{V}^{o})^{\otimes(m+n)}$ (moment form) $K_Q^{(m,n)}$, $m,n \in Z_+^1$, of a state Q, which have the obvious properties of sesquilinearity and positive definiteness. The functionals $K_Q^{(m,n)}$ with $m+n \leq 2$ are called the lower moment functionals (in the case of a quasi-free state they determine uniquely the whole state Q). In the fermion case the functional $K_Q^{(m,n)}$ may be extended to $\mathcal{V}^{\otimes(m+n)}$ and hence is bounded ($\|K_Q^{(m,n)}\| \leq 1$, $m,n \in Z_+^1$).

A convenient topology on a set of states is defined by means of convergence of the (m,n) -moment functionals $K_Q^{(m,n)}$. In the fermion case this is merely the w^* -topology.

In the sequel we shall intensively use "weak dependence" properties of a given state Q. For $m,n \in Z_+^1$ and two orthogonal subspaces $\mathcal{V}^{(1)}, \mathcal{V}^{(2)} \subset \mathcal{V}$ we set

$$\alpha_Q^{(m,n)}(\mathcal{V}^{(1)},\mathcal{V}^{(2)}) = \max^{(m_1,n_1)} \sup^{(\mathcal{V}^{(1)},\mathcal{V}^{(2)})}$$

$$\Big| K_Q^{(m,n)}(f_1^{(1)},\ldots,f_{m_1}^{(1)},f_1^{(2)},\ldots,f_{m-m_1}^{(2)};\, g_1^{(2)},\ldots,g_{n-n_1}^{(2)},g_1^{(1)},\ldots,g_{n_1}^{(1)}) -$$
$$-K_Q^{(m_1,n_1)}(f_1^{(1)},\ldots,f_{m_1}^{(1)};g_1^{(1)},\ldots,g_{n_1}^{(1)})K_Q^{(m-m_1,n-n_1)} \times \tag{4}$$
$$\times (f_1^{(2)},\ldots,f_{m-m_1}^{(2)};g_1^{(2)},\ldots,g_{n-n_1}^{(2)}) \Big|$$

The maximum in RHS of (4) is taken over all $m_1=0,\ldots,m$ and $n_1=0,\ldots,n$ with $m_1+n_1 > 0$, the supremum is taken over all vectors $f_1^{(1)},\ldots,f_{m_1}^{(1)}$, $g_1^{(1)},\ldots,g_{n_1}^{(1)} \in \mathcal{V}^{(1)}$, $f_1^{(2)},\ldots,f_{m-m_1}^{(2)},g_1^{(2)},\ldots,g_{n-n_1}^{(2)} \in \mathcal{V}^{(2)}$ with norm ≤ 1.

The following case is particularly interesting: $\mathcal{V}^{(1)} = \ell_2(I(x,r))$ or $L_2(I(x,r))$, $\mathcal{V}^{(2)} = \ell_2(\mathcal{C}I(x,s+r))$ or $L_2(\mathcal{C}I(x,s+r))$ where $x \in Z^\nu$ or R^ν, $s,r>0$ and $I(x,u)$ is the cube on Z^ν or in R^ν centered at x with the edges parallel to coordinate axes and the edge length u.

In this case we set

$$\alpha_Q^{(m,n)}(r,s) = \sup \alpha_Q^{(m,n)}(V^{(1)}, V^{(2)})$$

where sup is taken over all subspaces of the form mentioned above where s, r are fixed and $x \in Z^\nu$ or R^ν is arbitrary. The quantity $\alpha_Q^{(m,n)}(r,s)$ characterizes the decrease of the "space correlations" in the state Q.

3. Linear models (groups of Bogoliubov transformations): the fermion systems

Let $T^{(1)}$ be a bounded linear operator and $T^{(2)}$ a bounded antilinear operator in V satisfying the relations

$$(T^{(2)})^* T^{(1)} + (T^{(1)})^* T^{(2)} = T^{(1)}(T^{(2)})^* + T^{(2)}(T^{(1)})^* = 0 \qquad (5)$$

$$(T^{(2)})^* T^{(2)} + (T^{(1)})^* T^{(1)} = T^{(1)}(T^{(1)})^* + T^{(2)}(T^{(2)})^* = \mathbb{1} \qquad (6)$$

Then the formula

$$\mathcal{T}a^+(f) = a^+(T^{(1)}f) + a(T^{(2)}f), \quad \mathcal{T}a(g) = a^+(T^{(2)}g) + a(T^{(1)}g) \qquad (7)$$

defines a $*$-automorphism $\mathcal{T}$ of the C^*-algebra $\mathbb{A}^-$. This automorphism is called the Bogoliubov transformation, or linear canonical transformation (LCT).

Assume that $\mathbb{T}_t = (T_t^{i,j})$, $i,j = 1,2$, $t \in R^1$, is a group of bounded operator (2×2)-matrices with $T_t^{1,1} = T_t^{2,2} = T_t^{(1)}$, $T_t^{1,2} = T_t^{2,1} = T_t^{(2)}$, satisfying (5)-(6). Such a group generates the corresponding group of Bogoliubov transformations (LCT) $\{\mathcal{T}_t, t \in R^1\}$. This is the time dynamics under consideration. We shall suppose that the group $\{\mathcal{T}_t\}$ commutes with space translations S_x on $\mathbb{A}^-$, $x \in Z^\nu$ or R^ν. Given a state Q, we define

$$\mathcal{T}_t{}^*Q(A) = Q(\mathcal{T}_{-t}A), \quad A \in \mathbb{A}^- .$$

Our aim is to examine time-evolved states $\mathcal{T}_t{}^*Q$ as $t \to \pm\infty$.

Consider the following condition (A) on the initial state Q and the group of operator matrices $\{\mathbb{T}_t\}$ generating Bogoliubov transformations.

(A) For every $t \in R^1$ there exists a finite or countable orthogonal decomposition $V = V_0^{(t)} \oplus V_1^{(t)} \oplus \dots$ such that

A1) for all $m,n \in Z_+^1$ with $m+n \geq 2$

$$\lim_{t\to\pm\infty} \sum_{\ell\geq 1} c_t^{(m,n)}(\ell) = 0 ,$$

where

$$c_t^{(m,n)}(\ell) = \alpha_Q^{(m,n)}(V_\ell^{(t)}, \oplus_{\ell':\ell'\neq 0,\ell} V_{\ell'}^{(t)}) ,$$

A2) for every vector $h \in V^0$

$$\lim_{t\to\pm\infty} \sup_{\ell\geq 0} d_t(h,\ell) = 0 ,$$

where

$$d_t(h,\ell) = \max_{æ=1,2} \|\pi_{V_\ell^{(t)}} T_{-t}^{(æ)} h\| ,$$

$\pi_{V_\ell^{(t)}}$ is the orthogonal projection $V \to V_\ell^{(t)}$,

A3) for all $m,n \in Z_+^1$ with $m+n \geq 3$ and all $f_1,\dots,f_m, g_1,\dots,g_n \in V^0$

$$\lim_{t\to\pm\infty} \sum_{\ell\geq 1} e_t(f_1,\dots,f_m; g_1,\dots,g_n; \ell) = 0 ,$$

where

$$e_t(f_1,\dots,f_m; g_1,\dots,g_n; \ell) =$$

$$= \max_{\substack{æ_j, æ'_k \\ 1\leq j\leq m,\ 1\leq k\leq n}} \Pi_{j=1}^m \|\pi_{V_\ell^{(t)}} T_{-t}^{(æ_j)} f_j\| \Pi_{k=1}^n \|\pi_{V_\ell^{(t)}} T_{-t}^{(æ'_k)} g_k\| .$$

<u>Theorem 1</u> ([25]). Let an initial state Q and a group $\{\mathbb{T}_t\}$ satisfy condition (A). Then the states $T_t{}^*Q$, $t\in R^1$, converge to a quasifree state P as $t\to\pm\infty$ iff the lower moment functionals $K_{T_t{}^*Q}^{(æ_1,æ_2)}$, $æ_1+æ_2 = 1,2$, converge to the corresponding functionals $K_P^{(æ_1,æ_2)}$. □

This theorem reduces the problem of convergence of states $T_t{}^*Q$ to the more simple question about convergence of the lower moment functionals. The question of convergence for lower moment functionals may be solved separately (see Subsection 3.4). However the conditions of Theorem 1 are imposed on the pair $(Q, \{\mathbb{T}_t\})$, while from the physical point of view it is convenient to deal with separate conditions on Q and the group $\{\mathbb{T}_t\}$.

The group $\{\mathbb{T}_t\}$ is uniquely determined by its infinitesimal generators

$$B = \lim_{t\to 0} t^{-1}(T_t^{(1)} - \mathbb{1}) , \quad C = \lim_{t\to 0} t^{-1} T_t^{(2)} . \tag{8}$$

The Fourier transforms of these operators are given by the equalities

$$\hat{B}f(\theta) = b(\theta)f(\theta) , \quad \hat{C}f(\theta) = c(\theta)f(-\theta) \tag{9}$$

where b is a real and c an odd function on $[-\pi,\pi)^\nu$ or R^ν. Let

$$b_\pm(\theta) = \frac{1}{2}(b(\theta) \pm b(-\theta)) , \quad w(\theta) = (b_+^2(\theta) + |c(\theta)|^2)^{\frac{1}{2}} , \tag{10}$$

$$\omega_\pm(\theta) = b_-(\theta) \pm w(\theta) \tag{11}$$

The Fourier transform of the operator $T_t^{(1)}$ is the multiplication operator

$$\hat{T}_t^{(1)} f(\theta) = e^{itb_-(\theta)}\left[\cos(tw(\theta)) + \frac{ib_+(\theta)}{w(\theta)} \sin(tw(\theta))\right]f(\theta) ,$$

and that of $T_t^{(2)}$ is defined by the equality

$$\hat{T}_t^{(2)} f(\theta) = ie^{itb_-(\theta)} \frac{c(\theta)}{w(\theta)} \sin(tw(\theta))\bar{f}(-\theta) .$$

3.1. One-dimensional case

Let $\nu=1$. We shall suppose now that $b_+, c \in C^1$ and for some given $\mu_\varepsilon=2,3,\ldots$ the function $\omega_\varepsilon \in C^{\mu_\varepsilon+1}$ and the intersection

$$\beta(\omega_\varepsilon,\mu_\varepsilon+1) = \cap_{j=2}^{\mu_\varepsilon+1} \beta_j(\omega_\varepsilon) = \emptyset , \quad \varepsilon=\pm , \tag{12}$$

where

$$\beta_j(\omega) = \left\{\theta : \frac{d^j}{d\theta^j}\,\omega(\theta) = 0\right\} . \tag{13}$$

It is convenient to choose the minimal μ_ε with these properties. For $\mu_\varepsilon=2$ the above condition was formulated in [9]. As to the initial state, we shall suppose that for all $m,n \geq 0$ there exists $d = d(m,n) > 0$ such that

$$\lim_{s\to\infty} \sup_{r>0} s^d \alpha_Q^{(m,n)}(r,s) = 0 \tag{14}$$

Theorem 2 ([21]). Let $V = \ell_2(Z^1)$ or $L_2(R^1)$. If a group $\{\mathbf{T}_t\}$ and an initial state Q satisfy the above conditions, then the assertion of Theorem 1 holds. □

3.2. Multi-dimensional case: coranks of singularities

In this subsection we consider the general case $\nu\geq 1$. Suppose now that the functions $b,c \in C^\infty$. Set

$$\beta_2(\omega_\varepsilon) = \{\theta : \det d^2\, \omega_\varepsilon(\theta) = 0\} , \quad \varepsilon=\pm \quad . \tag{15}$$

Suppose that for all $y\in R^\nu$ the set $\{\theta:\operatorname{grad}\omega_\pm(\theta)=y\}$ consists of a finite number of points. Set

$$p = \max_{\varepsilon=\pm}\ \max_{\theta\in\beta_2(\omega_\varepsilon)}\ \operatorname{corank} d^2\, \omega_\varepsilon(\theta) \tag{16}$$

Theorem 3 ([21]). Suppose that $\nu>3p$ and the initial state Q satisfies the following condition: for all $m,n\geq 0$ and $j>0$

$$\lim_{s\to\infty} s^{\frac{3p\nu}{\nu-3p}}\, \alpha_Q^{(m,n)}\left(s, j\, s^{1-\frac{\nu p}{\nu-3p}}\right) = 0 \tag{17}$$

Then the assertion of Theorem 1 holds. □

Remark. In the lattice case $(V = \ell_2(Z^\nu))$ the set $\beta_2(\omega_\pm)$ is always non-empty. Hence, Theorem 3 is applicable only for $\nu\geq 4$. In the continuous case $(V=L_2(R^\nu))$ the assertion of the theorem remains true when the sets $\beta_2(\omega_\pm)=\emptyset$: in this case one sets $p=0$. If $\nu=1$ and $\beta_2(\omega_\pm)=\emptyset$, we can suppose that Q satisfies (14) with $d=0$ which is weaker than (17) with $p=0$.

3.3. Multi-dimensional case: local singularity types

In this subsection we shall use a more detailed information about the functions $\omega_\pm$ in a neighbourhood of the sets $\beta_2(\omega_\pm)$. Suppose that for every $\theta_o \in \beta_2(\omega_\pm)$ the function $F_\pm : \theta \mapsto -\theta \cdot \mathrm{grad}\, \omega_\pm(\theta_o) + \omega_\pm(\theta)$ has a simple or parabolic singularity at the point $\theta = \theta_0$ (the singularity of one of the types A_k $(k\geq 2)$, D_k $(k\geq 4)$, E_6, E_7, E_8, P_8, X_9, J_{10} (see, e.g., [5], [6] and the books [7-8])). It is convenient to introduce

$$\varepsilon(\omega_\pm) = \sup_{\theta_o \in \beta_2(\omega_\pm)} \varepsilon(\sigma(\theta_o)) . \tag{18}$$

Here $\sigma(\theta_o) = \sigma_\pm(\theta_o)$ is the singularity type of $F_\pm$ at θ_o , and $\varepsilon(\sigma)$ is the so-called index of singularity of type σ (see [4], [6] and the books [7-8]). For simple and parabolic singularities the index ε is given by the following table (see [6], [8], [14-15]):

σ	A_k	D_{k+1}	E_6	E_7	E_8	P_8	X_9	J_{10}
$\varepsilon(\sigma)$	$\frac{1}{2}-\frac{1}{k+1}$	$\frac{1}{2}-\frac{1}{2k}$	$\frac{5}{12}$	$\frac{4}{9}$	$\frac{7}{15}$	$\frac{1}{2}$	$\frac{1}{2}$	$\frac{1}{2}$

Theorem 4 ([21]). Suppose that $\nu > 6\varepsilon$ where $\varepsilon = \max \varepsilon(\omega_\pm)$ and an initial state Q satisfies (17) with p replaced by 2ε . Then the assertion of Theorem 1 holds. □

Remarks. 1. For $\nu \geq 6$ the above conditions on the functions $\omega_\pm$ describe a "generic" situation (in the class C^∞), see [6], [7].

2. If $\nu=2$ then the theorem is applicable only if the functions have a singularity of a type A_k , $2\leq k\leq 4$, at every $\theta_o \in \beta_2(\omega_\pm)$.

3.4. Convergence of the lower moment functionals

In this subsection we discuss briefly the question of convergence for lower moment functionals $K^{(m,n)}_{T_t{}^*Q}$, $m+n \leq 2$, as $t \to \pm\infty$. The reader can find the details in [21]. For the sake of simplicity we consider the one-dimensional case although there is no restriction on the space dimension.

First, suppose that the initial lower functionals $K_Q^{(m,n)}$, $m+n \le 2$, are S_x -invariant, $x \in Z^1$ or R^1. It is easy to see that in this case $K_Q^{(1,0)} = K_Q^{(0,1)} = 0$. By the Riesz theorem, one can write

$$K_Q^{(2,0)}(f,g) = \langle f, M_{1,1}^{(Q)} g\rangle , \quad K_Q^{(0,2)}(f,g) = \langle M_{2,2}^{(Q)} g, f\rangle$$
$$K_Q^{(1,1)}(f,g) = \langle f, M_{1,2}^{(Q)} g\rangle , \quad f,g \in V \tag{19}$$

where $M_{1,2}^{(Q)}$ is a bounded linear operator, $M_{1,1}^{(Q)}$ and $M_{2,2}^{(Q)}$ are bounded antilinear operators in V commuting with S_x, $x \in Z^1$ or R^1. In addition, it is convenient to use the linear operator $M_{2,1}^{(Q)} = \mathbb{1} - (M_{1,2}^{(Q)})^*$. Denote by $\mathbb{M}^{(Q)}$ the operator (2×2) -matrix with entries $M_{i,j}^{(Q)}$, $1 \le i,j \le 2$.

Let us introduce also the operator (2×2) -matrix $\mathbb{D} = (D_{i,j})$, $1 \le i,j \le 2$, with $D_{1,1} = D_{2,2} = B$, $D_{2,1} = D_{1,2} = C$ (this is the generator for the group $\{\mathbb{T}_t\}$, see (8)). Passing to the Fourier transform we write the matrix $\hat{\mathbb{D}}$ in the form

$$\hat{\mathbb{D}} = \hat{\mathbb{D}}_o + \hat{\mathbb{D}}_1 ,$$

where $\hat{\mathbb{D}}_o$ is the diagonal matrix which consists of the multiplication operators by the function b_+.

Suppose that for some given $\mu = 1,2,\ldots$ the function $w \in C^{\mu+1}$ and the intersection

$$\cap_{j=1}^{\mu+1} \beta_j(w) = \emptyset .$$

Then $\mathbb{M}^{(T_t{}^*Q)}$ converges weakly as $t \to \pm\infty$ to the limiting matrix $\mathbb{M}$ with the Fourier transform

$$\hat{\mathbb{M}} = \frac{1}{2}\hat{\mathbb{M}}^{(Q)} + \frac{1}{2w^2} (i\hat{\mathbb{D}}_1)^* \hat{\mathbb{M}}^{(Q)} (i\hat{\mathbb{D}}_1) .$$

Suppose now that initial lower functionals $K_Q^{(m,n)}$, $m+n \le 2$, have the following "periodicity" property: for some fixed $z \in Z_+^1$ or R_+^1

$K_Q^{(m,n)}(h) = K_Q^{(m,n)}(S_{zk}h)$, $m+n=1$, $h \in V$, $k \in Z^1$ and the operators $M_{i,j}^{(Q)}$, $1 \leq i,j \leq 2$, commute with the space translations S_{zk} (this also implies $K_Q^{(1,0)} = K_Q^{(0,1)} = 0$). Under some natural conditions of the non-degeneracy of the functions $\omega_\pm$ which are omitted from this text one can check that the matrix $\mathbb{M}^{(T_t*Q)}$ converges as $t \to \pm\infty$ to a limiting matrix $\mathbb{M}$. The limiting matrix $\mathbb{M}$ may be obtained from $\mathbb{M}^{(Q)}$ by some space-averaging procedures (which are closely related to periodicity properties of the functions $\omega_\pm$, see (11)). Compare with [9], theorem 4.2.

Finally, the periodicity of the initial lower functionals $K_Q^{(m,n)}$, $m+n \leq 2$, may be replaced by more general "almost periodicity" conditions. The same statements hold for the multi-dimensional case. In this case the functions $\omega_\pm$ should satisfy conditions which correspond to the conditions of Subsection 3.2.

In the one-dimensional case it is possible to construct examples of initial lower moment functionals $K_Q^{(m,n)}$, $m+n \leq 2$, for which

$$\lim_{t\to+\infty} K_{T_t*Q}^{(m,n)}(f,g) \neq \lim_{t\to-\infty} K_{T_t*Q}^{(m,n)}(f,g) \ , \ f,g \in V \ ,$$

(compare with [9], Theorem 4.1).

4. Linear models: the boson systems

The boson case is more difficult. First, a Bogoliubov transformation (7) created by a pair of operators $T^{(1)}$, $T^{(2)}$, in general, does not preserve the CCR C^* -algebra $\mathbb{A}^+$ (this is the case, whenever either $T^{(1)}V^o \not\subset V^o$ or $T^{(2)}V^o \not\subset V^o$ which is typical for physically interesting examples). However, a simple idea will allow us to introduce a linear time evolution $\{T_t*Q, t \in R^1\}$ of an initial state Q without using the time dynamics on $\mathbb{A}^+$.

Let us give some definitions. A C^∞ -state Q on $\mathbb{A}^+$ is called bounded if the moment functionals $K_Q^{(m,n)}$ may be extended to $V^{\otimes(m+n)}$ and hence are bounded, $m,n \in Z_+^1$. Let

$\mathbb{T}_t = (T_t^{i,j})$, $i,j=1,2$, $t \in R^1$, a group of bounded operator (2×2) -matrices be given where the operators $T_t^{1,1} = T_t^{2,2} = T_t^{(1)} : V \to V$ are antilinear and $T_t^{(1)}, T_t^{(2)}$ satisfy (5)-(6) . A bounded quasifree state P is called $\{\mathbb{T}_t\}$ -invariant if the lower moment functionals $K_P^{(m,n)}$, $m+n \le 2$, satisfy the relations

$$
\begin{aligned}
&K_P^{(1,0)}(f) = K_P^{(1,0)}(T_t^{(1)} f) + K_P^{(0,1)}(T_t^{(2)} f) , \\
&K_P^{(0,1)}(g) = K_P^{(1,0)}(T_t^{(2)} g) + K_P^{(0,1)}(T_t^{(1)} g) , \\
&K_P^{(2,0)}(f_1, f_2) = K_P^{(2,0)}(T_t^{(1)} f_1, T_t^{(1)} f_2) + K_P^{(1,1)}(T_t^{(1)} f_1, T_t^{(2)} f_2) + \\
&+ \langle T_t^{(1)} f_2, T_t^{(2)} f_1 \rangle + K_P^{(1,1)}(T_t^{(1)} f_2, T_t^{(2)} f_1) + K_P^{(0,2)}(T_t^{(2)} f_1, T_t^{(2)} f_2) ,
\end{aligned}
\tag{20}
$$

etc. These equalities express only the fact that the lower functionals do not change under the "transformations" (7). Such a definition seems to be reasonable since these relations imply that all (m,n) -moment functionals $K_P^{(m,n)}$ do not change under the "transformations" (7).

We say that a C^∞-state Q_1 is majorized by another C^∞-state Q_2 if for any $n \in Z_+^1$ and $f_1, \ldots, f_n \in V^\circ$

$$
K_{Q_1}^{(n,n)}(f_1,\ldots,f_n; f_n,\ldots,f_1) \le K_{Q_2}^{(n,n)}(f_1,\ldots,f_n; f_n,\ldots,f_1) \tag{21}
$$

A bounded state Q is called stable if it is majorized by a $\{\mathbb{T}_t\}$ -invariant state (which is called a $\{\mathbb{T}_t\}$ -invariant majorant for Q).

Let Q be a bounded state. We may define the family $\{(K_Q^{(m,n)})_t\}$, $m,n \in Z_+^1$, $t \in R^1$, of "time-evolved" moment functionals by using the formulas analogous to (20). In other words, the functionals $(K_Q^{(m,n)})_t$ describe the "time evolution" of the initial functionals $K_Q^{(m,n)}$ under the "transformations" (7). The precise definition is rather complicated and is given in [25]. The problem is to reconstruct the states Q_t , $t \in R^1$, for which $(K_Q^{m,n})_t = K_{Q_t}^{(m,n)}$. Such a family $\{Q_t, t \in R^1\}$ defines the linear time evolution of the initial state Q .

For the sake of simplicity we consider the particular case where $T_t^{(2)}=0$ (i.e., the matrix $\mathbb{T}_t$ is diagonal). We assume also that the operators $T_t^{(1)}$ commute with the space translations. Then any bounded t.i., g.i., quasi-free state P is $\{\mathbb{T}_t\}$ -invariant. Such a state is called regular. The only non-zero lower moment functional of the state P is $K_P^{(1,1)}$. The corresponding bounded linear operator in $\mathcal{V}$ is denoted, as in the fermion case, by $M_{1,2}^{(P)}$ (see (19)). Given a bounded $\Lambda\subset Z^\nu$ or R^ν, we denote by $\mathcal{V}(\Lambda)$ the subspace of $\mathcal{V}$ consisting of elements with support in Λ and by π_Λ the orthogonal projection $\mathcal{V}\to\mathcal{V}(\Lambda)$.

Theorem 5 ([25]). Let Q be a l.n., q.i., stable state with a $\{\mathbb{T}_t\}$ -invariant regular majorant P satisfying the following conditions: (a) $\|K_P^{(1,1)}\| < \frac{1}{2}$, (b) for any bounded Λ the operator

$$\sum_{n=1}^{\infty} 2^n(\pi_\Lambda\, M_{1,2}^{(P)}\, \pi_\Lambda)^n$$

is of trace class. Then for any $t\in R^1$ there exists a unique l.n., g.i., stable state Q_t such that $(K_Q^{(m,n)})_t = K_{Q_t}^{(m,n)}$, $m,n\in Z_+^1$. The mappings $\mathcal{T}_t^*\colon Q\mapsto Q_t$, $t\in R^1$, have the group property: $\mathcal{T}^*_{t_1+t_2} = \mathcal{T}^*_{t_1}\mathcal{T}^*_{t_2}$, $t_1,t_2\in R^1$. □

By construction, the states $\mathcal{T}_t^*Q$ depend continuously on $t\in R^1$.

Conditions (a) and (b) are sufficient, but not necessary. Roughly speaking, it means that the particle density in the regular state P is relatively small (for instance, in the case of the usual free time evolution we work outside the region of Bose condensation). An interesting question is to find weaker sufficient conditions.

Having the reconstruction theorem, one can repeat the analysis of convergence of the states $\mathcal{T}_t^*Q$ as $t\to\pm\infty$ given in the fermion case. We omit the details (see [25]).

5 Degenerated one dimensional models

In this section we discuss some quantum models with interaction. A natural conjecture is that in a "generic" situation the time-evolved state Q_t of a large particle system converges as $t \to \pm\infty$ to an equilibrium Gibbs state corresponding to the interaction potential (the Boltzmann-Gibbs postulate). For linear models, as we have seen in the preceding sections, such a conjecture is false: the set of limiting states consists not only of equilibrium Gibbs states but includes many other time-invariant states. This is related to the fact that a linear motion which corresponds to a quadratic Hamiltonian has many invariants (this is, in a sense, a simple example of quantum completely integrable systems).

Among the interacting systems there are models which may be reduced in some way to linear ones. We shall consider two classes of such models which are probably the simplest ones. The first class consists of the X-Y model and its generalizations, the second class is formed by hard-rod models in space and on lattice. Both classes are considered in one dimension. We believe that next examples of such type will be the one dimensional X-X-X model (isotropic spin $\frac{1}{2}$ Heisenberg model) and the one-dimensional continuous boson model with δ -interaction. All the models we have mentioned are examples of what is called now quantum completely integrable systems (moreover, ones are speaking about systems which are integrable by the (quantum) Lax method (the method of (L,A) -pair)). This list may probably be continued further on, but the corresponding models seem yet to be too hard for rigorous treatment.

5.1. The X-Y model and its generalizations

The C^* -algebra of a quantum spin system (spin $\frac{1}{2}$) on the lattice Z^1 is defined as the infinite tensor product

$$\mathfrak{M} = M^{\otimes Z^1}$$

where M is the complex (2×2) -matrix algebra. The local $*$ -subalgebra of $\mathfrak{M}$ is denoted by $\mathfrak{M}^o$. As usually, σ_j^x, σ_j^y, σ_j^z denote the Pauli matrices associated with the point $j \in Z^1$. Consider the derivation δ_{X-Y} of $\mathfrak{M}$ which is defined by

$$\delta_{X-Y} A = i[H_{X-Y}, A] , A \in \mathfrak{M}^o \tag{22}$$

where H_{X-Y} is the formal Hamiltonian of the X-Y model

$$H_{X-Y} = \sum_{j\in Z^1} (\alpha\sigma_j^x\sigma_{j+1}^x + \beta\sigma_j^y \sigma_{j+1}^y + h_o\sigma_j^z) , \alpha,\beta,h_o \in R^1 . \tag{23}$$

Although (23) is a divergent series, formula (22) is correct due to the locality of A. The group of $*$-automorphisms $\{\omega_t^{X-Y}, t\in R^1\}$ of $\mathfrak{M}$ generated by δ_{X-Y} describes the time dynamics in the X-Y model. If Q is an initial state of $\mathfrak{M}$, then its time evolution is given by

$$Q_t^{X-Y}(A) = Q(\omega_{-t}^{X-Y}A) , A\in\mathfrak{M} \tag{24}$$

The problem under consideration is: what happens with Q_t^{X-Y} as $t\to\pm\infty$?

Let ϕ be the automorphism of $\mathfrak{M}$ defined by the formula

$$\phi(A) = \Pi_{j=-\infty}^{\infty} \sigma_j^z A \Pi_{j=-\infty}^{\infty}\sigma_j^z , A\in\mathfrak{M}^o \tag{25}$$

(this definition is correct due to the locality of A). Let $\mathfrak{M}(+)$ be the $*$-subalgebra of $\mathfrak{M}$ consisting of ϕ-invariant elements.

In the analysis of the one-dimensional X-Y model the key role is played by the so-called Jordan-Wigner transformation. This transformation is formally written in the form

$$a_j^+ + a_j \leftrightarrow \Pi_{s<j} \sigma_s^z\sigma_j^x , \quad a_j^+ - a_j \leftrightarrow i\Pi_{s<j} \sigma_s^z\sigma_j^y , \quad j\in Z^1 ,$$

and generates the $*$-automorphism of C^*-algebra $\mathfrak{M}(+)$ and the even C^*-subalgebra $\mathcal{A}_{ev}^-$ of CAR C^*-algebra $\mathcal{A}^-$. This $*$-automorphism commutes with the space translations and establishes the one-to-one correspondence between the groups of $*$-automorphisms of both C^*-algebras as well as between their states. It is convenient to consider the induced correspondence between ϕ-invariant states of spin $\frac{1}{2}$ C^*-algebra $\mathfrak{M}$ and even states of CAR C^*-algebra $\mathcal{A}^-$. It is easy to check that the time-evolved state Q_t^{X-Y} (see (24)) is ϕ-invariant iff the initial state Q is ϕ-invariant. The transformation of Jordan and Wigner maps the X-Y dynamics into the LCT -dynamics which is defined by the infinitesimal generators B and C (see (7)-(8)) such that

$$b(\theta)=2[h_o-(\alpha+\beta)\cos\theta],\quad c(\theta)=2i(-\alpha+\beta)\sin\theta,\ \theta\in[-\pi,\pi)$$

We shall consider the more general situation. Let $\{\mathcal{U}_t,\ t\in R^1\}$ be any group of $*$-automorphisms of the C^*-algebra $\mathfrak{M}(+)$ which induces a group of LCT $\{T_t,\ t\in R^1\}$ of the C^*-algebra $\mathbb{A}^-$ under the Jordan-Wigner transformation. Given a ϕ-invariant state Q of the C^*-algebra $\mathfrak{M}$, we introduce the family of time-evolved states $\{\mathcal{U}_t^*Q,\ t\in R^1\}$.

Theorem 6. Suppose that a group $\{\mathcal{U}_t,\ t\in R^1\}$ of $*$-automorphisms of $\mathfrak{M}(+)$ satisfies the above condition and the corresponding group of LCT $\{T_t,\ t\in R^1\}$ of $\mathbb{A}^-$ satisfies conditions of Theorem 2. Let Q be a ϕ-invariant state of $\mathfrak{M}$ which corresponds to an even state Q^o of $\mathbb{A}^-$ satisfying conditions of the same theorem. Then the states $\mathcal{U}_t^*Q$ converge as $t\to\pm\infty$ to a ϕ-invariant state P of $\mathfrak{M}$ which corresponds to an even quasifree state P^o of $\mathbb{A}^-$ if for all $j_1,j_2\in Z^1$ with $j_1<j_2$

$$\lim_{t\to\pm\infty}\mathcal{U}_t^*Q\left(\sigma_{j_1}^{\cdot}\ \Pi_{s=j_1+1}^{j_2-1}\ \sigma_s^z\sigma_{j_2}^{\cdot}\right)=P\left(\sigma_{j_1}^{\cdot}\ \Pi_{s=j_1+1}^{j_2-1}\ \sigma_s^z\sigma_{j_2}^{\cdot}\right),\quad \cdot=x,y.\quad\square$$

The proof of this theorem is evident. Using the Jordan-Wigner transformation we reduce the problem of convergence for the state $\mathcal{U}_t^*Q$ to the problem of convergence for the state $T_t^*Q^o$. The latter is solved by Theorem 2.

Remark. For the proper X-Y model this theorem is applicable whenever $(\beta\alpha)^2+h_o^2\neq 0$.

In the same way we reduce the question of convergence for

$$\mathcal{U}_t^*Q\left(\sigma_{j_1}^{\cdot}\ \Pi_{s=j_1+1}^{j_2-1}\ \sigma_s^z\ \sigma_{j_2}^{\cdot}\right)$$

to the question of convergence for the lower moment functionals of the state $T_t^*Q^o$. See Subsection 3.4.

5.2. Hard rod model

In this subsection we discuss the one-dimensional continuous hard rod model. The lattice version of this model is simpler and is to be discussed later. The details and proofs of the theorems which follow may be found in [23].

Given $d>0$, we introduce the n-particle Hamiltonian of the model by the formula

$$H^{(n)}_{(d)} = -\frac{1}{2} \sum_{j=1}^{n} \frac{\partial^2}{\partial x_j^2} + \sum_{1\leq j_1<j_2\leq n} \Phi_{(d)}(|x_{j_1}-x_{j_2}|) \tag{26}$$

where $x_j \in R^1$, $j=1,\ldots,n$, and

$$\Phi_d(r) = \begin{cases} +\infty, & 0\leq r\leq d \\ 0, & r>d \end{cases} \tag{27}$$

is the pair interaction potential of hard rods of the length d. The operator (26) is considered in the subspace $H_n[d] \subset \subset H_n$ which consists of functions vanishing whenever $\min|x_{j_1}-x_{j_2}| < d$. We shall use the Dirichlet boundary conditions on the set

$$\{(x_1,\ldots,x_n) \in R^n : \min|x_{j_1}-x_{j_2}|=d\} \tag{28}$$

$H_{(d)}$ denotes the Hamiltonian of the system with arbitrary number of particles

$$H_{(d)} = \oplus_{n=0}^{\infty} H^{(n)}_{(d)} . \tag{29}$$

The first problem is to define the time evolution of a given initial state Q. For the model under consideration we are not able to construct the one-parameter group of $*$-automorphisms of C^*-algebra $\mathbb{A}$ nor its appropriate subalgebra (the time dynamics). As in Section 4, we proceed in a round-about way. For a given initial state Q which satisfies certain conditions we define a family of states $\{\mathcal{W}_t^{h.r.*}Q, t\in R^1\}$ and declare this family to be the time evolution of Q. A reason for this is that the states $\mathcal{W}_t^{h.r.*}Q$ give a (unique) solution of the Liouville equation. For precise statement we need some definitions and constructions.

Given $\Lambda \subseteq R^1$, we introduce the Fock space $H_{V(\Lambda)} = \oplus_{n=0}^{\infty} H_{n,V(\Lambda)}$ in the volume Λ (cf. with (1)). $H_{V(\Lambda)}[d]$ denotes the subspace of $H_{V(\Lambda)}$ consisting of vectors $f = (f_n, n \geq 0)$ such that

$$f_n(x_1, \ldots, x_n) = 0 \quad \text{whenever} \quad \min_{1 \leq j_1 < j_2 \leq n} |x_{j_1} - x_{j_2}| < d \; .$$

If $\Lambda = R^1$, we omit the index $V(\Lambda)$ in these notations.

It is convenient to identify both spaces $H^{\pm}_{V(\Lambda)}$ with the Hilbert space $L_2(C(\Lambda), \lambda_\Lambda)$ where $C(\Lambda)$ is the collection of finite subsets $X \subset \Lambda$ (including the empty set) endowed with the natural topology and the corresponding Borel σ-algebra $\sum_\Lambda$, λ_Λ is the Lebesgue-Poisson measure on $C(\Lambda)$. The subspace $H_{V(\Lambda)}[d]$ thereby coincide with the subspace of $L_2(C(\Lambda), \lambda)$ which consist of functions with support on the subset

$$C(\Lambda, d) = \{X \in C(\Lambda) : \min_{x, x' \in X : x \neq x'} |x - x'| \geq d\} \; .$$

Given $A \subset \sum_\Lambda$, we denote by $\Pi_{(A)}$ the orthogonal projector in $L_2(C(\Lambda), \lambda_\Lambda)$ onto the subspace of functions which vanish outside A.

The next object we need is the collection C of "locally finite" subsets of R^1, i.e., subsets $X \subset R^1$ having the property: $X \cap C$ is finite provided $C \subset R^1$ is a bounded set (it is clear that such X should be either finite or countable). The set C is endowed with the natural topology; the corresponding Borel σ-algebra is denoted by $\sum$. The σ-algebra $\sum$ is generated by the family $\{\sum^{(\Lambda)}\}$ where $\Lambda \subset R^1$ is a bounded (Borel) set and $\sum^{(\Lambda)}$ is the σ-subalgebra of $\sum$ generated by the map

$$X \in C \longmapsto X_\Lambda = X \cap \Lambda \; .$$

Clearly, $\sum^{(\Lambda)}$ is isomorphic to $\sum_\Lambda$ and we do not distinguish these σ-algebras further on.

Given $X \in C$, it is convenient to label the points $x \in X$ by integers j so that (a) index 0 is attached to the point of X with the smallest non-negative coordinate, (b) the indices of next neighbour points differ by 1,

(c) the indices of points increase with the coordinates. Correspondingly, we shall use the notation $x_j(X)$.

In the usual way one introduces the action of the space translation group $\{S_x, x\in R^1\}$ on C and the definition of translation invariant (t.i.) probability measure on $(C,\sum)$. Every such measure is uniquely determined by its restrictions on the σ-algebras $\sum_\Lambda$.

Now let Q be a l.n. state. Given a bounded $\Lambda\subset R^1$, we set

$$\mu_Q^{(\Lambda)}(A) = Q(\Pi_{(A)}) , \quad A \in \textstyle\sum_\Lambda .$$

This defines the probability measure $\mu_Q^{(\Lambda)}$ on $\sum_\Lambda$. The probability measure μ_Q on $(C,\sum)$ whose restrictions on $\sum_\Lambda$ coincide with $\mu_Q^{(\Lambda)}$ is called the diagonal measure of the state Q . It is clear that if a state Q is t.i., then the measure μ_Q is t.i., and if Q is ergodic, then μ_Q is ergodic. If Q is t.i., then the formula

$$k_Q^{(1)}(A) = \int_C \mu_Q(dX)\,\text{Card}\, X\cap A , \quad A\subset R^1 ,$$

defines the t.i. (Borel) measure $k_Q^{(1)}$ on R^1 which has the form

$$k_Q^{(1)}(dx) = \alpha_Q\cdot dx ,$$

where $\alpha_Q \in \bar{R}_+^1 = [0,+\infty]$. The number α_Q is called the particle density of the state Q .

A l.n. state Q is called d-admissible if the diagonal measure μ_Q is concentrated on the set $C[d]$ where

$$C[d] = \Big\{X\in C : \text{Card}\, X\cap R_+^1 = \text{card}\, X\cap R_-^1 = \infty ,$$
$$x_j(X)-x_{j-1}(x) > d \quad \text{for all} \quad j\in Z^1 \quad \text{and}$$
$$\textstyle\sum_{k<0}(x_k(X)-x_{k-1}(X)-d) = \sum_{k>0}(x_k(X)-x_{k-1}(X)-d) = \infty\Big\} .$$

It is clear that if a t.i. state Q is d-admissible, then $\alpha_Q<d^{-1}$.

Given a t.i. d-admissible ergodic state Q , one can define a new t.i. ergodic state $\mathbb{C}Q$ with $\alpha_{\mathbb{C}Q}=d\alpha_Q(1-d\alpha_Q)^{-1}$

which is called the contraction of the state Q . Conversely, if Q is a t.i. ergodic state with a finite particle density α_Q , then one can define a t.i. d-admissible ergodic state $\mathbb{D}Q$ with $\alpha_{\mathbb{D}Q} = d\alpha_Q(1+d\alpha_Q)^{-1}$ which is called the dilatation of the state Q . The maps $\mathbb{C}$ and $\mathbb{D}$ are mutual inverses:

$$\mathbb{C}\mathbb{D}Q = Q \ , \ \mathbb{D}\mathbb{C}Q = Q \ .$$

The precise definition of the maps $\mathbb{C}$ and $\mathbb{D}$ may be found in [23]. We want to emphasize that these maps are not induced by any transformation of the C^*-algebra $\mathbb{A}$.

In addition, one can introduce a natural map Θ which transforms states of CAR C^*-algebra $\mathbb{A}^-$ into states of CCR C^*-algebra $\mathbb{A}^+$ (see [23]). The definition of all the maps $\mathbb{C}$, $\mathbb{D}$ and Θ essentially uses the one-dimensional structure of the system.

In what follows we denote by $\mathbb{T}_t^{free}$ the operator (2×2)-matrix with $T_t^{(2)}=0$, $T_t^{(1)} = \exp(it\frac{1}{2}\Delta)$, $t\in R^1$, where $\Delta = \frac{\partial^2}{\partial x^2}$ is the Laplacian in $V = L_2(R^1)$. Correspondingly, T_t^{free} , $t\in R^1$, denotes the corresponding Bogoliubov transformations of CAR C^*-algebra $\mathbb{A}^-$. Let Q be a t.i., d-admissible ergodic state of $\mathbb{A}^-$. We set

$$W_t^{h.r.*}Q = \mathbb{D}T_t^{free*}\mathbb{C}Q \ , \ t\in R^1 \ ,$$

and call the family of states $\{W_t^{h.r.*}Q\}$ the hard rod time evolution of Q . If Q is t.i., d-admissible ergodic state of CCR C^*-algebra $\mathbb{A}^+$, the corresponding definition is as follows:

$$W_t^{h.r.*}Q = \Theta\mathbb{D}T_t^{free}\,\mathbb{C}\Theta^{-1}Q \ .$$

Theorem 7 ([23]). The mappings $W_t^{h.r.*}$, $t\in R^1$, transform the set of t.i., d-admissible, ergodic states into itself. They are continuous and have the group property:

$$W_{t_1+t_2}^{h.r.*} = W_{t_1}^{h.r.*}\,W_{t_2}^{h.r.*} \ , \ t_1,t_2 \in R^1 \ .$$

The states $W_t^{h.r.*}Q$ depend continuously on $t \in R^1$. □

Our definition is justified by the following fact. One can choose subsets $\mathfrak{D}^{\pm} \subset \mathfrak{A}^{\pm}$ having the property that any t.i., d-admissible, ergodic state is uniquely determined by its restriction onto $\mathfrak{D}^{\pm}$ and are such that for any $A \in \mathfrak{D}^{\pm}$ and an above-mentioned state Q the function

$$t \in R^1 \longmapsto \mathcal{W}_t^{h.r.*} Q(A)$$

is of class C^1 and satisfies the Liouville equation

$$\frac{d}{dt} \mathcal{W}_t^{h.r.*} Q(A) = \mathcal{W}_t^{h.r.*}([H_{(d)}, A]) .$$

Moreover, this is a unique (in some sense) solution (for the details see [23]).

Now we turn to the problem of convergence when $t \to \pm\infty$. A natural candidate for the role of invariant state is a state P of the form $\mathbb{D}P^o$ in fermion case and $\Theta^{-1}\mathbb{D}P^o$ in boson case where P^o is a t.i., g.i. quasifree state (which is invariant with respect to the free time dynamics $\{T_t^{free}\}$). Such a state P will be called d-equilibrium. It is completely determined by its $(1,1)$-moment functional $K_P^{(1,1)}$.

Theorem 8 ([23]). Let Q be a t.i., d-admissible, ergodic state. Suppose that the contraction $\mathbb{C}Q$ in fermion case and the state $\mathbb{C}\Theta Q$ in boson case satisfies condition (14). Then the states $\mathcal{W}_t^{h.r.*}Q$, $t \in R^1$, converge as $t \to \pm\infty$ to the d-equilibrium state P defined by the relation

$$K_P^{(1,1)} = K_Q^{(1,1)} .$$

Proceeding in the same way one can treat the lattice version of the hard rod model. The Hamiltonian (26) is replaced by its discrete analog here. The main feature of the lattice hard rod model is that the time evolution may be constructed in a "direct" way for any d-admissible state (although we are not able to construct any reasonable time dynamics). The details will appear in [22].

6. Concluding remark: the validity of conditions of Theorems 1-8

The problem to give reasonable examples of states which satisfy conditions of theorems formulated above is not only of academic interest. It is connected with the physical background of the approach we adopted in this paper. We shall not dwell on this problem here with all the details. The main class of such examples is constituted by the so-called Gibbs states. Up to now the theory of Gibbs states has been developed mainly for spin systems (see, e.g. [17-20]) using the KMS boundary condition. For the "locally infinite-dimensional" systems there are only particular results. However, even for particular Gibbs states constructed so far the problem to verify conditions of theorems under consideration is not trivial. This is especially true for theorems from Subsection 5.2. For the details see the references cited above.

References

[1] Anshelevich V. V. First integrals and stationary states of quantum spin dynamics of Heisenberg. (Russian) Teoret. Mat. Fiz. 43, no. 1, 107-110 (1980).

[2] Anshelevich V. V., Gusev E. V. First integrals of one dimensional quantum Ising model with diametrical magnetic field. (Russian) Teoret. Mat. Fiz. 47, no. 2, 230-242 (1981).

[3] Araki H., Barouch E. On the dynamics and ergodic properties of the X-Y model. J. Stat. Phys. 31, no. 2, 327-346 (1983).

[4] Arnold V. I. Integrals of quickly oscillating functions and singularities of projections of Lagrange manifolds. (Russian) Funktsional. Anal. i Prilozhen. 6, no. 3, 61-62 (1972).

[5] Arnold V. I. Normal forms of functions near degenerated critical points, Weyl groups A_k, D_k, E_k and Lagrange singularities. (Russian) Funktsional. Anal. i Prilozhen. 6, no. 4, 3-25 (1972).

[6] Arnold V. I. Remarks on stationary phase method and Coxeter numbers. (Russian) Uspekhi Mat. Nauk 28, no. 5, 17-44 (1973).

[7] Arnold V. I., Varchenko A. N., Gussein-Zadeh S. M. Singularities of differentiable mappings, vol. I. (Russian) "Nauka", Moscow, 1982.

[8] Arnold V. I., Varchenko A. N., Gussein-Zadeh S. M. Singularities of differentiable mappings, vol. II. (Russian) "Nauka", Moscow, 1984.

[9] Boldrighini C., Pellegrinotti A., Triolo L. Convergence to stationary states for infinite harmonic systems. J. Stat. Phys. 30, no. 1, 123-155 (1983).

[10] Botvich D. D. Spectral properties of fermion dynamical systems. (Russian). Diss. Thesis. Moscow State University (M. V. Lomonosov), 1983.

[11] Botvich D. D., Malyshev V. A. Unitary equivalence of temperature dynamics for ideal and locally perturbed Fermi-gas. Commun. Math. Phys. 91, 301-312 (1983).

[12] Bratelli O., Robinson D. W., Green's functions, Hamiltonians and modular automorphisms. Commun. Math. Phys. 50, 133-156 (1976).

[13] Bratelli O., Robinson D. W. Operator algebras and quantum statistical mechanics, vol. II. Springer-Verlag, New York - Heidelberg - Berlin, 1981.

[14] Colin de Verdière Y. Nombre de points entiers dans une famille homothétique de domaines de R^h. Ann. Scient. École Norm. Sup. 4ême ser. 10 no. 4, 559-576 (1977).

[15] Duistermaat J. T. Oscillatory integrals, Lagrange immersions and unfoldings of singularities. Commun. Pure Appl. Math. 27, no. 2, 207-281 (1974).

[16] Gusev E. V. First integrals of some stochastic operators of statistical physics. (Russian) Diss. Thesis, Moscow State University (M. V. Lomonosov), 1982.

[17] Lanford O. E., Robinson D. W. Statistical mechanics of quantum spin systems III. Commun. Math. Phys. 327-338 (1968).

[18] Lanford O. E., Robinson D. W. Approach to equilibrium of free quantum systems. Commun. Math. Phys. 24, 193-210 (1972).

[19] Robinson D. W. Statistical mechanics of quantum spin systems. Commun. Math. Phys. 6, 151-160 (1967).

[20] Robinson D. W. Statistical mechanics of quantum spin systems II. Commun. Math. Phys. 7, 337-348 (1968).

[21] Shuhov A. G., Suhov Yu. M. Ergodic properties of groups of Bogoliubov transformations of CAR C*-algebras. (Russian) Izv. Akad. Nauk SSSR, ser. Mat. (to appear).

[22] Shuhov A. G., Suhov Yu. M. in preparation.

[23] Suhov Yu. M. Convergence to equilibrium state for one dimensional quantum system of hard rods. (Russian) Izv. Akad. Nauk SSSR, ser. Mat. 46, no. 6, 1274-1315 (1982).

[24] Suhov Yu. M. Convergence to equilibrium for free Fermi--gas. (Russian) Teoret. Mat. Fiz. 55, no. 2, 282-290 (1983).

SYSTEMS WITH RANDOM COUPLINGS ON DIAMOND LATTICES

P. Collet

I. Introduction

It is by now well established that renormalization group techniques provide a very powerful tool for the study of models in statistical mechanics. These ideas were first introduced to study the problem of phase transitions, but they have proven to be useful in many other contexts like the high temperature phase in constructive quantum field theory, dynamical systems, turbulence, etc... The main difficulty in a renormalization group analysis is to derive the renormalization transformation (sometimes called the renormalization group for historical reasons). This is sometimes so difficult that one has to abandon the idea of using the simple general scheme which needs a lot of informations about this transformation. Migdal and Kadanoff tried to develop a general technique of approximation, the main idea being to obtain a tool to study gauge invariant problems. One of the advantage of their approximation is that the renormalization transformation can be written explicitly. This is very convenient for numerical and theoretical considerations (see for example [4]). Unfortunately, it is very difficult to appreciate the quality of the approximation.

Nevertheless, it was soon realised that one can construct models of statistical mechanics on so-called diamond lattices for which the Migdal Kadanoff renormalization transformation is exact. This idea has been applied to obtain rigorous results in various areas of physics where they are otherwise very hard to get. It is believed that regular lattices have properties which are not too different from those of diamond lattices. Without trying to be exhaustive, we can list a few applications. Percolation problems were considered in [3] , while problems associated with the Schrödinger equation were investigated in [19] , [23] , and [11]. The interested reader may also refer to the litterature about fractal lattices for very similar techniques. Another important area is the theory of spin glasses. Models with complicated (chaotic) phase diagrams were first constructed in [18] (see also [14] and [10]). It was then realised by different groups that one can also treat models with random couplings. Numerical computations were performed by Benyoussef and Boccara [2]. Their result suggests a lower critical dimension equal to 4. Pott's like models were considered in [1] , and Ising like models in [15] , [5] , [7] and [8] . In the remainder of the present paper we shall concentrate on the study of these Ising like models.

In chapter II we shall explain the construction of the diamond lattices and of the two models we have considered (model A and model B). Chapter III will be devoted to the renormalization analysis of model A, and chapter IV to the study of some spin observables. Chapter V contains some informations about model B.

II. Construction of the models

We shall first construct the diamond lattice (see [16]). We shall choose once for all an integer $n \geq 2$. The construction of the lattice is recursive (see the figure). To explain the basic induction step, we start with a lattice composed of one bond and two sites A and B. We transform this lattice into a lattice with n+2 sites $A, B, B_1, \ldots B_n$, and 2n bonds which connect the sites $(B_i)_{1 \leq i \leq n}$ to the sites A and B. We now apply this basic process to each bond previously constructed. We shall denote by L_N the lattice obtained after N successive applications of the basic construction.

If $n=2^d$, $d \in \mathbb{N}$, it is possible to consider the above lattice as a regular $\mathbb{Z}^d$ lattice where some of the sites have been identified. This is done as follows. Let $\underline{e}$ be a direction of bond in the $\mathbb{Z}^d$ lattice. We only keep those bonds which are parallel to $\underline{e}$. We consider a parallelipipedic box B_N of size 2^N in the $\underline{e}$ direction and 2^{N-1} in the other $\mathbb{Z}^d$ directions. Let i_1 be the $\underline{e}$ coordinate of a lattice point $\underline{i}$ inside the box. We shall assume that $0 \leq i_1 \leq 2^N - 1$. Let r be the largest integer such that 2^r divides i_1. Then the set of points in the box B_N with $\underline{e}$ coordinate i_1 is divided into $2^{(N-r)(d-1)}$ hypercubes of dimension d-1 and size 2^r, and the sites inside each hypercube are identified. In this way we obtain the lattice L_{N-1}. We now describe two models with random couplings which we shall call model A and model B

Model A

To each site $\underline{i}$ of the lattice L_N we associate an Ising variable $S_i = \pm 1$, and to each bond $(\underline{i}, \underline{j})$ we associate a random variable $\xi_{\underline{i},\underline{j}}$ which is an independant copy of a

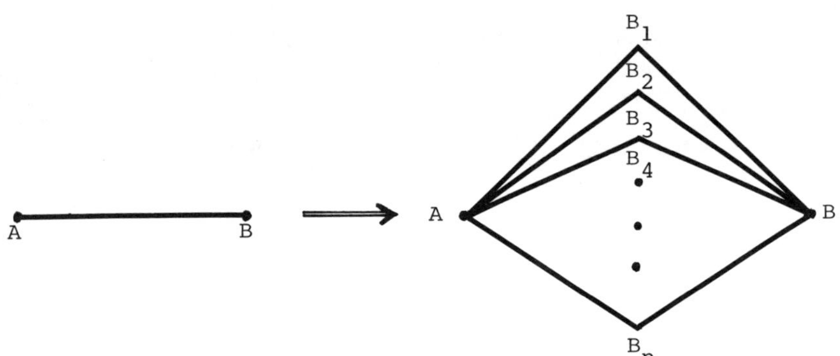

Recursive construction of the lattice

given random variable ξ . We shall consider the hamiltonian $H_N(\underline{\xi}, \underline{S})$ given by

$$H_N(\underline{\xi}, \underline{S}) = \sum_{\substack{\underline{i},\underline{j} \\ \text{nearest} \\ \text{neighbours}}} \xi_{\underline{i},\underline{j}} \, S_{\underline{i}} \, S_{\underline{j}}$$

Model B

This model is very similar to model A, except that we have imposed some relations between the random couplings. At each site $\underline{i}$, we have again an Ising variable $S_{\underline{i}}$, but the coupling $\xi_{\underline{i},\underline{j}}$ is defined as follows assuming $\bar{n}=2^d$. We have $\underline{i}=(i_1,\ldots, i_2)$ where i_1 is the coordinate of $\underline{i}$ in the $\underline{e}$ direction. For each value of i_1 we shall take the couplings $\xi_{\underline{i},\underline{j}}$ with $j_1 = i_1 + 1$ identical and equal to an independant copy of the given random variable ξ . It is easy to extend this construction to the case of a general n. The hamiltonian is the same as for model A. From now on we shall assume that the random variable ξ satisfies $E(\xi) = 0$ where E denotes the expectation (see [12] for the case $E(\xi) \neq 0$). For both models we first consider separately each possible configuration of random couplings. We then construct the Gibbs state with probability $Z^{-1} e^{-\beta H_N(\underline{\xi},\underline{S})}$ for the configuration $\underline{S}$ of spins. Next we consider the thermodynamic limit $N \to +\infty$, and should ask for informations about physical observables which are true almost surely, with respect to the distribution of couplings (see [13]). We shall only deal with average properties which are simpler to analyse.

The renormalization group program is now easy to implement. We described above the transition from a lattice L_N to a larger lattice L_{N+1} . The renormalization operation can be thought of as the inverse transition, i.e. replacing a model on lattice L_{N+1} by a similar model on lattice L_N. In this operation, the couplings will be modified, and their transformation is called the renormalization mapping. In the present case, the transition from lattice L_{N+1} to lattice L_N is performed by summing over the spins at the last level of L_{N+1}, i.e. the spins which are on the sites newly created when going from lattice L_N to lattice L_{N+1}. We shall consider for the moment that the inverse temperature β has been included in the coupling. We obtain an equivalent model on a lattice L_N, where the random couplings are copies of a new random variable ξ'. The transformation $\xi \to \xi'$ is the renormalization transformation. Obtaining the expression of ξ' as a function of ξ is an easy exercise in Ising spins algebra and we refer to [5] for the details. It is important to notice that the reduction of the renormalization operation (summing over the spins) to a simple transformation $\xi \to \xi'$ is a special feature of the particular lattices and models we are looking at. For $\mathbb{Z}^d$ lattices this operation can be very complicated (see [17] for a similar problem). We now give the rules for constructing ξ'.

Case of model A

Given ξ, one constructs 2n independent copies of ξ : $\xi_1, \xi_2, \ldots, \xi_n, \tilde{\xi}_1, \tilde{\xi}_2, \ldots, \tilde{\xi}_n$. ξ' is now given by

$$\xi' = R(\xi) = \sum_{j=1}^{n} \text{th}^{-1}(\text{th}\xi_i \ \text{th}\tilde{\xi}_i) \qquad (1).$$

Case of model B

Given ξ one constructs two independant copies ξ_1 and ξ_2. The formula for ξ' is

$$\xi' = R(\xi) = n \ \text{th}^{-1}(\text{th}\xi_1 \ \text{th}\xi_2). \qquad (2)$$

These formulas give the new couplings in term of the old ones. However, in our case, these couplings are random variables and it is sometimes useful to think of the renormalization transformation as an action on the probability law of the random variable ξ. The renormalization analysis can now proceed along the usual steps. We shall be interested in the behaviour of the sequence of random variable (ξ_p) given by $\xi_{p+1} = R(\xi_p)$ and $\xi_o = \beta\xi$. To do so, we shall determine the fixed points and local behaviour of the renormalization transformation. We shall then look at some adapted physical observables and determine their critical behaviour. Notice that if $E(\xi_o) = 0$, formulas (1) and (2) imply that $E(\xi_p) = 0$ for every p. We shall first consider model A which is better understood and then give some indications about model B.

III. Renormalization of model A, large d expansion.

As mentioned in the previous paragraph, one may interpret the quantity $\log_2 n$ as a dimension of the lattice. Looking at formula (1) it is tempting to consider the case of large n (i.e. large d) and to apply the central limit theorem. As we shall see, this simple idea turns out to be a very powerful method of investigation. We shall first give a heuristic argument in the case of large n. Assume first ξ is almost surely small. Then we can replace (1) by the approximate recursion relation

$$\xi' = \sum_1^n \xi_i \tilde{\xi}_i .$$

If $\sigma = E(\xi^2)^{1/2}$ and $\sigma' = E(\xi'^2)^{1/2}$ (we have assumed $E(\xi) = E(\xi') = 0$), we obtain

$$\sigma' = n^{1/2} \sigma^2 .$$

Therefore, if σ_p is the variance of ξ_p we have the following behaviour

i) If $\sigma_o < n^{-1/2}$ (high temperature), $\sigma_p \to 0$ ($\xi_p \to 0$ a.s).

ii) If $\sigma_o > n^{-1/2}$, $\sigma_p \to +\infty$ and we cannot draw any conclusion since our approximation (ξ_p small) breaks down.

iii) If $\sigma_o = n^{-1/2}$, $\sigma_p = n^{-1/2}$ $\forall p$, i.e. we have a fixed point. Moreover

$$\left.\frac{d\sigma'}{d\sigma}\right|_{\sigma = n^{-1/2}} = 2 .$$

If we reintroduce the temperature dependence we obtain

$\sigma_o = \beta E(\xi^2)^{1/2}$, and there is a phase transition at inverse temperature $\beta = n^{-1/2} E(\xi^2)^{-1/2}$ with critical index to equal 2. These heuristic results are confirmed (up to small corrections) by the rigorous analysis. Our first result concerns the behaviour at large temperature.

<u>Theorem 1</u> : <u>Given ξ symmetric and of variance</u> 1, <u>for every</u> $\beta < (50n)^{-1/2}$ <u>we have</u>

$$E[(R^p(\beta\xi))^2] \leqslant 2^{-p}/50n .$$

<u>Proof</u> : We observe that

$$E(R(\xi_p)^2) = n\ E[(th^{-1}(th\xi_p\ th\xi'_p))^2]$$

where ξ'_p is an independant copy of ξ_p.

From $|th^{-1}(thx\ thy)|^2 \leqslant 25x^2y^2 \quad \forall\ x,y$ in $\mathbb{R}$, we deduce

$$E(R(\xi_p)^2) \leqslant 25\ E(\xi_p^2)^2$$

and the result follows easily.
Theorem 1 tells us that if the temperature is large enough, the renormalization drives the coupling almost surely to zero, i.e. to the infinite temperature fixed point.

We now investigate the low temperature behaviour.

<u>Theorem 2</u> : <u>Let</u> ξ <u>be a symmetric random variable of variance</u> 1. <u>Let</u> $Q = E(\mathrm{Inf}(\xi^2, \xi'^2))/8$ <u>where</u> ξ' <u>is an independent copy of</u> ξ <u>(note that</u> $Q \neq 0$ <u>since</u> ξ <u>has a non zero variance)</u>. <u>There is a constant</u> $M_o = M_o(Q) > (320)^2$ <u>such that for</u> β <u>large enough, and for</u> $n > M_o$, <u>the sequence</u> $\xi_p = R^p(\beta\xi)$ <u>has the following properties</u> :

1) $\sigma_p = E(\xi_p^2)^{1/2} > (n^{1/2}/320)^p\, \beta\, Q^{1/2}$ (i.e. $\sigma_p \to +\infty$).

2) <u>If the probability density of</u> ξ <u>is in</u> $L\infty$, <u>the density</u> f_p <u>of</u> ξ_p <u>satisfies</u>

$$|f_p(x) - (2\pi\sigma_p^2)^{-1/2} \exp(-x^2/2\sigma_p^2)| \leqslant K\sigma_p^2 n^{-1/2}/(\sigma_p + |x|)^3$$

<u>where</u> K <u>is a universal constant</u>.

<u>Proof</u> : Let η and η' be two independant and identical random variables. We define a new random variable $\tilde{\eta}$ by

$$\tilde{\eta} = th^{-1}(th\eta \; th\eta')$$

and we shall estimate the variance $\tilde{\sigma}$ of $\tilde{\eta}$. It will turn out that we shall also need the skewness $\tilde{\alpha}$ of $\tilde{\eta}$ which is given by

$$\tilde{\alpha} = E(|\tilde{\eta}|^3)/\tilde{\sigma}^3.$$

From the inequality $|th^{-1}(thx \; thy)| \leqslant |xy|^{1/2}$, we have

$$\tilde{\alpha} \leqslant E(\eta^2)^{3/2} \tilde{\sigma}^{-3} .$$

We now use the fact that η will be identified with ξ_p. This is a random variable which is the sum of n independant random variables. Let η be the random variable $\sum_1^n \hat{\eta}_i$, where $\hat{\eta}_i$, $2 \leqslant i \leqslant n$ are independent copies of $\hat{\eta}_1'$. We shall denote by F the distribution function of η. We have the Berry - Essen inequality [22]

$$|F(xn^{1/2} \hat{\sigma}) - \Phi(x)| \leqslant \hat{\alpha} n^{-1/2} \quad \forall x \in \mathbb{R},$$

where Φ is the normal distribution, $\hat{\sigma}$ and $\hat{\alpha}$ are the variance and skewness of $\hat{\eta}_1$. Assume now the variance σ of η satisfies $\sigma > 9$ (notice that $\sigma = n^{1/2} \hat{\sigma}$), and also

$$|F(x\sigma) - \Phi(x)| \leqslant 1/40 \quad \forall x \in \mathbb{R}.$$

Then standard probability estimates imply that $\tilde{\sigma} < \sigma/320$ and $\alpha < (320)^3$. Therefore, if n is so large that $40n^{-1}(320)^2 < 1$, and

$n^{1/2} > 320$, we can use recursively the above estimates. This proves 1) provided the initial random variable $\xi_o = \beta\xi$ satisfies the recursive hypothesis. This is insured by taking β large enough. Notice that $\sigma_p \to +\infty$ does not imply that the random variables ξ_p tend to infinity almost surely. This result follows however from an estimate on the probability laws like 2) which is proven using a more precise version of the Berry-Esseen inequality due to Sahaidarova (see [22]). We refer the reader to [5] for a complete proof.

We now come to the more delicate case of the critical behaviour. We shall only deal with random variables ξ which have a density of the form $(2\pi s)^{-1/2} \exp(-x^2/2s) + \varphi(x)$ where s belongs to $\mathbb{R}^+$ and φ belongs to a function space H' defined by

$$H' = \{\varphi \in L_\infty \mid \int x^8 \ |\varphi(x)|^2 \ dx < +\infty \ , \text{ and } \int\varphi(x) \ dx = \\ = \int x^2 \ \varphi(x) \ dx = 0\} \ .$$

We shall denote by H the set of couples (s,φ) as above. We define $\|x^k\varphi\|_p$ to be the L_p norm of the function $x \to x^k\varphi(x)$, $\|\varphi\|_{H'}$ and $\|(s,\varphi)\|_H$ are defined by

$$\|\varphi\|_{H'} = \|\varphi\|_\infty + \|x^4\varphi\|_2 \ , \qquad \|(s,\varphi)\|_H = |s| + \|\varphi\|_{H'} .$$

We shall think of the renormalization transformation as a map from H into itself, and we shall look for a fixed point near $(n^{-1/2}, 0)$. We now need the expression of the renormalization tranformation in the space H. We define first a map S_1 from H to the set of probability densities by

$$S_1((s,\varphi))(x) = (n/s)^{1/2} \ (2\pi)^{-1/2} \ \exp(-nx^2/2s) + \varphi(n^{1/2}x) .$$

Notice that the above expression defines a function in L_1. However this function is a probability density only if φ satisfies some special conditions. Let $f = S_1((s,\varphi))$, we shall denote by η the random variable with probability density f. $S(f)$ will denote the probability density of the random variable $th^{-1}(th\eta\, th\eta')$ where η' is an independent copy of η. The formula for $\tilde{f} = S(f)$ is easily seen to be

$$S(f)(x) = (1 - th^2 x) \int_{|thx| \leq |y| \leq 1} dy\, |y|^{-1} (1-y^2)^{-1} (1-y^{-2}(thx)^2)^{-1} \times$$

$$\times f(th^{-1}y)\, f(th^{-1}(thx/y)). \qquad (3)$$

We now define a map T among probability densities by $\tilde{f} \to \hat{f} = T\tilde{f}$ given by

$$T\tilde{f} = \tilde{f} * \tilde{f} * \ldots * \tilde{f} \quad (n \text{ fold convolution}).$$

Finally there is a map T_1 from probability densities back to H : $(s,\varphi) = T_1\hat{f}$ which is given by

$$s = n \int x^2 \hat{f}(x)\, dx, \text{ and } \varphi(x) = (s/n)^{1/2} [\hat{f}(xn^{-1/2})$$

$$- (2\pi)^{-1/2} \exp(- x^2/2ns)].$$

Notice that T_1 is roughly the inverse of S_1. It is easy to verify that as a mapping from H to H the renormalization transformation R is given by the product $M = T_1 \circ T \circ S \circ S_1$. We are now able to state the result for the critical behaviour.

<u>Theorem 3</u> : <u>Let</u> $B = \{(s,\varphi) \in H \mid \|(s-1,\varphi)\|_H \leq (\log n)^{10}/n\}$. <u>The map</u> M <u>from</u> B <u>to</u> H <u>has a unique fixed point</u> (s_o,φ_o) <u>in</u> B <u>for</u> n <u>large enough</u>. <u>Moreover</u>

$$s_o = 1 + \mathcal{O}((\text{Log } n)^{10}/n^{1/4}),$$

$$\varphi_o = \mathcal{O}((\text{Log } n)^{10}/n^{1/4}),$$

<u>and the function</u> $S_1(s_o,\varphi_o)$ <u>is a positive function</u>. <u>The linear operator</u> $DM_{(s_o,\varphi_o)}$ <u>has in</u> H <u>a simple</u>

<u>eigenvalue equal to</u> $2 + O(n^{-1/2})$. <u>The rest of the spectrum is contained in the open unit disk</u>.
Using this result, one can repeat the usual renormalization group analysis in the space H i.e. one first constructs the local stable and unstable manifolds at the fixed point and proves the universal scaling behaviour (see [6] or [9]). The 1/n expansion was investigated in [15].

We now give the main ideas for the proof of theorem 3 (see [5] for more details). As mentioned before, if n is large, we expect the fixed point to be nearly gaussian, i.e. near to (1,0). This suggests to use the inverse function theorem. We implement this ideas as follows. We have

$$DM_{(s,\varphi)} - \mathbb{1}_H = M^* + M'_{(s,\varphi)} \quad \text{where} \quad M^* = \begin{pmatrix} 1 & 0 \\ 0 & -\mathbb{1}_{H'} \end{pmatrix},$$

and $M'_{(s,\varphi)}$ is a linear operator of norm less than $(\log n)^{-2}$ for (s,φ) in B. We define a map Φ from B to H by

$$\Phi = \mathbb{1}_H - M^{*-1}(M - \mathbb{1}_H).$$

It turns out that Φ is a contraction from B to B ($D\Phi$ has norm smaller than 1/2 in B), and the fixed point of Φ in B is a fixed point of M.
There are two technical steps in the above argument. The first one is to prove that $\Phi((1,0))$ is near to (1,0) for large n. The second one is to show that $D\Phi$ has norm less than 1/2 in B. The proofs of these two assertions are very similar and we shall only give some indications about the first one. We first observe that only T and S give non-trivial contributions to the operator M(T_1 and S_1) are trivial linear operators). The good part is T, although this is a very nonlinear operator (n convolution). T is almost a projection onto gaussian functions, the correction being of order $n^{-1/2}$. We have the following lemma

<u>Lemma 4</u> : <u>Let</u> $f = n^{-1/2}\,\tilde{f}^{*n}(n^{-1/2}\cdot)$, <u>where</u> $\tilde{f}$ <u>satisfies</u>

$$\|\tilde{f}\|_1 + \|\tilde{f}\|_2 + \|x^4\tilde{f}\|_1 < \infty \quad \underline{\text{and}} \quad \int_{-\infty}^{+\infty} x\,\tilde{f}(x)\,dx = 0\,.$$

Then, for n *large enough*

$$f(x) = (2\pi\, s^{-1})^{-1/2} \exp(-\, x^2\, s/2) + \varphi(x)$$

and $\|\varphi\|_{H'} \leqslant O(n^{-1/2})$.

This lemma is very much in the flavor of the results about corrections to the central limit theorem. Unfortunately it does not seem to have already appeared in the litterature under this functional form. The proof consists in estimating the Fourier transform of f (and of its first few derivatives). For $|t| \leqslant O(n^{1/2})$ one performs the usual expansion and cancellations of the usual proof of the central limit theorem. For $|t| \geqslant O(n^{1/2})$ one uses a general result of Statulyavichus (see [5] for details). In order to be able to apply this lemma, we have to show that if (s,φ) is in H, $SS_1(s,\varphi)$ satisfies the hypothesis. This is the main reason for the choice of the space H. The proof relies on various L_p estimates about formulas (3) (see [5]). It is worth mentioning that the function $SS_1(1,0)$ is not at all like a gaussian. It is infinite at $x = 0$. However, the operator T is so strongly contracting toward gaussians that for n large enough this bad behaviour desappears.

IV Spin observables

We now come to the more concrete part of the renormalization analysis, i.e. the analysis of those physical observables which behave in a particular way under the renormalization. Notice that various boundary conditions can be obtained by fixing the two extreme spins. We shall give here some details about the averaged spin, the same ideas can be used to analyse some correlations. The results are summarized in the following theorem.

Theorem 4 : Under the conditions of Theorem 2, we have in the thermodynamic limit and for any boundary condition

i) $E(\langle s\rangle) = 0$ at every temperature

ii) $E(\langle s\rangle^2) = \begin{cases} 0 & \text{if } \beta \text{ is small enough} \\ \neq 0 & \text{if } \beta \text{ is large enough.} \end{cases}$

This theorem proves that the Edwards-Anderson order parameter has a phase transition. Using the results of chapter III, it is also possible to compute the critical exponent. Recent works based on the replica symmetry breaking ([21] , [20]) have shown that in the Sherrington-Kirkpatrick model, the random variable $V^{-1} \sum_{\underline{i}} S_{\underline{i}}^{(1)} S_{\underline{i}}^{(2)}$ (where $S_{\underline{i}}^{(1)}$ and $S_{\underline{i}}^{(2)}$ are the spins of two independant replicas and V the number of spins) has a non trivial distribution in terms of the couplings. In the present case, it was shown in [15] that this is not the case. It was also shown in [15] that the replica technique cannot be applied to this model (at least in it's initial form). We now give some ideas about the proof of theorem 4. We shall obtain for the thermal average of a spin a recursion relation associated to the renormalization of the model. Let S_o be a spin which is on a lattice site of L_{N+1} which does not belong to L_N. S_o has two neighbouring spins which belong to L_N. We shall denote by S_1 that neighbour which is not on a site of L_{N-1} and by S_1' the other one. Let S_o' be equal to S_1 or to S_1'. If A and B are some constants, it is easy to verify that the renormalization acts as follows

$$< AS_o + B S_o' >_{L_{N+1}} = <A'S_1 + B'S_1' >_{L_N}$$

where

$$A' = B + AX(1-X'^2)(1-X^2X'^2)^{-1}, \quad B' = AX'(1-X^2)(1-X^2X'^2)^{-1} \quad \text{if}$$
$S_o' = S_1$.
$$A' = AX(1-X'^2)(1-X^2X'^2)^{-1}, \quad B' = B + AX'(1-X^2)(1-X^2X'^2)^{-1} \quad \text{if}$$
$S_o' = S_1'$,

$X = \mathrm{th}\,\xi$, $X' = \mathrm{th}\,\xi'$, where ξ and ξ' are the couplings between S_o and S_1 and S_o and S_1' respectively.

It should be obvious that iterating infinitely many times the above relations, one should be able in principle to compute the thermal average of a spin in the thermodynamic limit. Starting with a lattice L_N, the recursion ends with a lattice L_o of two sites, where the thermal average can be computed explicitly. We summarize the argument by the following formula

$$< S_o >_{L_N} = A_N S_N + B_N S_N' \; .$$

A_N and B_N are random variables which have (complicated) expressions in terms of the couplings. The behaviour of these random variables can be controlled as follows. Given a starting site S_o on lattice L_N, we shall denote by $(A_p)_{0 \leqslant p \leqslant N}$ $(B_p)_{0 \leqslant p \leqslant N}$ the sequences of random variables produced by the above recursion and with initial conditions $A_o = 1, B_o = 0$. The following lemma summarizes some useful properties of A_p and B_p.

<u>Lemma 5</u> : <u>Under the hypothesis of Theorem</u> 2, <u>and for</u> $0 \leqslant p \leqslant N$

we have

i) $E(A_p) = E(B_p) = E(A_p\, B_p) = 0$

ii) $|A_p| + |B_p| \leqslant 1$

iii) If $\sigma_p = E((R^p(\beta\xi))^2)^{1/2}$ is large enough, we have

$$E(A_{p+1}^2 + B_{p+1}^2) \geqslant (1 - \sigma_p^{-1/2})\; E(A_p^2 + B_p^2) - \sigma_p^{-1/2} \quad .$$

i) and ii) are easy consequences of the gauge symmetry and of the recursion formulas. iii) is more technical and we refer to [5] for the proof. Theorem 4 is now an easy consequence of Lemma 5. In the case of high temperature, we have $X_p \to 0$ exponentially fast by Theorem 1. We get smaller and smaller coefficient in the recursion relations and it is easy to show that A_N and B_N tend to zero if N tends to infinity. For the low temperature, we have to use Theorem 2 instead which implies that $\sigma_p \to +\infty$ exponentially fast. Therefore, if σ_o is large enough, using iii) of Lemma 5, we get a uniform lower bound $\gamma > 0$ for $E(A_p^2 + B_p^2)$. This implies

$$E(< S_o >^2) = E(A_N^2 + B_N^2) + S_N S_N'\; E(A_N B_N) \geqslant \gamma$$

which is ii) of Theorem 4. At this writing no information has been derived in the presence of a magnetic field. This is due to the fact that no simple recursion relation with simple estimates (like in Lemma 5) has been derived up to now.

V. Some results about model B

As we shall see, this model is less understood than model A. We shall use the hyperbolic tangents of couplings to describe the system. The recursion relation takes the following form. Let $X_o = \text{th}(\beta\xi)$, and define recursively a sequence of random variables (with values in $[-1,1]$) $X_p = R^p(X_o)$, $p \in \mathbb{N}$, by

$$X_{p+1} = q(X_p X'_p)$$

where X'_p is an independent copy of X_p and $q(s) = \text{th}(n\,\text{th}^{-1}(s))$. We shall assume that $n \geq 2$, and show that for certain distributions of ξ the model has a phase transition.

<u>Theorem 5</u> : <u>Assume that for some number</u> $c > 0$, <u>we have</u>

$$c < |\xi| < c^{-1} \quad \text{almost surely.}$$

<u>Then</u>

i) <u>for</u> β <u>small enough</u> $X_p \to 0$ a.s.

ii) <u>for</u> β <u>large enough</u> $|X_p| \to 1$ a.s. .

<u>Proof</u> It is easy to verify that if $|X_p| \subset [a,b]$ a.s., then $X_{p+1} \subset [q(a^2), q(b^2)]$ almost surely. The map $t \to q(t^2)$ has two attractive fixed points on $[0,1]$: $t = 0$ and $t = 1$, and there is also a repulsive fixed point t_o. Moreover the map is monotone. Therefore, if $|X_o| \subset [0, t_o[$ a.s., then $X_p \to 0$ a.s. and if $|X_o| \subset]t_o, 1]$ a.s., we have $|X_p| \to 1$ a.s.

We also observe that these convergences to 0 or 1 have an exponential rate. Therefore one can repeat the argument of chapter IV to analyze the behaviour of the spin observables (see [7]). The following result implies that some initial random couplings do not give rise to models with phase transition.

Theorem 6 : If for some $\delta \in [0, 1/128[$, the event $|X_o| < \delta$ has probability larger than 4δ , then $X_p \to 0$ a.s. This result also indicates that numerical computations must be done with extreme care since a small error near 0 can destroy a phase transition.

The critical behaviour seems to be much harder to analyze, and we have not yet obtained rigorous results. Instead we have studied an approximate version of the renormalization transformation which is built as follows. We replace the probability density by a new one which is constant on the sets $[-2^{-q}, -2^{-q-1}[\cup]2^{-q-1}, 2^{-q}]$. We shall denote by $f(z)$ the generating function of the sequence of numbers obtained this way. f is a function which is holomorphic in the unit disk, has positive (or zero) Taylor coefficients in $z=0$, and satisfies $f(1)=1$. The approximate renormalization transformation R is given by

$$(Rf)(z) = f(0)^2 + (f(z)^2 - f(0)^2)/z.$$

It is easy to verify that Rf has the same properties as f. R has only three fixed points : $f(z)=1$ (low temperature), $f(z)=z$ and $f(z)=z^\infty$ ($=0$ if $|z|<1$, ∞ if $|z|>1$) which is the high temperature fixed point. The global behaviour of R in the set of admissible f's is completely given by the following theorem [8].

Theorem 7 : Assume f is holomorphic in the unit disk, with positive (or zero) coefficients at $z=0$ and satisfies $f(1)=1$. Then

i) If f does not extend to a function holomorphic up to $z=2$, or if $f(2)-2f'(2)<0$, then $R^p f(z) \to z^\infty$ if $p \to +\infty$

ii) If $f(2)-2f'(2)>0$, or $f(2)-2f'(2)=0$, but f is not the function $z \to z$, then $R^p f \to 1$ if $p \to +\infty$.

It is easy to verify that $f(2)-2f'(2)=0$ is an invariant surface which plays the role of a critical surface in the usual renormalization analysis. However, except the fixed point

$f(z) = z$, all the models on this surface are in the law temperature phase. In particular it does not seem possible to define critical indices. We conjecture that model B has a similar behaviour.

References

[1] D. Andelman, A. N. Berker, Scale-invariant quenched disorder and its stability criterion at random critical points. Preprint MIT (1983).

[2] A. Benyoussef, N. Boccara, Real space renormalization group investigation of three-dimensional Ising spin glasses. Phys. Lett. 93A, 351-353 (1983). Existence of spin-glass phases for three and four dimensional Ising and Heisenberg model. Preprint CEA-Saclay (1983).

[3] V.P. Bovin, V.V. Vas'kin, I. Ya. Shneiberg. Recursive models in percolation theory. Journ. Theor. and Math. Phys. 54, 175-181 (1983).

[4] P.M. Bleher, E. Zylas, Existence of long range order in Migdal's recursion relation. Commun. Math. Phys. 67, 17-42 (1979).

[5] P. Collet, J-P. Eckmann, A spin glass model with random couplings. Commun. Math. Phys. 93, 379-407 (1984).

[6] P. Collet, J-P. Eckmann, A Renormalization Group Analysis of the Hierarchical Model in Statistical Mechanics. Lecture Notes in Physics 74, Springer, Berlin, Heidelberg, New York (1978).

[7] P. Collet, J-P. Eckmann, Y. Glaser, A. Martin, A spin glass with random couplings. Preprint IHES (1983).

[8] P. Collet, J-P. Eckmann, Y. Glaser, A. Martin, Study of the iterations of a mapping associated to a spin glass model. Commun. Math. Phys. 94, 353-370(1984).

[9] P. Collet, J-P. Eckmann, O.E. Lanford III, Universal properties of maps on an interval. Commun. Math. Phys. 76, 211-254 (1980).

[10] B. Derrida, L. De Seze, C. Itzykson, Fractal structure of zeroes in hierarchical models. Journ. Stat. Phys. 33 559 (1983).

[11] E. Domany, S. Alexander, D. Bensimon, L.P. Kadanoff, Solutions to the Schrödinger equation on some fractal lattices. Phys. Rev. B28, 3110-3123 (1983).

[12] B. Derrida, E. Gardner. To appear.

[13] S.F. Edwards, P.W. Anderson, Theory of spin glasses. J. Phys. F5, 965 (1975).

[14] B. Derrida, J-P. Eckmann, E. Erzan, Renormalization group with periodic and aperiodic orbits. Journ. Phys. A16, 893 (1983).

[15] E. Gardner, A spin glass model on a hierarchical lattice. Preprint spht-84-44 CEA-Saclay (1984).

[16] M. Kaufman, R.B. Griffiths, Spin systems on hierarchical lattices. Introduction and thermodynamic limit. Phys. Rev. B26, 5022-5032 (1982).

[17] A. Kupiaïnen, These proceedings.

[18] S. Mc Kay, A.M. Berker, S. Kirkpatrick, Amorphously packed, frustrated hierarchical models, chaotic rescaling and spin glass behaviour. J. Appl. Phys. 53, 7974-7976 (1982).

[19] J-M. Langlois, A-M. S. Tremblay, B.W. Southern, Chaotic scaling trajectories and hierarchical lattice models of disordered binary harmonic chains. Phys. Rev. B28, 218-231 (1983).

[20] M. Mézard, G. Parisi, Self averaging correlation functions in the mean field theory of spin glasses. Preprint LPTENS 87-4, Paris (1984).

[21] M. Mézard, G. Parisi, M. Sourlas, G. Toulouse, M. Virasoro, Nature of the spin glass phase. Phys. Rev. Lett. 52, 1156-1159 (1984).

[22] V.V. Petrov, Sums of independent random variables Springer-Verlag Berlin, Heidelberg-New York (1975).

[23] A-M. S. Tremblay, B.W. Southern, Generating function and scaling for the density of states of fractal lattices. Preprint (1983).

P. Collet

Centre de Physique Théorique
Ecole Polytechnique
91128 Palaiseau France

ON THE DIFFUSION IN DYNAMICAL SYSTEMS

L. A. Bunimovich

In the recent years there has been a considerable progress in understanding the equilibrium mechanisms of statistical mechanics. This progress has been achieved with help of methods of the ergodic theory. For some realistic systems there have been proved such properties as ergodicity, mixing, K and Bernoulli property. At the same time the same progress has not been obtained in the field of non-equilibrium statistical mechanics. It is connected with the following well-known fact: stochasticity may be strong or it may be weaker. In other words, the onset of stochasticity is not sufficient in itself to explain the hydrodynamical laws from those of statistical mechanics. So dealing with the problems of transport theory one must consider the most delicate property in the hierarchy of statistical properties of dynamical systems - the speed of mixing (or the rate of correlation decay). Knowing this characteristics one can calculate the values of transport coefficients for the considered dynamical system by Green-Kubo formulae [14], can obtain some information about its energy spectrum etc.

Quite recently this property was formulated as "exponential decay of correlations". Really there was a general belief that, in stochastic systems, correlations $b_f(n) = \int_M f(T^n x) f(x) d\mu$ must decay exponentially fast. Here M is a phase space of a dynamical system under consideration, x - its point, T - a transformation which defines this system, μ - an invariant measure on M and n a (discrete) time. Of course, it was clear that this property in such a general formulation fails even for the "most stochastic" classical dynamical systems as the hyperbolic automorphisms of tori [7]. One must consider not all functions $f(x)$ on the phase space of the considered dynamical system but only sufficiently "nice" (for instance, continuously differentiable or Hölder) functions $f(x)$. The physically natural functions arising in realistic models belong, as a rule, to these classes.

But the numerical computations carried out in the late sixties have shown that the integrands in Green-Kubo formulae can exhibit long-time algebraic decay [14]. These results have forced to reconsider previous intuitive ideas that the relaxation processes in dynamical systems of statistical mechanics are controlled by processes localized in space (on scales of mean free path) and in time (on scales of relaxation time).

From the general point of view, the description of dynamical processes in many-body systems by kinetic equations can be considered as approximating a non-Markovian stochastic process by a Markovian one. The fast decrease of correlations indicates that the system is close in some sense to a Markov chain. The main tool for the rigorous investigation of this approximation is the method of Markov partitions [16].

In [5] such a partition was constructed for the two-dimensional Lorentz gas with a periodic configuration of scatterers and a bounded free path. With the help of this partition it was shown in [6] that, in this system, the correlations of smooth functions decay quasi-exponentially. We shall consider below the problem of approximation of such dynamical system by the process of random motion.

Let us consider the two-dimensional Lorentz gas consisting of a single point particle moving in a triangular array of immobile disk scatterers (Fig. 1). Outside all scatterers the point particle moves with a constant velocity and at the moments of reflections it changes its velocity according to the usual law of elastic collision. Without any loss of generality one can assume that the radius of the scatterers equals 1 while the lattice spacing is

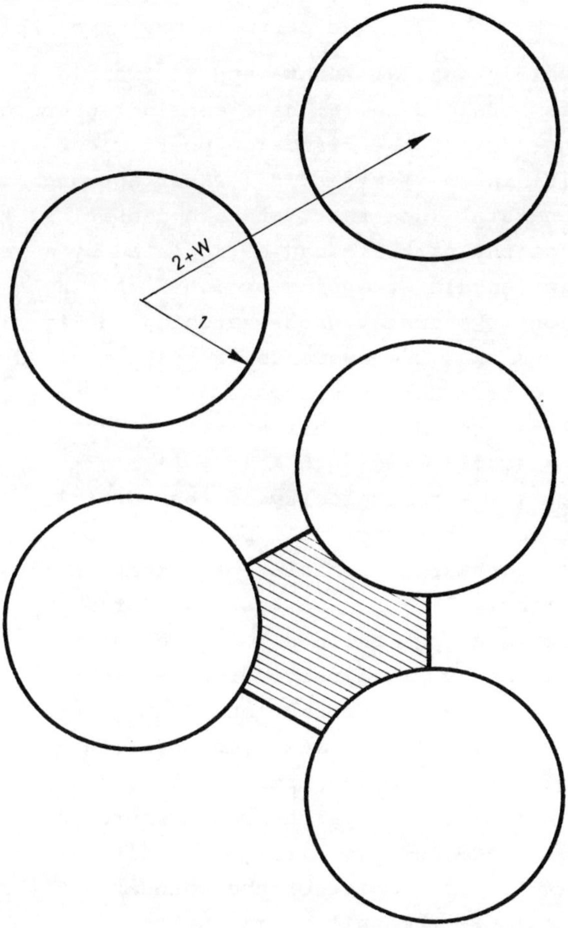

Figure 1.

$2+W$. If $W=0$ the moving particle is trapped in a bounded triangular region formed by three touching disks. If $0 < W < \frac{4}{\sqrt{3}} - 2$, the free path of the moving particle is bounded by the value $2\sqrt{3}$. In this case, the corresponding dynamical system is called a Lorentz gas with a finite horizon. For $W > \frac{4}{\sqrt{3}} - 2$, the particle may move arbitrary far between collisions. J. Machta and R. Zwanzig [11] basing on results obtained in [6] have considered at physical level the behaviour of the diffusion coefficient D for this system [9] in the limit $W \to 0$. Their approach was based on the natural idea that, at high densities, the exact motion of the particle can be replaced by a random walk between triangular trapping regions. (A single such trapping region is shown by cross-hatching in Fig. 1.) This assumption means that the sequence of traps visited by the moving particle is a Markov process.

The phase space M of this dynamical system consists of points $x = (q,v)$, where $q = (q^1,q^2)$ is the position and $v = (v^1,v^2)$ is the velocity of the particle. The flow corresponding to this system will be denoted by $\{S^t\}$. We shall consider a natural special representation [10] of the flow $\{S^t\}$. Namely let M_0 be the space of points $x = (q,v)$ such that q belongs to the boundary of one of the scatterers and v is directed inside the scatterers. We denote by T_0 the transformation of M_0 into itself which arises when the point $x \in M_0$ moves along its trajectory till the next reflection from a scatterer and $T_0 x = (q_1,v_1)$, where $q_1 \in \mathbb{R}^2$ is the point where the next reflection takes place and v_1 is the velocity after the reflection. If q is a point of the boundary of a scatterer then $n(q)$ is the unit normal vector directed outwards the scatterer and $\cos\phi = (n(q),v)$, $\frac{\pi}{2} \le \phi \le \frac{3\pi}{2}$, where $(\cdot,\cdot)$ denotes scalar product. Thus, for every scatterer K_i , $i=1,2,\ldots,$ we can introduce natural coordinates r , ϕ on the set $K_i' \subset M_0$ of points $x = (q,v)$, $q \in K_i$, where r is a cyclic coordinate along the boundary K_i and ϕ measures the angle between $n(q)$ and v . Let

U_0 be the trapping region with center in the point $(0,0)$. Let us consider the differential $d\mu_0$ of the measure on the set $U_0 \cap (\bigcup_i K_i)$ such that its restriction to K_i is proportional to $|\cos\phi|\,dr\,d\phi$.

Let $f_i(x) = q^i$, $x = (q,v)$, $i=1,2$. We denote by $g_i(x)$, $i=1,2$, the function which equals to the i-th coordinate of center of a trapping region containing the point x. If x belongs to the boundary of two neighbouring trapping regions, then, by definition, the value of $g_i(x)$, $i=1,2$, equals to the i-th coordinate of center of a trapping region with the less second coordinate.

Let us consider the following expression

$$\mathcal{D}_n^{(i)} = \int_{M_0} (f_i(T_0^n x) - f_i(x))^2 d\mu_0 . \tag{1}$$

Theorem. For any W, $0 < W < \frac{4}{\sqrt{3}} - 2$, there exists the limit $\lim_{n\to\infty} \frac{\mathcal{D}_n^{(i)}}{n} = \mathcal{D}^{(i)}(W)$ and as $W\to 0$ the value $\frac{\mathcal{D}^{(i)}(W)}{W}$ tends to a limit $\bar{\mathcal{D}}^{(i)} > 0$.

It is easy to see that $\mathcal{D}_n^{(i)}$ can be replaced (for large n) by

$$\tilde{\mathcal{D}}_n^{(i)} = \int_{M_0} (g_i(T_0^n x) - g_i(x))^2 d\mu_0 . \tag{2}$$

We shall consider the function $h_i(T_0^m x) = g_i(T_0^{m+1}x) - g_i(T_0^m x)$. So we can write

$$\frac{\tilde{\mathcal{D}}_n^{(i)}}{n} = \int_{M_0} h_i^2(x)\,d\mu_0 + 2 \sum_{2\le j\le n} \left(1+\frac{1-j}{n}\right) \int_{M_0} h(T_0^{j-1}x)h(x)\,d\mu_0 . \tag{3}$$

In what follows we will denote all constants which do not depend on W simply by const.

The new point arising in this problem comparing with the common situation in the periodic Lorentz gas with a finite horizon is the following. The dynamical system considered here is not uniformly hyperbolic [13] in the limit $W\to 0$. Indeed, in this limit the free path of the moving particle is not bounded away from zero and the components of the boundary of the trapping region touch each other.

(The last property makes this system essentially different from billiard systems considered in [4].) Therefore in the phase space of our system there is a "bad" subset sitting in the neighbourhood of straight segments belonging to the boundary of a trapping region (Fig. 1). Such segments will be called boundary segments.

It is easy to see that the first term in the right-hand side of (3) equals to const $\cdot$ W . So the main point is to estimate the second term.

The set of all trajectories which intersect on the given step one of the boundary segments can be decomposed into two subsets. The first subset consists of all such trajectories which intersect the same scatterers as the corresponding boundary segment. The second subset consists of all other trajectories. From elementary geometrical considerations one can see that the measure of the second subset equals to const$\cdot W^2$. So only the first subset must be considered.

All trajectories can be decomposed into series of reflections from the boundaries of scatterers taking place in one and the same trapping region. According to this decomposition, for any point $x \in M_0$, the sequence $h(x)$, $h(T_0 x), \ldots, h(T_0^n x), \ldots$ is decomposed into such segments that all elements in any of them besides the last one are equal to 0 .

Now one must take into account that in each such series there is at least one reflection with a free path not shorter than $\alpha > 0$, where α does not depend on W . Therefore from the general theory of dispersed billiards [15] a corresponding coefficient of expansion of neighbouring trajectories in phase space is bounded from below by $1+\gamma$, where γ does not depend on W , too. So an application of methods derived in [6] allows us to obtain for the second term an estimation from above by const. W . This already gives the assertion of the theorem formulated above.

The last estimation can be improved if we use the fact that a typical trajectory is sitting in a trapping region during a sufficiently long time. Really it follows from the ergodicity of the dynamical system under consideration that the average rate for leaving a trap is const. W^{-1} . It can be shown that the second term in (3) has an order $o(W)$ as $W \to 0$. In order to obtain this estimation one must use some sufficiently subtle properties of the Markov partition of this dynamical system. The approach based on replacing the system under consideration by a random motion between trapping regions [11] allows to obtain estimation of the first term in (3), only. Computer simulations [11] have shown that it is relevant only for very small values of W . The rigorous approach based on Markov partitions allows us to obtain correction to this estimation, in principle, in the whole segment $0<W<\frac{4}{\sqrt{3}}-2$.

The new method for the construction of a Markov partition for billiard systems proposed in [3] can be applied to the Lorentz gas in arbitrary (but finite) dimension. Besides this Markov partition has the same properties as the one constructed in [5] for the two-dimensional case. It allows us to prove the same theorem for the periodic Lorentz gas with triangular lattice of scatterers in higher dimensions, too.

It will be interesting to compare the limiting behaviour, as $W \to 0$, of the Lorentz gas when the sets of centers of the scatterers form triangular and square lattices on the plane $\mathbb{R}^2$. These two dynamical systems essentially differ due to the unboundedness of a free path length in the second one. The rate of correlation decay for the Lorentz gas with an infinite horizon and a periodic configuration of scatterers was treated in [2]. In particular, it was proved that, in this system, the decrease in the correlations is polynomial but, nevertheless, the diffusion coefficient exists and is positive.* Besides in [2] it was shown that the diffusion coefficient D in the low density limit depends on the density of scatterers nonanalitically. Earlier this result was obtained at a physical

---

*Quite recently the numerical simulation of this system which confirms an algebraic decay has been performed [8].

level for the hard spheres gas and the Lorentz gas [14]. It seems that, in the high density limit, the diffusion coefficient D depends on the density analytically. Therefore it may be the case that, in spite of the general opinion, the expansion of the diffusion coefficient into a power series with respect to the density does not exist for small densities but does exist for large densities of particles.

In conclusion, we shall discuss the question of the integrability of the velocity autocorrelation function $b_v(n)$. Recently some authors have obtained for some models by considerations at a physical level or numerical computations that $|b_v(n)| \sim \frac{1}{n}$ [12, 17]. Then it has been concluded that the diffusion coefficient does not exist. But the symmetry properties of these systems which provide the existence of D were not taken into account. Let us consider these systems.

The first system is the so called "stadium", i.e. the billiard in the region in $\mathbb{R}^2$ bounded by two semi-circles of one and the same radius and two straight segments tangent to them parallel [1]. It was shown [17] analytically (at a physical level) and numerically that, in this system, $|b_f(n)| \sim \frac{1}{n}$. It can be proved rigorously that $|b_f(n)| < < \text{const.}\ n^{-1+\varepsilon}$, where $\varepsilon>0$. Thus, for a general function in the phase space of this system, $b_f(n)$ is nonintegrable. But for the velocity it can be proved that $|b_v(n)| < < \text{const.}\ n^{-2+\varepsilon}$. Really the main contribution to $b_f(n)$ is made by trajectories which spend a long time in neighbourhoods of the family of periodic orbits which are perpendicular to the boundary line segments. But it is easy to see that, in the case $f(x) = v$, after two consequtive reflections from the boundary the moving particle has almost opposite velocities. It gives the desired estimation. The same arguments can be applied to the billiard system considered in [12].

References

[1] Bunimovich L. A. On the ergodic properties of nowhere dispersing billiards. Comm. Math. Phys. 65 (1979), 295-312.

[2] Bunimovich L. A. Statistical properties of the Lorentz gas with infinite horizon. Proc. of 3rd Int. Vilnius Conf. on Probability Theory, 1981, 1, 85-86.

[3] Bunimovich L. A. Some new advancements in the physical applications of ergodic theory. In "Ergodic Theory and Related Topics" (ed. by H. Michel), Akademie-Verl., Berlin, 1982, 27-33.

[4] Bunimovich L. A., Sinai Ya. G. On a fundamental theorem in the theory of dispersed billiards. Math. USSR Sbornik 19 (1973), 407-423.

[5] Bunimovich L. A., Sinai Ya. G. Markov partitions for dispersed billiards. Comm. Math. Phys. 78 (1980), 247-280.

[6] Bunimovich L. A., Sinai Ya. G. Statistical properties of Lorentz gas with periodic configuration of scatterers. Comm. Math. Phys. 78 (1981), 479-497.

[7] Crawford J., Cary J. Decay of correlations in a chaotic measure-preserving transformation. Physica 6D (1983), 223-232.

[8] Friedman B., Martin R. F., Jr. Decay velocity autocorrelation function for the periodic Lorentz gas, Preprint, 1984, 13 p.

[9] Hauge E. H. What can one learn from Lorentz models? In "Transport Phenomena" (ed. by L. Garrido), Springer, Berlin, 1974, 337-367.

[10] Kornfeld I. P., Sinai Ya. G., Fomin S. V. The ergodic theory. Moscow, Fizmatgiz press, 1980.

[11] Machta J., Zwanzig R. Diffusion in a periodic Lorentz gas. Phys. Rev. Letters 50 (1983), 1959-1962.

[12] Machta J. Power law decay of correlations in a billiard problem. J. Statistical Phys. 32 (1983), 555-564.

[13] Pesin Ya. B., Sinai Ya. G. Hyperbolicity and stochasticity of dynamical systems. Soviet Math. Surveys 2 (1981), 53-116, Gordon and Bridge.

[14] Resibois P., DeLeener M. Classical kinetic theory of fluids. John Wiley and Sons, 1977.

[15] Sinai Ya. G. Dynamical systems with elastic reflections. Russian Math. Surveys 25 (1970), 137-190.

[16] Sinai Ya. G. Markov partitions and C-diffeomorphism. Functional Anal. Appl. 2 (1968), 61-82.

[17] Vivaldi F., Casati F., Guarneri I. Origin of long-time tails in strongly chaotic systems. Phys. Rev. Letters 51 (1983), 727-730.

ϕ_4^4 with negative coupling

A. Kupiainen

1. Introduction

Construction of a Quantum Field Theory model in four space-time dimensions, which has a non-trivial S-matrix, has for a long time been one of the foremost problems of mathematical physics. In Constructive Field Theory this program has been carried out for space-time dimensions two and three for the superrenormalizable scalar ϕ^4 model [1], whereas the four dimensional renormalizable case has so far defied all attempts (see however [2] and [3] for the construction of a planar theory). The ϕ^4 interaction in four dimensions is perturbatively renormalizable, i.e. the S-matrix has an (divergent) expansion with finite coefficients in powers of the physical coupling constant: a non-trivial result of perturbative renormalization theory. In particular this expansion, if asymptotic to some "true" theory, predicts non-trivial scattering. However, already three decades ago, it was speculated by Landau and others ([4], [5], for a review, see [6]) that the theory might not be consistent in the ultraviolet (UV), and after the advent of the renormalization group (RG), it was suggested [7] that the theory be free for non-perturbative reasons. The <u>conjecture</u>, that the continuum limit of the (Euclidean) Schwinger functions, determined by the action

$$S_\Lambda(\phi) = \frac{1}{2}Z(\Lambda)\int(\nabla\phi)^2 + \frac{1}{2}Z(\Lambda)m_0^2(\Lambda)\int\phi^2 + \frac{1}{4!}Z(\Lambda)^2\lambda(\Lambda)\int\phi^4 \tag{1}$$

where the field ϕ has some UV cutoff Λ, are those of the (generalized) free field, no matter how the bare parameters $Z(\Lambda)>0$, $\lambda(\Lambda)>0$, $m_0(\Lambda)^2\in\mathbb{R}$ are chosen, is "almost" proven in the context of lattice cutoff with ∇ the nearest neighbour lattice gradient, [8], [9] . These negative results still leave open the question of the meaning of the formal perturbative construction. In fact, the same perturbative RG which suggests the triviality of (1) for $\lambda(\Lambda)>0$ indicates, that a non-trivial continuum limit might exist, if $\lambda(\Lambda)$ is taken small <u>negative</u>: $\lambda(\Lambda)\to -0$ as $\Lambda\to\infty$, the theory then being <u>asymptotically free</u> in the UV (see [10]). However, taking $\lambda(\Lambda)<0$ we run into the problem of how to stabilize the action (1): the functional integral doesn't make sense even for the cutoff theory.

In this talk I would like to discuss some investigations on the negative coupling ϕ^4 theory done in collaboration with K. Gawedzki. We difine the cutoff negative coupling ϕ^4 model by analytic continuation in terms of a <u>stable</u> (convergent) functional integral and show, that the continuum limit of the Schwinger functions exist and are <u>non-gaussian</u>. The resulting theory is UVAF. Moreover, we expect the renormalized perturbation theory to be asymptotic and <u>Borel-summable</u> to our construction. This result thus answers to the question concerning the meaning of the formal construction. Our non-perturbative construction provides euclidean Schwinger functions. However, they probably have no analytic continuation to Minkowski space since the crucial property of Reflection Positivity is most likely lacking, due to the analytic continuation involved in the construction. Nevertheless, we consider our result showing, how the continuum limit of an asymptotically free theory may be rigorously established. The Reflection Positivity problem is peculiar to ϕ_4^4 and is essentially an "infrared" problem, already present in the unit lattice negative λ theory.

The method used in the construction is very similar to the one used by us in the proof of <u>infrared AF</u> property of the <u>positive</u> coupling <u>critical</u> lattice ϕ^4 theory, [11] . It consists of a rigorous application of the Kadanoff-Wilson block spin transformation to perform the functional integral, originally over field fluctuations in all distance scales, each dealing with fluctuations in a fixed scale. These are performed iteratively, each step of lowering the UV-cutoff producing an effective theory of approximately

the ϕ^4-type, but the coupling replaced by a running coupling constant, in accordance with the perturbative RG. Each of these fixed scale integrations has, in proper dimensionless variables, both an infrared and ultraviolet cutoff of order unity and we evaluate them by performing perturbation theory in the running coupling constant to third order and bounding rigorously the remainder.
We sould like to stress then, that the question of renormalizability and existence of AF theories may be reduced to an IR and UV finite perturbation theory, up to only a low order, and bounding rigorously the remainder. Thus we believe, that AF is a sufficient condition for the construction of continuum limits, contrary to some recent speculations [12].

2. Negative λ theory

We consider the action (1) where ϕ is taken to be on a hypercubic lattice of spacing Λ^{-1} ($\int$ is replaced by a Riemann sum) and restricted to a finite box V, to be removed in the end. The correlations are given by

$$G_N(x_1,\ldots,x_N) = \int \Pi\phi(x_i)e^{-S}D\phi / \int e^{-S}D\phi \equiv N/D \tag{2}$$

where $D\phi = \prod_{x\varepsilon V} d\phi(x)$. Both N and D in (2) are analytic in Z, m_0^2, λ, in the region Z, $m_0 \varepsilon \mathbb{C}$, $\mathrm{Re}\, Z^2\lambda>0$. Moreover, we may rotate the contour $\phi \to e^{i\alpha}\phi$, provided $\mathrm{Re}(e^{4i\tilde{\alpha}}Z^2\lambda)>0$, $0\leq\tilde{\alpha}\leq\alpha$ (take $\alpha>0$), obtaining

$$N(Z,m_o^2,\lambda) = \exp(i\alpha(N+\Lambda^4|V|))N(e^{2i\alpha}Z,m_o^2,\lambda) \tag{3}$$

which furnishes an analytic continuation of the left hand side to the whole region Θ_α given by $\mathrm{Re}(e^{4i\alpha}Z^2\lambda)>0$. Obviously we may continue rotating, (3) giving an analytic continuation to an arbitary Θ_α. In particular, for Z, $m_0^2 \varepsilon \mathrm{IR}$ and λ <u>negative</u> N is given in terms of a functional integral ($\alpha = \pi/4$) with <u>imaginary</u> mass and $(\nabla\phi)^2$-term and a <u>positive</u> ϕ^4 coupling. We take this stable integral as our definition of the negative λ theory: a different one (complex conjugate of the above) is obtained by $\alpha = -\frac{\pi}{4}$. If the partition function D in 2 doesn't have zeros, (as we will show) the negative λ correlations are given by

$$G_N(x_i,Z,m_o^2,\lambda) = e^{Ni\alpha}G_N(x_i,e^{2i\alpha}Z,m_o^2,\lambda) \tag{4}$$

for $\mathrm{Re}(Z^2\lambda e^{4i\alpha}) > 0$.

<u>Remarks</u> 1. Due to complex action in (4), reflection positivity is lost. This occurs already on unit lattice and is not related to the UV-limit.

2. Formal perturbation theory obviously satisfies (4): it corresponds to multiplying each propagator by $e^{-2i\alpha}$, each vertex by $e^{4i\alpha}$ and external legs by $e^{i\alpha}$. Thus our theory has the same formal perturbation theory as the negative coupling theory. In particular the complex nature of the measure is a non-perturbative effect.

3. The result

We take as our bare parameters in (1)

$$Z(\Lambda) = \exp(\pm i(\tfrac{\pi}{4}-\varepsilon)) \tag{5}$$

$$\lambda(\Lambda)^{-1} = g^{-1} - \beta_2 \log\Lambda + (\beta_3/\beta_2)\log(1-g\beta_2\log\Lambda) \tag{6}$$

where β_2 and β_3 are computable numbers (the first two coefficients of the β-function), ε is .1, say. We prove the

Theorem For the renormalized coupling g small, negative and $m^2 = O(1)$, positive, there exists a bare mass $m_0^2(\Lambda) = m_0^2(m^2, \lambda(\Lambda))$ so that the continuum (and $V \to \mathbb{R}^4$) limit of G_N exist (eg. in $S'(R^4)^N$) and are non-gaussian with m the physical mass.

Remarks 1. The wave function renormalization Z in (5) is taken finite. This is indeed what perturbative RG predicts too, see below.

2. The construction in fact works for g complex in the region depicted in picture 1 where near the positive real axis $(\mathrm{Re}g)^2 < g_0 |\mathrm{Im}g|$, $g_0 > 0$ small.

3. We get detailed information on correlations. E.g. the usual physical coupling is

$$\Gamma^4(p_i=0) = g + O(g^2) \tag{7}$$

and the theory is UVAF:

$$\Gamma^4(tp_i) \underset{t\to\infty}{\simeq} (\log t)^{-1} \tag{8}$$

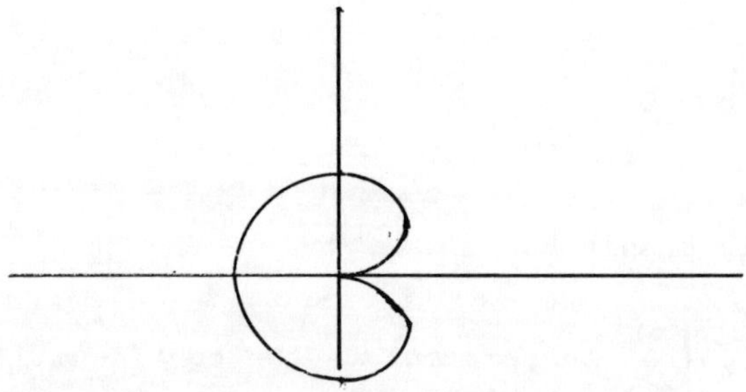

Picture 1. The analyticity region in g

4. We expect to prove analyticity in g in the region of Picture 1 and combined with (7) to prove asymptoticity and <u>Borel-summability</u> of the standard renormalized perturbation series to our solution.

5. ε in (5) is for technical reasons, in order to have a covariance with positive real part.

4. The Block spin RG

It is convenient to reformulate the continuum limit in terms of a <u>scaling limit</u> of a system of <u>Statistical Mechanics</u>. For ϕ a $\frac{1}{\Lambda}$-lattice field, introduce the dimensionless field Φ on <u>unit lattice</u>

$$\phi(x) = \Lambda\Phi(\Lambda x) \tag{9}$$

and the (bare) "Hamiltonian"

$$H_\Lambda(\Phi) = S_\Lambda(\Lambda^{-1}\Phi(\Lambda^{-1}\cdot)) \tag{10}$$

whence the continuum limit is given as a scaling limit

$$G_N(x_i) = \lim_{\Lambda\to\infty} \Lambda^N e^{Ni(\frac{\pi}{4}-\varepsilon)} \langle \prod_i \Phi(\Lambda x_i) \rangle_{H_\Lambda} \tag{11}$$

H_Λ has the same form as (1), except $m_0^2(\Lambda)\to\Lambda^{-2}m_0^2(\Lambda)$. We study the continuum limit of (1) by "integrating out" fluctuations on scales $1\leq\bar{\Lambda}\leq\Lambda$. In terms of (11) this corresponds to the <u>block spin transformation</u>. Explicitly, consider the "effective theory" for "low momenta" resulting from such an integration, that is, the action S_Λ^{EFF} determining the distribution of the average continuum fields in unit cubes $\square(x)$, centered at $x\varepsilon Z^4$. It is given by

$$\exp(-S_\Lambda^{EFF}(\Psi)) = \int \exp(-S_\Lambda(\phi)) \prod_x \delta(\Psi(x) - \int_{\square(x)} \phi) D\phi \tag{12}$$

where x runs through $V\cap\mathbb{Z}^4$. In terms of the Hamiltonian (10):

$$S_\Lambda^{EFF}(\Psi) = (R_\Lambda H_\Lambda)(\Psi) \qquad (13)$$

where R_Λ is the block spin transformation mapping the "space of unit lattice Hamiltionians" to itself:

$$\exp(-R_\Lambda H(\Psi)) = \int e^{-H(\Phi)} \Pi\delta(\Psi(x)-(C_\Lambda\Phi)(x))D\Phi \qquad (14)$$

the $C_\Lambda\Phi$ being the block spins:

$$(C_\Lambda\Phi)(x) = \Lambda(\Lambda^{-4} \sum_{|y_\mu|<\Lambda} \Phi(\Lambda x+y)) \qquad (15)$$

The existence of the scaling limit (11) may thus be reformulated in terms of R_Λ: find a sequence of unit lattice Hamiltonians H_Λ, which after an infinite (as $\Lambda\to\infty$) flow under R have a limit

$$\lim_{\Lambda\to\infty} R_\Lambda H_\Lambda = S^{EFF} \quad , \qquad (16)$$

the effective unit scale action of the field theory. In particular, H_Λ will approach the <u>critical manifold</u> in the "space of Hs".

5. Perturbation theory for R_Λ

The idea is to perform R_Λ in many steps. R_Λ has a semigroup property

$$R_\Lambda = R_L^N \tag{17}$$

for $\Lambda = L^N$. We take $L = O(1)$ (i.e. $N\to\infty$) and analyze R_L. This is given in terms of a functional integral with UV cutoff 1, IR cutoff L^{-1} and is thus well suited for a rigorous analysis. Explicity, we split the gaussian measure $d\mu_{Z^{-1}G}$ where the covariance is $Z^{-1}G = Z^{-1}(-\Delta)^{-1}$ to the block spin and fluctuation parts:

$$d\mu_{Z^{-1}G}(\Phi) = d\mu_{Z^{-1}G'}(\Psi) \otimes d\mu_{Z^{-1}}(Z) \tag{18}$$

$$\Phi(x) = L^{-1}\Psi'(L^{-1}x) + Z(x) \tag{19}$$

(here $\Psi'$is linearly and locally related toΨ and $G' = G$ in the IR). (18) and (19) now give, that for

$$H(\Phi) = -\tfrac{1}{2}Z(\Phi,\Delta\Phi) + V(\Phi) \quad , \tag{20}$$

$$\exp(-R_L H(\Psi)) = \text{const.}\exp(-\tfrac{1}{2}Z(\Psi,G'^{-1}\Psi)-TV(\Psi)) \tag{21}$$

and the interaction is transformed to

$$e^{-TV(\Psi)} = \int \exp(-V(L^{-1}\Psi'(L^{-1}\cdot)+Z))d\mu_{Z^{-1}\Gamma}(Z) \tag{22}$$

Crucially, the fluctuation covariance Γ has exponential falloff (with mass $O(\frac{1}{L})$).

The rigorous analysis of (22) is built upon its perturbative analysis in powers of V. For V local as in (1), TV is given in terms of connected graphs with $Z^{-1}\Gamma$, as lines $L^{-1}\psi$, on legs and as vertices the vertices of V. Each term in this perturbation series is well defined due to the massiveness of Γ(the series itself of course is divergent). This also assures that TV, although not local, still is approximately local, up to exponential tails. For V as (we Wick order with G)

$$V(\Phi) = \frac{1}{2}Zr\Sigma:\Phi(x)^2: + \frac{1}{4!}Z^2\lambda\Sigma:\Phi(x)^4: \tag{23}$$

the leading part of TV will be given by a similar expression, with r and λ replaced by

$$r_1 = L^2 r + O(\lambda^2, r\lambda) \tag{24}$$

$$\lambda_1 = \lambda - \beta_2(L)\lambda^2 + O(\lambda^3, r\lambda) \tag{25}$$

where the first term in (24) reflects the "relevant" nature of the Φ^2 term whereas by (25) Φ^4 is "mariginal" in the leading order. The second term in (25) is due to the graph  and for λ negative, we see that $|\lambda|$ increases in the iteration. There are of course plenty of other terms in TV, all of "irrelevant" nature upon subsequent iteration, except for a mariginal $\Sigma(\nabla\Psi)^2$ term of order λ^2(from the graph ), leading to

$$Z \to Z_1 = (1+O(\lambda^2))Z \tag{26}$$

Already this heuristic second order analysis suggests, that if we succeed "fine turning" the bare mass r so as to counter its increase (see below), the bare coupling $\lambda(\Lambda)$ should be chosen approximatively as

$$\lambda(L^N)^{-1} \simeq -N\beta_2(L) \qquad (27)$$

to counter its expansion. Then (26) would lead to

$$Z_N \simeq Z \prod_{M=0}^{N-1} (1+O(\lambda_M^2)) \simeq Z\Pi(1+O(M^{-2})) \qquad (28)$$

which stays bounded in Λ as claimed.
To fix $\lambda(\Lambda)$ completely a third order analysis exhibiting the loglogΛ dependence is required. It may be shown, still perturbatively, that the iteration of T quickly stabilizes T^nV to a fixed scaling form where λ flows as

$$\lambda_{M+1} = \lambda_M - \beta_2(L)\lambda_M^2 - \beta_3(L)\lambda_M^3 + O(\lambda_M^4) \qquad (29)$$

Our choise of the bare coupling is dictated by (29). The reader may verify, that for the differential equation

$$d\lambda(\bar{\Lambda})/d\log\bar{\Lambda} = \beta_2\lambda(\bar{\Lambda})^2 + \beta_3\lambda(\bar{\Lambda})^3 + O(\lambda(\bar{\Lambda})^4) \qquad (30)$$

with our choise of $\lambda(\Lambda)$ as the initial value, $\lambda(1)$ will be $g+o(g^2)$, no matter what the $O(\lambda^4)$ in (30) is. This is essential for our rigorous analysis, where (29) will be established rigorously.
Finally, the fine tuning of the bare mass is achieved by iteratively restricting its range so that the running mass flows as desired, at the end becoming m^2. This succeeds, since the expansion of r_m at each step gives us room so readjust r. The idea is from [13] .

6. The rigorous analysis

Our rigorous analysis is based on the heuristic picture presented above. Crudely speaking, we wish to establish bounds for the corrections to (24)-(26) as well as to the irrelevant terms in R_L^M. These bounds need to be such as to allow us to iterate the transformation (22). Since the covariance Γ in (22) has exponential falloff and we expect the "couplings" in V to be small, it is natural to analyze the non-locality of the Z integration using a high temperature (cluster) expansion. The problem is not standard, however, since the external field Ψ' in (22) may take arbitary large values, for which V is not a small perturbation of a gaussian. Also, we need a very detailed knowledge of TV for iteration, in particular a knowledge of the "couplings"

$$\partial^n TV(0)/\partial\Psi'(x_1)\ldots\partial\Psi'(x_n) \quad . \tag{31}$$

The solution to these problems is the following. A natural devise to go from bounds on TV to properties such as bounds of (31) is analyticity. Indeed, it is easily seen that T preserves analyticity of e^{-V} in a large region, a natural one being a (poly-) strip around Ψ' real. Thus in particular we expect analyticity in a neighbourhood of $\Psi' = 0$, which allows to study (31). The size of the analyticity domain is determined by asking for which Ψ', V in (22) is a small perturbation of $d\mu_{Z^{-1}\Gamma}$: essentially, this occurs for

$$|\Psi'| < \quad 0|\lambda|^{-1/4}) \tag{32}$$

a large region for $|\lambda|$ small.

Given a field configuration Ψ', we expect the cluster

expansion to be effective in the region $\bar{Y}$ of our lattice where Ψ' satisfies (32) allowing to obtain TV there as

$$TV(\Psi'|_{\bar{Y}}) = \sum_{Y\subset\bar{Y}} V_Y(\Psi') \tag{33}$$

where the multibody potential V_Y depends on $\Psi'|_Y$ and has exponential falloff in Y. We use now the analyticity idea to extract from (33) the quadratica and quartic pieces and prove bounds for the remainder. Moreover, it is possible to show that the third order perturbative computation of these objects is asymptotic, thus rigorously establishing the estimates of section 5.

What remains, is the control of TV in the large field region

$$|\Psi'| > O(|\lambda|^{-1/4}) \tag{34}$$

Note, that the large field part of V enters to our small field analysis above: even for Ψ'small, $L^{-1}\Psi'+Z$ in (22) may take large values (34). This, however, involves large Z, having a neglible probability

$$\exp(-O(|\lambda|^{-1/2})) \tag{35}$$

due to the gaussian $d\mu_{Z}$-1_Γ, (here the ε in (5) is used) <u>provided</u> $\exp(-V(\Phi))$ won't get too big for Φ large. If this is the case, (33) makes the above small field discussion feasible.

Thus, some stability bound for $\exp(-T^M V(\Phi))$ is needed. A good one turns out to be

$$e^{-T^M V(\Phi)} < \exp(\sum_x(-|\lambda_M|^{1/2}|\Phi(x)|^2 + |\lambda_M|(\mathrm{Im}\Phi(x))^4 + C) \tag{34}$$

(36) is easily verified for (23) and its iteration succeeds since the relevant $|\Phi|^2$ term tends to increase, part of which increase may be used to kill dangerous constants.

Of course, the small and large field regions are actually coupled by $d\mu_{z^{-1}\Gamma}$ and we need a representation for a general Φ configuration, a representation which involves the Hamiltonian (33) in the small field region and only the Gibbs factor (36) in the large field region. This naturally emerges from the cluster expansion and is the following.

Let D_M be the region where $|\Phi| > (|\lambda_M|^{-\frac{1}{4}})$. $T^M V$ is analytic in $||\Im m\Phi||_\infty < (|\lambda_M|^{-1/4})$ and has the representation

$$\exp(-T^M V(\Phi)) = \sum_{\{X_j\}} \prod_j g^M_{X_j}(\Phi)$$

$$\exp(-\sum_{Y \subset X^c} V^M_Y(\Phi)) \qquad (37)$$

where X_j are deisjoint sets and $g^M_{X_j}$ depends on $\Phi|_{X_j}$. X_j are "islands" around D_M: $X = \cup X_j \supset D_M$ and $X_j \cap D_M$ is a union of components of D_M.

V^Y_M have a representation in terms of the 3rd order perturbation theory in λ_M and a remainder with proper bounds. g^M_X satisfy a bound analogous to (36). The form (37) is preserved by the iteration of T.

This analysis establishes the limit lim $R_\Lambda H_\Lambda = S^{EFF}$ as a well behaved unit lattice theory. The analysis of the effective action may be extended to a full analysis of the continuum limit correlations G_N.

References

[1] J.Glimm,A.Jaffe, Quantum Physics, Springer 1981

[2] V.Rivasseau, Preprint, Ecole Polytechnique, Jan.1984

[3] G.´t Hooft, Commun.Math.Phys.86,449 (1982), Commun.Math Phys.88,1 (1983)

[4] L.D.Landau, Coll.Papers, Gordon and Breach 1965

[5] I.Pomeranchuk,V.Sudakov,K.Ter Martirosyan, Phys.Rev. 103,784 (1956)

[6] A.Sokal, Ann.Inst.H.Poincaré,37,317 (1982)

[7] K.Wilson,J.Kogut, Phys.Rep.12C,75 (1974)

[8] J.Fröhlich,Nucl.Phys.B200,281 (1982)

[9] M.Aizenman, Phys.Rev.Lett.47 (1981), Commun.Math.Phys. 86,1 (1982)

[10] K.Symanzik, Lett.Nuovo Cim.6,no.2 (1973)

[11] K.Gawedzki,A.Kupiainen, IHES preprint, July 1984

[12] G.´t Hooft, Phys.Rep.104,129 (1984)

[13] P.M.Blecher,Ya.G.Sinai, Commun.Math.Phys.33,23 (1973)

A.Kupiainen
Helsinki University of Technology
Department of Technical Physics
Espoo 15 Finland

THE DYNAMICS OF A PARTICLE INTERACTING WITH A SEMI-INFINITE IDEAL GAS IS A BERNOULLI FLOW

C. Boldrighini, A. De Masi, A. Nogueria,
E. Presutti

Abstract

A one-dimensional semi-infinite system of point particles moving in $[0,\infty)$ is considered. The particles have mass m and are neutral except for the first one which has mass $M>m$ and charge 1. The particles undergo elastic collisions with each other and with the wall at the origin; between collisions the m-particles move with constant velocity and the M-particle with constant acceleration, E/M. A Gibbs measure exists for those values of the density and the temperature for which the thermodynamical pressure P of the free gas (of only m-particles) is larger than E.

The so defined dynamical system is a Bernoulli scheme when $E<\frac{1}{2}P$. This extends a result of [1], namely that for E=0 the system is a k-system.

0. Introduction

The ergodic properties of an infinite one-dimensional system of point particles moving on the half line $[0,\infty)$ have been studied in a recent paper, [1]. The system is interacting but the interaction is localized, as it happens in

[5], cf. also [4]. More precisely the system has infinitely many point particles which only interact through elastic collisions among each other and against a wall at the origin. All the particles have the same mass, m, except the first one which has mass M, M>m. The Gibbs measures are of course stationary and in [1] it has been proven that the dynamical system associated to any of them is a k system.

In this paper we consider a more general case, namely when a constant force E acts on the first particle. We prove that if E is not too large the system is Bernoulli, thereby strengthening the result of [1]. We hope that our paper will provide more insight into the ideas and the techniques introduced in [1] which might be relevant in solving a number of interesting open problems. We discuss some such problems in the concluding section of the present paper.

Be shall now give a heuristic outline of our results. Precise definitions and statements will be given in Section 1.

Let $q_o(t)$ denote the position of the first particle at time t, such particle will be called the "heavy particle", h.p., throughout the paper. Since all the particles with non-zero velocity will collide with the h.p. in the past or in the future (with probability 1) then it is clear that the infinite particle configuration x at time zero is (modulo zero) identified by the history of the h.p. $\{q_o(t), t\in \mathbb{R}\}$. This implies that if we consider the partition ζ generated by the "past" history of the h.p., i.e. by $\{q_o(t), t\leq 0\}$ then $\bigvee_{t\geq 0} T_t \zeta$ is the partition of the phase space into points (modulo zero) with respect to the Gibbs measure considered. The main point in [1] for proving the k-poperty, was to show that $\zeta_\infty = \bigwedge_{t\leq 0} T_t\zeta$ is, modulo zero, the trivial partition.

Proving triviality of ζ_∞ amounts to prove that the "behaviour" of $q_o(t)$ for $T\leq t\leq +\tau$, conditioned to its past (i.e. to $\{q_o(t), t\leq 0\}$) is close to the unconditioned one for any fixed $\tau>0$ and for "most" specifications of the past, when T is large enough. We shall prove here that one can take $\tau=\infty$, namely the whole future becomes independent of the past for T large. In other words we prove that the process $q_o(t)$

is α-mixing, which implies the weak Bernoulli property, and hence that the dynamical system is Bernoulli [8] (see Section 1 for the details).

A main point in [1] as well as in the present paper, is the use of a "copying procedure" by which the behaviour of the process $q_o(t)$, $t \geq T$ taken in any "atom of the past", was copied within "any other" atom of the past; i.e. it was shown that a copy of it appears in "most" other atoms of the past. This was used in [1] to prove that the conditional probabilities for different past histories when restricted to the tail σ-algebra of the far future, are equivalent. By a general result, proved in [1], this implies the k-property. The procedure used here is more straightforward. We can show that the discrepancy between the conditional processes is small, by making the copying measure preserving, and get directly the α-mixing condition and the Bernoulli (B-)property.

The copying procedure, here as in [1], is based on the occurence of a "nice event", which "reduces" the memory of the past history of the h.p. This event consists of a "cluster of fast incoming particles" which collide with the h.p. in such a way and for so long time that the h.p. never collides again with the particles with which it collided before the cluster arrived. So the future of the h.p. depends only on its position and velocity at the times when it starts colliding with the cluster ("cluster time", c.t.) and on the incoming particles. In [1] the future behaviour, after a c.t. of the h.p. with some given past was reproduced in an atom corresponding to a different past by suitably inserting a finite number of particles, and leaving the incoming particles (at c.t.) unchanged.

In this paper we show that we can choose such cluster that completely "cancels" the past. Namely, if the particle configuration in $[0,L]$ is in a compact set C, there is a configuration is $[L,+\infty)$ (cluster) such that the motion of the h.p. for large times is "focalized", i.e. is driven towards a fixed attractive cycle, independent of the choice of the configuration in C. This is the loss of memory re-

required in the proof of the B-property. We think that this property is interesting by itself and we present it separately in Sect.2.

Some modifications have to be introduced when the h.p. is subject to an external force E, which we assume for simplicity to be constant. First of all for the Gibbs measure to be invariant we must have $E<P$, where P is the thermodynamic pressure of the gas of light particles. If density and temperature are such that this inequality holds, it is easy to show that there are "focalizing clusters" similar to that for $E=0$. Troubles arises with the last part of the proof which we have not mentioned yet. Namely we need to prove that after focalization the h.p. is very unlikely to catch up with the outgoing particles of its past. We need an explicit estimate for that, since that equilibrium estimate on the displacement of the h.p. cannot be used in the untypical situation created by the cluster. This can be done only if $E<\frac{1}{2}P$. In fact our bound comes from an estimate on the momentum trasferred by the incoming (negative velocity) particles to the "moving barrier" of the h.p.. We know that their velocities eventually turn positive (after collision with the h.p.), but we do not know much about their outgoing values. So we know only that the momentum transfer due to a particle of velocity v is $m|v|$, and hence that the average momentum transfer per unit time is larger than $\frac{1}{2}P$. After this paper was completed Carlo Boldrighini and Anna De Masi have obtained a proof by a different method which extends the result to any $E<P$.

1. Definitions and main results

The one particle phase space is $\mathbb{R}_+^2 = \{(q,v)\in\mathbb{R}^2 : q\geq 0\}$ where q denotes the position of a particle and v its velocity. Let X denote the phase space of the locally finite configurations in $\mathbb{R}_+^2$, and M the σ-algebra of the Borel sets of X . If $x\in X$ and $A\subseteq\mathbb{R}_+^2$ is a measurable set we denote by x_A

the configuration $x \cap A$. $\mathcal{M}_A$ denotes the σ-algebra generated by x_A. By $\mathcal{X}_A$ we denote the phase space of a system of particles in A. Sometimes we shall not distinguish between $\mathcal{X}_A$ and the subset $\hat{\mathcal{X}}_A = \{x \in \mathcal{X} : x_{\mathbb{R}_+^2/A} = \emptyset\}$ and we shall identify the subsets of $\mathcal{X}_A$ with the corresponding subsets of $\hat{\mathcal{X}}_A$. A point $x \in \mathcal{X}$ can be identified by a sequence q_k, v_k, $k \in \mathbb{N}$, where the paticles are labelled in order of increasing position and, for equal position, of increasing velocity. M will be the mass of the first particle (with coordinate q_o) and m<M the mass of any other particle.

We fix the inverse temperature β and the chemical potential λ and we denote by μ_o the Gibbs measure (with the above parameters) for a gas of free identical particles in $0,\infty$ all having mass m. We deonte by P the corresponding thermodynamical pressure, namely

$$\beta P = \left(\frac{2\pi}{\beta m}\right)^{1/2} \exp(\beta\lambda) \tag{1.1a}$$

$$\beta P = \rho \tag{1.1b}$$

where ρ is the density of the gas.
We then fix E<P and we denote by μ the Gibbs measure defined as follows. The position q_o of the h.p. is distributed with Poisson law of parameter $\beta(P-E)$, i.e.

$$\mu(q_o > x) = [\beta(P-E)]^{-1} \int_x^\infty dy \exp[-\beta(P-E)y] \tag{1.2}$$

Given $q_o = x$ the distribution of the positions of the other particles is Poisson with parameter βP, as in μ_o. Velocities are independently distributed with Maxwellian law.

The time evolution is defined as follows. Light particles cross each other without collision. The h.p. moves with constant acceleration E/M, until a collision occurs. When it collides with the wall at the origin it just inverts its velocity, when it collides with a light particle the outgoing velocities V' of the h.p. and u' of the light particle are given in terms of the corresponding incoming velocities V,u by the elastic collision law,

$$(V,u) \longrightarrow (V',u')$$
$$V' = \alpha V + (1-\alpha)u \tag{1.3a}$$
$$u' = -\alpha u + (1+\alpha)V$$

with

$$\alpha = \frac{M-m}{M+m} \tag{1.3b}$$

The main result in this paper is

Theorem 1.1. If $E<\frac{1}{2}P$ the dynamical system $(\mathcal{X},\mu,T_t)$ is Bernoulli.

The proof of the theorem is based on the α-mixing property for the process $q_o(t)$. Namely let $\zeta_-[\zeta^\tau]$ denote the partition generated by $q_o(t)$, $t\leq 0$ $[t\geq\tau]$ and consider the quantity

$$\alpha(\tau) = \sup_{A\in\zeta^-} \sup_{B\in\zeta^\tau} |\mu(A\cap B)-\mu(A)\mu(B)| \tag{1.4a}$$

(if Π is a partition by $A\in\Pi$ we mean that A is Π-measurable) (1.4b)

The following results holds.

Theorem 1.2. The process $q_o(t)$ is α-mixing, i.e.

$$\lim_{\tau\to\infty} \alpha(\tau) = 0 \tag{1.4c}$$

Proof of Theorem 1.1. The proof follows easily from Theorem 1.2. As mentioned in the introduction $\{q_o(t),\ t\in\mathbb{R}\}$ generates the partition into points, cf. also [1]. Let π be any finite partition coarser than that generated by $q_o(t)$ for $t\in(-1,0]$. Then by (1.4) the $\bigvee_{n=-\infty}^{+\infty} T_n\pi$ of the shift T_1 is weakly Bernoulli. From this the B-property follows, [8], since we can choose a sequence of increasing partitions π_n which generate the whole $q_o(t)$, $t\in(-1,0]$ and the corresponding factors are all B-shifts.

Proof of Theorem 1.2. The rest of this section will contain an outline of the proof of the Theorem.
To prove eq. (1.4c), we introduce another coefficient, namely $\gamma(L,T)$, as follows.
For $L>0$ let

$$\mathbb{R}^2_+ = \mathbb{R}_+ \times \mathbb{R} \tag{1.5a}$$

$$R_L = \{(q,v)\in\mathbb{R}^2_1 : q\leq L\}\ \{(q,v)\in\mathbb{R}^2_+ : v\geq 0\} \tag{1.5b}$$

$$C_L = \mathbb{R}^2_+\setminus R_L \tag{1.5c}$$

and let ξ_L be the partition of X into points of R_L, namely the atom of ξ_L containing y is

$$\xi_L(y) = \{x\in X : x_{R_L} = y_{R_L}\} \tag{1.6}$$

We define

$$\gamma(L,\tau) = \sup_{\xi_L\ni A}\ \sup_{\zeta^\tau\ni B} |\mu(A\cap B)-\mu(A)\mu(B)| \tag{1.7}$$

We will prove in Section 4 that ξ_L, for L large refines on a subset of the phase space of measure close to 1.

Lemma 1.1. For any $\varepsilon>0$ there is a L_ε such that for any $L>L_\varepsilon$ and any τ

$$\alpha(\tau) \leq \gamma(L,\tau) + \varepsilon \tag{1.8}$$

The proof of Theorem 1.2. is completed by

Lemma 1.2. For any $\varepsilon>0$ there is a L_ε such that for any $L>L_\varepsilon$

$$\limsup_{\tau\to\infty} \gamma(L,\tau) < \varepsilon \tag{1.9}$$

Proof. The proof is based on coupling arguments and the copying procedure of [1], which we now outline.

Definition 1.1. Let L be fixed. We denote by ∂_L the class of the measures Q on $(X \times X)$ with the following properties.

1) the marginals of Q are μ, namely if $\hat{f}$ is a bounded measurable function on X let $f(x,x') = \hat{f}(x)$ then $Q(f)=\mu(\hat{f})$ and if $g(x,x')=\hat{f}(x')$ then $Q(g)=\mu(\hat{f})$. We say that Q is a joint representation of (X,μ), (X,μ).

2) For Q almost all atoms of $\xi_L \times \xi_L$ the following happens: fix $\xi_L(x) \times \xi_L(x')$ Q almost surely, then the marginals of $Q(\cdot|\xi_L(x) \times \xi_L(x'))$ are $\mu(\cdot|\xi_L(x))$ and $\mu(\cdot|\xi_L(x'))$, namely, using the same notation as above

$$Q(f|\xi_L(x) \times \xi_L(x')) = \mu(\hat{f}|\xi_L(x))$$

$$Q(g|\xi_L(x) \times \xi_L(x')) = \mu(\hat{f}|\xi_L(x))$$

3) The relativization of Q to $\xi_L \times \xi_L$ is $(\mu\ \xi_L)\times(\mu\ \xi_L)$. It is easy to see that if $Q\in\partial_L$ then

$$\gamma(L,\tau) \leq Q(\{(x,x'):\exists t\geq\tau \text{ and } q_o(T_t x)\neq q_o(T_t x')\}) \qquad (1.9)$$

In fact call 1_A, 1_B the characteristic functions of A and B. If $Q\in\partial_L$ then from (1.7) we get

$$|\mu(A\cap B)-\mu(A)\mu(B)| = |Q(1_B(x)[1_A(x)-1_A(x')])|$$
$$\leq Q(\{\exists t\geq\tau: q_o(T_t x)\neq q_o(T_t x')\})$$

The proof of Lemma 1.2 now follows from the following result proved in Section 4 (by Statement 4.4)

<u>Lemma 1.3.</u> For any $\varepsilon>0$ there is an L_ε such that for $L>L_\varepsilon$ there is a $Q\in\partial_L$ for which for soma $\tau>0$

$$Q(\{\exists t\geq\tau: q_o(T_t x)\neq q_o(T_t x')\}) < \varepsilon \qquad (1.10)$$

2. Loss of memory

In this section we show that any "reasonable" configuration of particles in R_L, cf. eq.(1.5b), can be completed by adding a special configuration y of particles in C_L, cf. eq. (1.5.c), in such a way that $q_o(T_t(x\ y))$ approaches asymptotically a periodic motion independently of x. This is a key point in our proof and we believe that such a property can be found for many systems consisting of a "localized" interacting subsystems in a free particle bath, to which our result can be extended.

To prove Thm. 1.1 we will have to consider small random perturbations of y. We will first examine the case E=0 then E≠0. Reasonable configurations are those in

$$\mathcal{D}(L,N,V) =$$

$$= \{x : x = x_{R_L}, g_o(x) < L, \sum_{(g,v)\in x} 1(g<2L) \le N, \max_{(g,v)\in x} 1(g<2L)|v| \le V\}, \quad (2.1)$$

where L,N,V are positive numbers, we will be interested in the case when they are all "very large".

For sake of notational simplicity in the next theorem we shall agree to have fixed a rule for the collisions which is compatible with energy and momentum conservation and which allows to define the evolution even triple or more collisions occur. This will not generate ambiguities since such events occur with probability zero.

Theorem 2.1. (E=0). For any positive L,N,V there is $y \in \mathcal{X}_{C_L}$, $\bar{\tau} \in (0,1)$, $\bar{v} = \frac{L}{\bar{\tau}}$ so that for any $x \in \mathcal{D}$ (L,N,V)

$$\lim_{t\to\infty} |q_o(T_t(xuy)) - \bar{q}_o(t)| = 0, \quad (2.2)$$

$\bar{q}_o(t)$ is a periodic function of period $\bar{\tau}$. It describes a motion with constant speed $\bar{v}$ which takes place between 0 and ½L, such that $q_o(\bar{\tau}) = \frac{1}{2}L$.

Proof. The configuration y. The first particle in y is the w_1-particle which is at L and has velocity $-w_1$. w_1 will be chosen to be much larger than V. After the w_1-particle there is a "cluster" of n-particles, the n-cluster, each particle in the n-cluster has velocity -1. All the particles in the n-cluster are at time zero in an interval of length ℓ with right end point 3/2L-1. n is much larger than N and ℓ is very small. After the n-cluster there is the w_2-particle which, at time zero, is to the right of 3/2 L (its actual position will be specified later). To the right of the w_2-particle there is the w_3-cluster, made of infinitely many particles: they have velocity $-w_3$ and they are equally spaced, the spacing is $(1-\alpha)^{-1}(1+\alpha)L$ and the first particle of the w_3-cluster is at some suitable distance from the w_2-particle, as we shall see below.

<u>Choice of the parameters.</u> After the collision with the w_1-particle, the h.p. takes a velocity between $-(1-\alpha)w_1-\alpha V$ and $-(1-\alpha)w_1+\alpha V$. We choose w_1 so large that the velocity is always negative.

The h.p. then collides with the wall at the origin, takes a positive velocity and then starts colliding with the light particles of x. After u collisions its speed will be larger than

$$\alpha^u(1-\alpha)w_1 - \sum_{i=0}^{u-1} \alpha^i(1-\alpha)V \geq \alpha^u(1-\alpha)w_i-V$$

We chose w_1 so that

$$\alpha^{N+n}(1-\alpha)w_i > V \tag{2.4}$$

As a consequence the h.p. will travel without being stopped by the particles of x and by those which are in the n-cluster. We set

$$\tau' = \frac{3/2L+L}{\alpha^N(1-\alpha)w_1-L} + \frac{\ell}{\alpha^{N+n}(1-\alpha)w_1-V}$$

so that in any case at τ' the h.p. has already interacted with all the particle of the n-cluster. We choose w_1 so large that

$$V\tau' < L/4 \tag{2.5}$$

and

$$(1+\alpha)[\alpha^{N+n}(1-\alpha)w_1 - V] - \alpha(V+1) > V \tag{2.6}$$

Therefore all the particles of x which were initially in $[0,L]$ have at time τ' positive velocity larger than V. We then take n large and ℓ small so that

$$\tfrac{3}{2}L - 1 + \tau'\alpha^n(1-\alpha)w_1 \leq \tfrac{3}{2}L \tag{2.7}$$

hence the h.p. at time τ' is still to the left of 3/2 L. We place the w_2-particle in such a way that at time τ' it is at 3/2 L. At that time the h.p. is in (3/2L - 2, 3/2 L) with velocity between $\alpha^{N+n}(1-\alpha)w_1 - V$ and $\alpha^n(1-\alpha)w_1$. We chall compare the actual evolution of the h.p. with that corresponding to the "ideal case" when at time τ' its position equals 3/2L and its speed is zero. By choosing w_2 very large we can make the actual and the ideal coordinates very "close" to each other.

Let $\bar{v}$ and $\bar{\tau}$ be the parameters which define the periodic motion $\bar{q}_o(t)$. We impose that $\tau'/\bar{\tau} = k$, k being a positive integer. We choose $w_2 = \alpha^{-}$ so that in the ideal case at time $\tau' + 2\bar{\tau}$ the h.p. will be at L/2 with positive velocity $\bar{v}$. The first w_3-particle is at time τ' at $L/2 + 2\bar{\tau}w_3 > 5/2$ L, since $w_3 = (1+\alpha)(1-\alpha)^{-\frac{1}{\bar{v}}}$ and $\bar{v}\bar{\tau} = L$.

In the ideal case therefore the h.p. moves like $q_o(t)$ after time $\tau' + 2\bar{\tau}$.

Actually this is not the case because of the "small" error with respect to the ideal case which is present at time τ'. We shall now see that the collisions with the w_3 particles are of contractive nature and lead the h.p. toward a periodic motion, i.e. $\bar{q}_o(t)$.

To understand this let us first consider the case where there are only two particles, the h.p. and a light particle, both on the line. There are two cases we consider: (a) their velocity and position are respectively (q,v) and (q',w), with $v>0$ and $w<0$, $q>q'$. (b) position and velocity are $(q+dq, v+dv)$ and (q',w), with $v+dv>0$ and $q+dq<q'$. In (a) the collision occurs at time T and in (b) at S. At time $t>\max\{T,S\}$ we have, $(\bar{q},\bar{v})$ and $(\bar{q}+d\bar{q}, \bar{v}+d\bar{v})$ denote, respectively, the coordinates of the h.p. at time t in (a) and (b),

$$
\begin{aligned}
d\bar{v} &= \alpha dv \\
d\bar{q} &= \alpha(dg+tdv)
\end{aligned}
\tag{2.8}
$$

We now compare our case with the ideal one, i.e. when at τ' the h.p. is at 3/2 L with zero speed. We can apply the above formulas if the h.p. does not recollide with the light particles and it collides when it is moving with positive velocity. In this case the collisions against the barrier at the origin do not change our analysis. We shall now set conditions for this to hold, we proceed by assuming we can apply eq. (2.8), we will then see that the result is indeed consistent with such assumption.

We will look at the coordinates of the h.p. when, in the ideal motion, it is at L/4, i.e., at the times $T_k = \tau' + \frac{3}{4}\bar{\tau}+k\bar{\tau}$, k=1,1,... At a time t, $T_k<t<T_{k+1}$, the position and velocity of the h.p. are $(L/4, \bar{V})$ in the ideal motion and $(L/4 + dq(k), V+dV(k))$ in the actual motion. By eq.(2.8) we have that for any $t>\max\{T_k,S_k\}$, $t<T_{k+1}$

$$
dq(k+1) = \alpha(t\, dV(k) + dg(k)) \tag{2.9a}
$$

$$
dV(k+1) = -\alpha dv\,(k). \tag{2.9b}
$$

As a consequence

$$|dq(n+1)| \leq \bar{\tau} \sum_{i=1}^{n} \alpha^{n+1-i}|dv(i)|+\alpha^{n}|dq(1)|.$$

Therefore

$$|dq(n+1)| \leq \bar{\tau}n \alpha^{n}|dv(1)| + \alpha^{n}|dq(1)|. \tag{2.10}$$

To apply eg. (2.8) we need that

$$|dv(1)| < \bar{V}$$
$$|dq(n)| < L/4, \quad n =1,2,\ldots .$$

By choosing w_3 very large we make $\bar{\tau}$ very small. This takes care of the first term on the r.h.s. of eq. (2.10), since the sequence $n\alpha^n$, $n=1,2,\ldots$, is bounded. For the second term it is enough to have

$$|dq(1)| < L/4.$$

<u>Definition 2.1.</u> We define here small perturbations of the configuration y introduced in Thm. 2.1. We order the particles of $x \in X_{C_L}$ in terms of the time they take to arrive at L under free motion. So $x \in X_{C_L}$ is described by the sequence $(\tau_i(x), v_i(x))$, $i \geq 1$, where $\tau_1(x) \leq \tau_2(x) \leq \ldots$ and $(\tau_i|v_i|+L,v_i)_{i\geq 1}$ is the set of he coordinates of all the particles of x. We choose δ very small, for instance $\delta < \frac{\bar{\tau}}{10}$, and we define, given L,N,V hence y,

$$N(y,\delta)=\{x\in_{C_L}X:\tau_i(y)\le\tau_i(x)\le\tau_i(y)+\delta,$$

$$|v_i(x)-v_i(y)|<\delta,\ \forall\, i\ge 1\}. \tag{2.11}$$

As a corollary of Thm.2.1, (and of its proof), we have that given any ε there is $\bar{n}_\varepsilon$ and $\delta_\varepsilon<\frac{\bar{\tau}}{10}$ so that for any $x\in X\,(L,N,V)$, $z\in N(y,\delta)$ $\delta<\delta_\varepsilon$,

$$|q_o(T_t(x\ z)-\bar{q}_o(t)|<\varepsilon \qquad \forall\ t\ge\bar{n}_\varepsilon\ \bar{\tau}$$

$$|v_o(T_{(n-1/4)\bar{\tau}}(x\ z)-\bar{v}_o(t)|<\varepsilon \quad \forall\ n\ge\bar{n}_\varepsilon$$

where $\bar{v}_o(t)$ denotes the derivative of $\bar{q}_o(t)$. We choose $\varepsilon<\frac{L}{10}$ and fix $\delta<\delta_\varepsilon$, so that at time $(n-1/4)\bar{\tau}$, $n\ge\bar{n}_\varepsilon$, the h.p. has already collided against the wall, but not yet with the next incoming particle of $N(y,\delta)$. We denote by (q_n,v_n) the position and velocity of the h.p. at time $(n-1/4)\bar{\tau}$. We call $\underline{\mathbb{P}}$ the probability on $N(y,\delta)$ and $\mathbb{P}_x$ the law of (q_n,v_n) induced by $\underline{\mathbb{P}}$ if $x\in D\,(L,N,V)$ is a configuration in X_{R_L}. The process (q_n,v_n), $n\ge n_\varepsilon$, is Markoffian since the law of (q_{n+1},v_{n+1}) is completely specified by (q_n,v_n). Actually we have

$$\underline{\mathbb{P}}_x\ (dq_{n+1}dv_{n+1}|q_n,v_n)\ =\ \rho(q_{n+1},\ _{n+1})dq_{n+1}dv_{n+1}$$

where ρ is a bounded function of its arguments uniformly in n and x. Furthermore there are positive constants $\bar{\rho}$ and a such that

$$\rho(q_{n+1},v_{n+1})\ \ge\ \bar{\rho},\ \text{if}\ \ ||(q_{n+1},v_{n+1})-\Gamma(q_n,v_n)||\ \le\ a,$$

where $||\cdot||$ is the Euclidean norm in $\mathbb{R}^2$ and $\Gamma(g_n, v_n)$ is the value of (q_{n+1}, v_{n+1}) when the first light particle which interacts after time $(n-1/4)\bar{\tau}$ has parameters as in y. We have

$$\Gamma(q_n, v_n) = (\frac{L}{4} + dq(n+1),\ \bar{V}+dv(n+1)), \tag{2.12}$$

where dq(n+1) and dv(n+1) are given by eq.(2.9).

The following is a classical theroem, see for instance [2], which applies directly to our case: let $U \subset \mathbb{R}^n$ be an open bounded neighborhood of the origin. Let $T: U \to U$, with $\overline{TU}$ U, be a continuous map having the origin as an attractive fixed point. Assume that for each $x \in TU$ there is a measurable function $y \to \rho(y, Tx)$ with support in U and that there exist positive constants $\bar{\rho}$ and a such that:
(i) $\rho(y, Tx) \geq \bar{\rho}$, if $||y-Tx|| \leq a$, (ii) $\mathbb{P}(x, dy) = \rho(y, Tx)dy$. Then P defines the transition probability of a Markov chain with state space U. The chain has only one stationary measure ν and this is absolutely continuous with respect to the Lebesgue measure and furthermore

$$\lim_n ||P^n(x,dy) - \nu(dy)|| = 0 \ ,$$

uniformly in x and exponentially fast. A similar argument was used in [3]. In our case the state space is in $\mathbb{R}^2$, i.e. the position and velocity of the h.p. observed at the times $(n-1/4)\bar{\tau}$, $n \geq n_\varepsilon$. The map T refers to the transformation (2.12).

Thm.2.1 and the above result extend easily to the case $E \neq 0$. In fact the action of the electric field can be regarded as a small perturbation if we choose the velocity of the incoming particles large enough. In this way the contractive properties of the map T remain unchanged and the previous conclusions keep their validity. We report the above considerations in Thm. 2.2. below in a form which will be suitable for the construction of the coupling.

Theorem 2.2. ($E \neq 0$). For any choice of the positive numbers L,N,V there is a configuration y in C_L and $\delta>0$ so that for any $z \in N(y,\delta)$, cf. eq.(2.11), and $x \in \mathcal{D}(L,N,V)$ the path $q_o(T_t(x\,z))$ has the following properties:

(1) there is $\tau'<1$ such that $q_o(T_{\tau'}(x \cup z)) \in (3/2L-2,\ 3/2\ L)$ and for any $t<\tau'$, $q_o(T_t(x \cup z))\ q_o(T_{\tau'}(x \cup z))$. Furthermore all the particles of x which at time zero are to the left of L have at time τ' positive velocity larger than 1.

(2) After τ' and before reaching $3/2\ L$ the h.p. undergoes a collision. Because of that the velocity of the h.p. becomes negative and without further collisions the h.p. reaches position $3/4\ L$ at time $\tau^* \leq \tau' + \bar{\tau}$, $\bar{\tau}<1$.

(3) After τ^* there are collisions with new light particles which keep the h.p. in the interval $[0,\ 3/4\ L]$ with speed less than $2\bar{v}$. The particles which collide with the h.p. after τ^* interact only once with the h.p. and after that they get some positive velocity larger than 1.

(4) Given any $\varepsilon>0$ there exists n_ε and for all $n>n_\varepsilon$ the following holds. For $S=(n-1/4)\bar{\tau}$ let

$$N(y,\delta,S) =$$

$$=\{z \in C_L(0,S):\ \exists z' \in N(y,\delta),\ z' \cap C_L(0,S) = z \cap C_L(0,S)\}$$

Let $P^{(S)}$ be the measure μ_o conditioned to $N(y,\delta,S)$. Then for any x and x' both in $\mathcal{D}(L,N,V)$ there is an isomorphism $\phi_{S,x,x'}$ of $N(y,\delta,S)$ onto itself such that the probability $P_{x,x'}$ on $N(y,\delta,S)^2$ which has support on

$$\{z,\ \phi_{S,x,x'}(z)\}$$

and with marginals equal to $P^{(S)}$, is such that

$$P_{x,x'}(\{(z,z'):q_o(T_s(x\cup z)) \neq q_o(T_s(x'\cup z'))\})<\varepsilon \quad (2.13)$$

3. Construction of the coupling

In this section we construct a coupling $Q\in\partial_L$, cf. Def. 1.1, which depends on the parameters L,N,V,..., introduced in Section 2. In the next section we will determine the values of the parameters so that Q will enjoy the properties required by Lemma 1.3 of Sect.1.

Definition 3.1. The sets J_1, J_2, J_3, . L,N,V, ,T are fixed, cf. Def.2.1. for notation. We introduce the sets

$$F_1 = \{(q,v)\in C_L(0,T): v<-2\}$$

$$F_2 = \{(q,v)\in C_L(0,T): -2 \leq v \leq 0\},$$

where

$$C_L(s,t) = \{(q,v)\in C_L:|v|^{-1}(q-L)\in(s,t)\} \quad s<t \qquad (3.1)$$

and C_L is defined in (1.5).

One should think that L is very large and $T^{-1}L$ is very small. Here T denotes the length of the "nice cluster", hence $T=(n-1/4)\bar{\tau}$ for a certain fixed n.

We set

$$J_1 = \{x\in X : x_{F_1} \in N(y,\delta,T)\} , \qquad (3.2)$$

c.f. Theorem 2.2 for the notation used here. At time T the cluster is over so we need to slow down the h.p. this is accomplished by putting particles in F_2. We define the set J_2 as follows. It has a large number of particles in F_2 with velocity in $[-2,-1]$. They would be in $[\frac{7}{4}L,+\infty]$ at time zero and in $(\frac{3}{4}L,L)$ at time T, if the motion were free. By (2) of Theorem 2.2 the h.p. will interact with the above cluster only after time T. In fact at time $\bar{\tau} + \bar{\tau}'$ the first particle will be to the right of $7/4\, L - 2(\tau'+2\bar{\tau}) > 3/2\, L$ (at least for large L). The h.p. will not cross 3/4 L before time $n\,\bar{\tau}$ (remember that $T = (n-1/4)\bar{\tau}$ because of condition (3) of Theorem 2.2. We also require that

$$\sum_{(q,v)\in x_{F_2}} m|v| > 4(2M\bar{v} + EL)$$

Notice that by Theorem 2.2 $r\bar{v}$ is an upper bound for the speed of the h.p. The motivation for the above requirement is the following: if no new light particle arrives, then the h.p. will cross L only after it has gone back to the origin from where it bounces off with speed less than 2. To have the situation described in theorems 2.1 and 2.2 we must require that the only particles in the interval L,2L which have non-positive velocity are those introduced so far. Such condition implies that no new particle interacts with the h.p. in the time interval O,T In fact the other particles are both in $C_L(T,\infty)$, cf. eq. (3.1), and to the right of 2L. A collision might occur before time T only if a particle reaches 3/2 L before time $\tau'+\bar{\tau}$, cf. Condition (2) of Theorem 2.2. Since at time zero such particle was to the right of 2L this means that under free motion it would reach L at time $2(\tau'+\bar{\tau})$. This is less than 4, we choose T large, certainly larger than 4, so the particle can only stay in $C_L(0,T)$. Besides the sets J_1 and J_2 introduced above, we will need to consider in Section 4 also the sets:

$$F_3 = \{(q,v)\in\mathbb{R}^2_+ : L\leq q\leq 2L,\ v\leq 0,\ q+vT\geq L\} \tag{3.3a}$$

$$J_3 = \{x\in X : x_{F_3} = \emptyset\} \tag{3.3b}$$

We set

$$J = J_1 \cap J_2.$$

We define

$$\mathcal{J}^{(k)} = T^o_{-t_k}\ \mathcal{J},\ k = 0,1,2,\ldots$$

where T^o_t denotes the free evolution, and $t_k = kT$. Analogously we define

$$J_i^{(k)} = T^o_{-t_k}\ J_i\ ,\quad i=1,2,3.$$

We then shorthand

$$G^{(k)} = J^{(k)} \cap C^{(k-1)} \tag{3.4a}$$

$$C^{(k)} = \bigcap_{i=0}^{k} (X \setminus J^{(i)}) \tag{3.4b}$$

so that $G^{(0)},\ldots,G^{(s)},\ C^{(s+1)}$ is a partition of X.

A final remark on the notation we are employing: in the definition of the previous sets the position of the h.p. is not mentioned. So, in principle, it is not correct to say that in $x\in J$, for instance, a cluster is arriving, because the first particle in x_{F_1} might be the h.p. itself. We will consider, however, $\mathcal{J}$ intersected with $\mathcal{D}(L,N,V)$ so, in particular, the condition $q_o(x)<L$ is fulfilled.

Definition 3.2. The coupling Q.Q is the product measure $\mu\upharpoonright \mathcal{X}_{R_L} \times \mu\upharpoonright \mathcal{X}_{R_L}$ on $\mathcal{X}_{R_L} \times \mathcal{X}_{R_L}$ cf. Def.1.1. So it remains to define the conditional probabilities of Q given $\mathcal{X}_{R_L} \times \mathcal{X}_{R_L}$. We set

$$Q(dx_{C_L}\ \ dx'_{C_L} | x_{R_L}, x'_{R_L}) = \mu(dx_{C_L} | x_{R_L})\mu(dx'_{C_L} | x'_{R_L}),$$

whenever either $q_o(x_{R_L}) \geq L$ or $q_o(x'_{R_L}) \geq L$. For notation simplicity $\mu(\cdot | x_{R_L})$ denotes the conditional probability to the atom of ξ_L which contains x_{R_L}. So we have to define Q when $q_o(x_{R_L})<L$ and $q_o(x'_{R_L})<L$. We will define a map $\Psi : \mathcal{X}_{C_L} \to \tilde{\mathcal{X}}_{C_L}$ in such a way that $Q(dx_{C_L}\ dx'_{C_L} / x_{R_L}, x'_{R_L})$ will have support on $\{(x_{C_L}, \Psi(x_{C_L}))\}$ The map Ψ depends on x_{R_L}, x'_{R_L}. Assume $x_{C_L} \in G^{(k)}$, see (3.4), $k \leq S$. Then

$$x_{C_L} \cap C_L(0,t_k) = \Psi(x_{C_L}) \cap C_L(0,t_k) := z$$

There are two cases:

(1) $y_1 := T_{t_k}(z \cup x_{R_L}) \in \mathcal{D}(L,N,V)$, $y_2 := T_{t_k}(z \cup x'_{R_L}) \in \mathcal{D}(L,N,V)$.

(2) its complement.

In case (2) we pose $x_{C_L} = \Psi(x_{C_L})$.

In case (1) we pose

$$\Psi(x_{C_L}) \cap F_1^{(k)} = T^o_{-t_k} \Phi_{y_1,y_2} (T^o_{t_k}(x_{C_L} \cap F_1^{(k)}))$$

where Φ is the map defined in Thm. 2.2. Finally

$$x_{C_L} \cap C_L \backslash F_1^{(k)} = \Phi(x_{C_L}) \quad C_L \backslash F_1^{(k)}.$$

If $x_{C_L} \in C^{(s+1)}$, see (3.4), we set $\Psi(x_{C_L}) = x_{C_L}$ therefore Ψ defines an isomoprhism of (X_{C_L}, μ_o) onto itself. We complete the definition of $Q(dx_{C_L} dx'_{C_L} / x_{R_L}, x'_{R_L})$, by requiring that its marginals are μ_o which is compatible with the above position because Ψ is an isomoprhism. By its very definition $Q \in \partial_L$ (see Def. 1.1).

4. Proof of Theorem 1.2

In this section we prove Lemma 1.1, Lemma 1.3 and then Theorem 1.2. We use extensively estimates proven in [1] which extend essentially unchanged to the case $E \neq 0$ (in such instances we will omit reporting the corresponding proofs).

We proceed as follows: we first choose the actual values of the parameters $L, N, V\delta, T, s$ introduced in the previous sections. At this stage we just want to make explicit their mutual interdependence. After that we define the coupling Q and the reason for the above choice will become clear. We first fix $\varepsilon > 0$, we are interested in the limit when ε goes to zero. We define L_ε so that for all $L \geq L_\varepsilon$

$$\mu\,(\xi_L \quad A_L) > 1-\varepsilon \tag{4.1a}$$

$$\mu\,(\tilde{\xi}_L^{(k)} \cap \tilde{A}_L^{(k)}) > 1-\varepsilon \tag{4.1b}$$

The sets appearing in eq. (4.1) will be introduced later on and, at that time, we will also prove the existence of such L_ε.

In the following we should think of ε as fixed and correspondingly L will always be taken larger than L_ε . We then fix N and V so that

$$\mu(\mathcal{D}(L,N,V)) > 1-\varepsilon \tag{4.2}$$

where $\mathcal{D}(L,N,V)$ has been define in eq.(2.1). It is easily seen that eq.(4.2) is satisfied in N and V are chosen to be large enough.

Given L,N.V as above we define y as in the proof of Theorem 2.1 and then δ, which depends on L,N,V,y, like stated in Theorem 2.2.

After that we will fix T which denotes the time-length of the cluster, and then, finally, the time we are going to wait for the cluster to arrive, this will be the time τ to which referred in Section 1.

The fist condition concerns the atoms of the past, we pose, as in [1],

$$\xi_L = \{x : q_o(T_t R_L) \le \sqrt{L}\log_+|t|,\ \forall t\le 0\} \tag{4.3a}$$

$$\Gamma_L = \{(q,v)\in C_L : q\ge L,\ 0\ge v\ge -\sqrt{L}e^{-(\frac{q}{\sqrt{L}}+1)}\} \tag{4.3b}$$

$$A_L = \{x : x_{\Gamma_L} = \emptyset\} \tag{4.3c}$$

C_L and R_L have been define in eq.(1.5). The following statement easily follows from the stationarity of the Gibbs measure, its proof is similar to that when E=0, which is reported in [1], so we do not give it

Statement 4.1. Let $x\in \xi_L \cap A_L$ then the path $q_o(T_t x)$, $t\leq 0$, is such that the h.p. has never collided with any particle which at time zero was in C_L. Hence $q_o(T_t x_{R_L}) = q_o(T_t x)$ for all $t\leq 0$.

Furthermore for any $\varepsilon>0$ there is L_ε so that eq. (4.1) holds.

Proof of Lemma 1.1. Let A be ζ-measurable and take $L\geq L_\varepsilon$. Define

$$\bar{A} = \{x : x_{R_L} \in A\}$$

hence $\bar{A}$ is ξ_L-measurable. Then, Δ below denotes symmetric difference,

$$\mu(A\Delta\bar{A}) \leq \mu((A\Delta\bar{A})\ \xi_L \cap A_L) + \varepsilon$$

On the other hand by Statement 4.1 $(A\Delta\bar{A})\ (\xi_L \cap A_L) = \emptyset$, hence Lemma 1.1. is proven.

Given L,N,V we take y and δ according to Theorems 2.1 and 2.2. We then define $T=(n-1/4)\bar{\tau}$ with n so large that (1)

$$\mu(J_3) > 1-\varepsilon \tag{4.4a}$$

where J_3 has been define in eq.(3.3). Notice in fact that by the defintion of F_3 it easily follows that

$$\lim_{T\to\infty} \mu_o(J_3) = 1-\varepsilon$$

The second requirement on T is (2)

$$P_{x,x'}(\{q_o(T_T x) \neq q_o(T_T x')\}) < \varepsilon \qquad (4.4b)$$

where $P_{x,x'}$ is the coupling defined in Theorem 2.2, cf. eq. (2.13), and both x and x′ are in $\mathcal{D}(L,N,V)$.

We will consider $x \in J^{(k)}$ for some given k, cf. Def.3.1 for notation; then everything is "prepared" for the nice cluster to arrive at time t_k. However for this to really occur we need to impose the following:

Requirement 1 The particles specified by $J^{(k)}$ should not interact with the h.p. before time t_k.

Requirement 2 They should find a configuration in R_L which belongs to $\mathcal{D}(L,N,V)$.

To fulfill the requirements we pose:

$$R_L(k) = R_L \cup C_L(0,t_k) \qquad (4.5a)$$

$$\xi_L^{(k)} = \{x : q_o(T_{t+t_k}\, x_{R_L(k)}) \leq \sqrt{L}\, \log_+|t|,$$

$$\forall t \leq 0;\ T_{t_k}\, x_{R_L(k)} \subset R_L\} \qquad (4.5b)$$

$$\Gamma_L^{(k)} = T^o_{-t_k}\, \Gamma_L = \{(q,v) \in \mathbb{R}^2_+ : (q+t_k, v) \in \Gamma_L\} \qquad (4.5c)$$

$$A_L^{(k)} = \{x : x \cap \Gamma_L^{(k)} = \emptyset\} \qquad (4.5d)$$

Statement 4.2 Let $x \in \xi_L^{(k)} \cap A_L^{(k)}$ then $T_{t_k} x = T^o_{t_k}\, x_{C_L(t_k,\infty)} \cup T_{t_k}\, R_L(k)$ and $T_{t_k}\, x_{R_L(k)} \subset R_L$. Furthermore eq. (4.1) holds for any $L \geq L_\varepsilon$.

Statement 4.2 is proven in [1], in fact the proof does not differ from the one when E=0.

We define

$$\mathcal{D}^{(k)} = \{x \in \mathcal{X} : T_{t_k} R_L(k) \in \mathcal{D}(L,N,V)\} \qquad (4.6a)$$

and we have that

$$\mu(\quad{}^{(k)}) \geq \mu(\mathcal{D}^{(k)} \cap \quad{}_L^{(k)} \cap A_L^{(k)}) > 1-\varepsilon \qquad (4.6b)$$

since by Statement 4.2

$$\mathcal{D}^{(k)} \cap \xi_L^{(k)} \cap A_L^{(k)} = \{x \in \xi_L^{(k)} \cap A_L^{(k)} : T_{t_k} x \in \mathcal{D}(L,N,V)\} \subseteq$$

$$\subseteq T_{-t_k} \mathcal{D}(L,N,V)$$

and then eq.(4.6b) follows from eq.(4.2).

We remark that the above requirements, (1) and (2), are verified in the set $\xi_L^{(k)} \cap A_L^{(k)} \cap \mathcal{D}_L^{(k)}$.

We will now specify the value of s, sT will then equal the time τ introduced in Lemma 1.3. s will be a positive integer, large enough so that the probability that a cluster arrives within sT will be close enough to one. It is possible and convenient to assume that the choice of our parameters is such that

$$\mu_o(J) < 1/2$$

s is chosen so that

$$\mu(C^{(s+1)}) = (1-\mu_o(J))^{s+1} < \varepsilon \qquad (4.7)$$

Our last concern refers to the recollisions of the h.p. with the light particles which had interacted before the arrival of the cluster.
We first define

$$\Gamma_L^{(+)} = \{(q,v)\in\mathbb{R}^2_+ : q\geq L,\ 0\leq v\leq 4q^{3/4}\, e^{-q^{1/4}}\}$$

$$\tilde{A}_L^{(k)} = \{x \in X : T_{t_k}\, x_{R_L(k)} \cap \Gamma_L^{(+)} = \emptyset\}$$

We then have the following

<u>Statement 4.3</u> For $x\in \tilde{A}_L^{(k)} \cap \xi_L^{(k)} \cap A_L^{(k)}$

$$T_{t_k} x \cap \Gamma_L^{(+)} = \emptyset$$

and for $L\geq L_\varepsilon$ $\mu(A_L^{(k)}) > 1-\varepsilon$.

The proof of the above satement easily follows from Statment 4.2 and the definition of $\tilde{A}_L^{(k)}$.

Notice that by using Theorem 2.2 we also have

$$x\in A_L^{(k)} \cap \xi_L^{(k)} \cap J^{(k)} \cap J_3^{(k)} \rightarrow T_{t_{k+1}} x \cap \Gamma_L^{(+)} = \emptyset$$

<u>Definition 4.1</u> Let N_L be thet set of configurations in R_L such that: (1) the speed of the incoming particles of x_{R_L} is less than 2, (2) if $x\in N_L$ then $T_t x_{R_L}$ evolves for $t\geq 0$ in such a way that the h.p. will cross L only after having gone on the origin, from where it bounced off with speed less than 2.

Notice that by Def. 3.1. at time t_k, after the cluster has passed away, the configuration is in N_L.

We need then

<u>Proposition 4.1.</u> Let $E<\frac{1}{2}P$. We pose

$$\tilde{\xi}_L^{(k)} = \{x\in X: \forall y\in N_L,\ q_o(T_t z) \leq \max(L,(\log_+ t)^4), \forall t\geq 0,$$

$$z = y \cup T^o_{t_k} x_{C_L(t_k,\infty)} \} ,$$

where N_L is defined in Def.4.1. Then for all $L>L_\varepsilon$

$$\mu(\tilde{\xi}_L^{(k)}) > 1-\varepsilon,\ \forall\ k\geq 1.$$

We postpone the proof Prop.4.1., which is similar to the proof of Prof.A.2 in [1], with the assumption that $E<\frac{1}{2}P$ (this is the only part in the paper where we need $E<\frac{1}{2}P$ rather than $E<P$. The choice of $\tilde{\xi}_L^{(k)}$ is such that together $\tilde{A}_L^{(k)}$, it will imply that the h.p. does not collide after t_k with outgoing particles on the right of L.

<u>Statement 4.4</u> Given any $\varepsilon>0$, take any $L>L_3$ so that (4.1) holds. Take then N,V, according to (4.2), δ as in Def.2.1, T as in (4.4), s as in (4.7). For such values of L,N,V,δ,T,s let $Q\in \partial_L$ be the coupling define in Def. 3.2. Then there is a function $\varphi(\varepsilon)$ which vanishes as ε goes to zero such that for $\tau=sT$

$$Q(\{\exists\ t\geq\tau\ :\ q_o(T_t x) \neq q_o(T_t x')\})<\varphi(\varepsilon).$$

This proves Lemma 1.3 of Section 1, which completes the proof of Theorem 1.2.

Proof of Statement 4.4. We will call $\chi(x,x')$ the characteristic function of the set $\{\exists\, t\geq\tau : q_o(T_t x)\neq q_o(T_t x')\}$. By (4.2), we have

$$\int dQ\chi(x,x')\leq 2\varepsilon+\int dQ\chi(x,x')1(q_o(x)<L)1(q_o(x')>L)$$

$$\leq 2\varepsilon+\sum_{k=0}^{s}\int dQ\chi(x,x')1(q_o(x)<L)1(q_o(x')<L)1(x\in G^{(k)})$$

$$+\int dQ\,(q_o(x)<L)1(q_o(x')<L)1(x\in{}^{(s+1)}),$$

where $G^{(k)}$ and $C^{(s+1)}$ are defined in (3.4).
WE use the following shorthand notation for $0\leq k\leq s$, where s is given by (4.7),

$$d\tilde{Q} = dQ1(q_o(x)<L)1(q_o(x')<L).$$

Let $\tilde{\chi}^{(k)}(x)$ be the characteristic function of $\xi_L^{(k)}\cap A_L^{(k)}\cap\tilde{\xi}_L^{(k+1)}\cap\tilde{A}_L^{(k)}\cap D^{(k)}\cap J_3^{(k)}$, where $J_3^{(k)}$ is defined in (3.3) and the other sets in (4.5) and (4.6). Then

$$\int d\tilde{Q}\;\chi(x,x')1(x\in G^{(k)})\leq\int d\tilde{Q}\;\chi(x,x')1(x\in G^{(k)})\tilde{\chi}^{(k)}(x)\tilde{\chi}^{(k)}(x')$$

$$+\int d\tilde{Q}1(x\in G^{(k)})\{[1-\tilde{\chi}^{(k)}(x)]+[1-\tilde{\chi}^{(k)}(x')]\}.$$

We have that

$$\chi(x,x')1(x\in G^{(k)})\tilde{\chi}^{(k)}(x)\tilde{\chi}^{(k)}(x')$$

$$\leq 1(x\in G^{(k)})\tilde{\chi}^{(k)}(x)\tilde{\chi}^{(k)}(x')1(q_o(T_{t_{k+1}}x)\neq q_o(T_{t_{k+1}}x'))$$

$$Q\text{-a.s.}\qquad(4.8)$$

In fact by the choice of $\xi_L^{(k)}$, $A_L^{(k)}$, $\mathcal{D}^{(k)}$, $\mathcal{J}^{(k)}$, we know that at time t_k the cluster is arriving both in x and x′ and that it leaves a configuration in N_L at time t_{k+1}. By the choice of $\tilde{\xi}_L^{(k+1)}$ and $\tilde{A}_L^{(k)}$ the h.p. does not interact after time t_{k+1} with outgoing particles which at time t_{k+1} are on the right of L, as well as with the particles in $F_1^{(k)}$, cf. Def. 3.1.

The argument is completed by noticing that in $F_2^{(k)}$ and $C_L(t_{k+1},\infty)$ the configurations x and x′ are the same with Q-probability one.

This r.h.s. term of (4.8) is bounded by dropping the condition that x and x′ are in $\tilde{\xi}_L^{(k+1)} \cap \tilde{A}_L^{(k)}$. We then take the conditional expectation with respect to fixing the configurations x and x′ in $R_L(k)$ (4.5a). The condition $q_o(T_{t_{k+1}} x) \neq q_o(T_{t_{k+1}} x')$ becomes the condition (cf.(4) of Thm.2.2)

$$z,z':q_o(T_s(x\ z)) \neq q_o(T_s(x'\ z'))\ ,\ s \text{ as in eq.(2.13)},$$

where $x = T_{t_k}(x_{R_L(k)})$ and $x' = T_{t_k}(x'_{R_L(k)})$ (we have used here that the remaining particles in $F_2^{(k)}$ are the same in x and x′ and that "fast" particles in F_1 do not recollide with the h.p. until time t_{k+1}). By using (4) of Thm. 2.2, we conclude that

$$\int dQ\ 1(x \in G^{(k)})\chi(x,x')\tilde{\chi}^{(k)}(x)\tilde{\chi}^{(k)}(x') \leq \varepsilon \int d\tilde{Q}\ 1\ (x \in G^{(k)}).$$

We have

$$\int dQ\, 1\,(x\in G^{(k)})[1-\tilde{\chi}^{(k)}(x)]$$

$$\leq \int d\mu\, 1\,(q_o(x)<L)1\,(x\in J^{(k)})\,1\,(x\in C^{(k-1)})$$

$$\{\,1\,(x\notin \tilde{\xi}_L^{(k)}) + 1\,(x\notin A_L^{(k)}) +$$

$$+\,1\,(x\notin \tilde{\xi}_L^{(k+1)}) + 1\,(x\notin \tilde{A}_L^{(k)}) + 1(x\notin \mathcal{D}^{(k)}) + 1(x\notin J_3^{(k)})\}$$

$$= \mu(\,J^{(k)})\mu(\,C^{(k-1)}\cap[\tilde{\xi}_L^{(k)^c} \cup \tilde{\xi}_L^{(k)^c}\,\tilde{A}_L^{(k+1)^c} \cup$$

$$\mathcal{D}^{(k)^c} \cup A_L^{(k)^c} \cup\ J_3^{(k)^c})\,,$$

Where we have used that

$$J^{(k)}\cap\ A_L^{(k)^c} = J^{(k)}\cap\ \{x\cap(T_L^{(k)}\backslash C_L(t_k,t_{k+1}))=\emptyset\}$$

$$\mu(\,J^{(k)}\cap A_L^{(k)^c} \leq \mu(\,J^{(k)})\mu(\,A^{(k)^c})$$

and analogously for $J^{(k)}\cap\ J_3^{(k)^c}$. So we get

$$\mu(C^{(k-1)}\ \xi_L^{(k)^c} \cup \tilde{\xi}_L^{(k+1)^c} \cup A_L^{(k)^c} \cup \mathcal{D}^{(k)^c} \cup\ A_L^{(k)^c} \cup\ J_3^{(k)^c}) \leq$$

$$\leq 6\sqrt{\varepsilon}(1-\mu(\,J))^{\frac{k-1}{2}}$$

and the proof is concluded by remarking that

$$\int dQ\ \chi(x,x') \leq 2\varepsilon + \varepsilon + \frac{\mu(J)12\sqrt{\varepsilon}}{1-\sqrt{1-\mu(\,J\,)}} + \mu(\,C^{(s+1)})$$

Proof of Proposition 4.1 The proof only requires a few modifications with respect to the case E=0, which is treated in [1]. For $x\in\mathcal{X}$, $q\in\mathbb{R}_+$, $t\geq 0$, $s>0$, let

$$\pi_q(t,t+s;x) = m \sum_{(q,v)\in C_q(t,t+s)\ x} |v|$$

which is the absolute value of the total momentum of the particles of x which cross q between times t and t+s. We will simply write in the following $\pi_q(t,t+s)$ instead of $\pi_q(t,t+s;x)$. We want to estimate the probability that within time T the h.p. reaches q ($\geq$L). For this to happen it is necessary that before T the h.p. collides against the wall at the origin; this is so because we know that the configuration in R_L is, at time zero, in $\mathcal{R}_L$, cf. Def. 4.1. Let t be the time of such collision and v the speed of the h.p. at that time.
Let t+s be the time when the h.p. reaches q, then, necessarily

$$MV + Es > \pi_q(t,t+s) \tag{4.9}$$

Since the average value of $\pi_q(t,t+s)$ is $\frac{1}{2}$Ps and E<$\frac{1}{2}$P, then the event in eq.(4.9) occurs only when there is a large fluctuation either of V or of $\pi_q(t,t+s)$. One then understands why the event defined by the inequality (4.9) has small probability: to prove Prop. 4.1. we simply need to make the above argument more precise. We start fixing L and then we consider the positions L, L+1, L+2, L+3, ... For notational simplicity let us suppose that L is an integer. Given $k\geq L$ we set

$$s_k = \exp((k+1^{(1/4)})$$

and

$$s_k = \{x \in C_L : \forall y \in \mathcal{N}_L \quad q_o(T_t(x\ y)) \le k \quad \forall t \le s_k\}$$

so that

$$\tilde{\xi}_L^{(o)} \subseteq \bigcap_{k \ge L} R_k$$

We define

$$R'_k = \{x \in \mathcal{X}_{C_L} : \min_{(q,v) \in \Sigma(k,s_k)} v > -v_k\}$$

where

$$\Sigma(k,s_k) = \{(q,v) \in \mathbb{R}^2_+ : L \le q \le k\} \quad C_k(0,s_k)$$

$$v_k = k^{3/8}$$

It is easy to see that

$$\lim_{k_o \to \infty} \mu_o \left(\bigcap_{k \ge k_o} R'_k \right) = 1$$

We then introduce $\tilde{s}_k$ so that

$$(2M)^{-1} E\tilde{s}_k^2 + {}_k\tilde{s}_k = k$$

and we define

$$R''_k = \{x\in C_L : \pi_k(t,t+s-2) > Mv_k + Es,\ \forall (t,s)\in \mathbb{N}^2,\ t\le s_k, s\ge s_k\}$$

We have that $R'_k \cap R''_k \cap R_k$ and it is a simple computation the proof that

$$\lim_{k_o\to\infty} \mu_o \left(\bigcap_{k\ge k_o} R''_k\right) = 1$$

cf. [1]. Proposition 4.1 is therefore proven.

5. Concluding remarks, open problems

In this section we discuss some open problems. The first and the most interesting one, in our opinion, is to prove that a particle which interacts with a large system at equilibrium approaches a Brownian motion in the "usual" space-time scaling, without the limiting assumption that the mass ratio M/m should diverge. The one dimensional version of this problem (the large system is an ideal gas at equilibrium which interacts with the test particle by elastic collisions) has been solved in the case when all masses are equal, cf. [9] for the equilibirum case and [7] for the analysis of more general situations.

The problem is still open when $M\neq m$. Even the ergodic properties are not known. Notice that a dynamical system can be defined also in the infinite case: the Palm measure which describes the ideal gas as seen from the h.p. is stationary. Such system is conjectured to be k. In fact in the infinite case we do not expect the process $q_o(t)$, $t\in\mathbb{R}$ to be α-mixing, as it is in the semi-infinite case, cf. Thm.1.1. The reason is that the analogue of Lemma 1.1 does not hold. If the h.p.

moves like $t^{1/2}$ then it will eventually collide with particles of the remote past. More precisely consider the particles which have coordinate (q,v) in the region

$$\{q \geq L,\ 0 \geq v \geq -\frac{1}{q}\} \cup \{q \leq -L,\ 0 \leq v \leq \frac{1}{|q|}\}$$

the space coordinate are chosen in the frame when the h.p. is at the origin.

Then for any choice of L the above region will contain infinitely many particles, since its Lebesgue measure is unbounded. Such particles will interact in the future with the h.p. and if the h.p. moves like $|t|^{1/2}$ there will be interactions also in the very remote past, if L is large. The copying procedure we have used in the present paper as well as that employed in [1] would therefore be destroyed by collisions with such very slow particles which are specified by the past history of the h.p.

Even assuming the k-property holds for the "real" infinite case, then, even in that case, the last step (i.e. to prove convergence to Brownian motion) might be troublesome.

There is an analogous problem also in the model we have considered here. In such case of course $t^{-1/2}\ q_o(t) \to 0$. There are however other variables for which a central limit theorem could be stated. For instance the fluctuations on the positions of the h.p. at the times when it gets a negative velocity (we are thinking now of the case E=0). Convergence to a Wiener process could be proven by obtaining a fast decay of the α-mixing coefficient [6]. Our estimate are too rough for such a purpose. We discussed this problem with Magda Peligrad, and we agreed on the following strategy. The ρ-mixing coefficient is defined as, [6],

$$\sup_{\substack{A \in \zeta \\ B \in \zeta^{\tau}}} |\mu(A\ B)-\mu(A)\mu(B)|[\bar{L}\mu(A)\mu(B)]^{-1/2} =: \rho(\tau)$$

Convergence to the normal law holds if the ρ coefficient goes to zero. No way to prove that in our case, at least with the techniques we used. So we tried the following. (1) to prove that if the supremum is taken over all subsets of a large set then it vanishes as $\tau \to \infty$ (2) to prove that in the remaining the α coefficient has fast decay properties; finally (3) to prove that when (1) and (2) both hold then there is a convergence to the normal law. We attacked this problem with Magda Peligrad, she actually accomplished her duty and solved her part (3); we did not do ours, unfortunately.

Acknowledgements

We are indebted to A. Pellegrinotti, Ya. G. Sinai and R.M. Soloveitchik for many helpful discussions and comments. We are also indebted to J.L. Lebowitz for proposing the problem and for many helpful comments. We are indebted to M. Peligrad for many helpful comments (c.f. the remarks in Setion 5). One of us (A.N.) would like to thank Francesco Guerra for the kind hospitality at the Dept. of Mathematics of the University of Rome. Three of us (C.B., A.N. and E.P.) also acknowledge the kind hospitality of the University of L'Aquila, during the workshop "Hydrodynamical Behavior of Many Particles Systems" in October of 1983.

References

[1] Boldrighini, C., Pellegrinotti, A., Presutti, E., Sinai, Ya.G., Soloveitchik, R.M. Ergodic Properties of a One Dimensional Semi-Infinite System of Statistical Mechanics. Preprint (1984).

[2] Doob, J.L. Stochastic Processes. Joh Wiley & Sons (1953).

[3] Goldstein, S., Ianiro,N., Kipnis, C. Stationary States for a Mechanical System with Stochastic Boundary Conditions (in preparation).

[4] Goldstein, S., Lebowitz,J.L., Presutti,E. Ergodic Properties of Dynamical Systems Coupled to a Thermal Reservoir. Colloquia Mathematica Societatis Janos Bolyai 27, Random Fields, Esztergom, pp. 403-419 (1979).

[5] Goldstein,S., Lebowitz,J.L., Ravishankar,K. Ergodic Properties of a System in Contact with a Heat Bath: A One Dimensional Model. Commun. in Math. Phys. 85 p. 419 (1982).

[6] Ibragimov, T.A., Linnik, Yu.V. Independent and Stationary Sequences of Random Variables. Wolters-Noordhoff Netherlands (1971).

[7] Major,P., Szasz,D. On the Effect of Collisions on the Motion of an Atom in R^1. Annals of Probability 8, p. 1068 (1980) .

[8] Ornstein, D.S. Ergodic Theory, Randomness and Dynamical Systems. Yale University Press, New Haven and London (1974).

[9] Spitzer,F. Uniform Motion with Elastic Collision of an Infinite Particle System. J. of Math. Mech. 18, p. 973 (1969).

A GENERALIZATION OF CARATEODORY'S CONSTRUCTION FOR DIMENSIONAL CHARACTERISTIC OF DYNAMIC SYSTEMS

Ya. B. Pesin

In this paper we propose a generalization of the classical Carateodory's construction of various characteristics of dimension type. Our approach allows us to obtain well-known dimensions (for example, the Hausdorff dimension) as well as new ones. In particular, for invariant sets of dynamical systems we define a class of dimensions which depend on the dynamics of the system. This class includes such well-known characteristics as topological pressure and topological entropy, and also a new notion: the dimension with respect to the map. It seems to me that the latter one can be used for the description of the topological and geometrical structure of the invariant set as well as the Hausdorff dimension. In multidimensional case we obtain the formulae, connecting this dimension with the characteristics of trajectory instability of the dynamical system (such as Lyapunov exponents). These results were obtained for the two-dimensional case in [7]. Our construction deals with continuous maps of non-compact subsets of compact metric spaces. Therefore, we can also consider discontinuous maps (they are continuous on the non-compact set which does not contain preimages of the discontinuity set), for example, one-dimensional piecewise monotonic maps, Lorentz attractor (cf. [6]) and so on.

1°. Let X be a compact topological space, $Y \subset X$. Let F be a collection of open subsets in Y. Suppose that there are three functions

$$\xi, \eta : 2^{Y} \times F \times \mathbb{R} \to \mathbb{R}^{+}, \quad \psi : F \to \mathbb{R}^{+}$$

(2^Y is a set of all subsets of Y), satisfying the following conditions: for any $Z\subset Y$, $U\in F$, $\alpha,\beta\in\mathbb{R}$.

A1) $\xi(Z,U,\alpha+\beta) \le \eta(Z,U,\beta)\xi(Z,U,\alpha)$,

A2) for any $\varepsilon>0$ there is $\delta>0$ such that for every $U\in F$ with $\Psi(U) \le \delta$ we have

$$\eta(Z,U,\beta) \le \varepsilon \quad \text{if} \quad \beta>0 \quad \text{and}$$

$$\eta(Z,U,\beta) \ge \varepsilon^{-1} \quad \text{if} \quad \beta<0 \ .$$

We put

$$M(F,\alpha,Z,\varepsilon) = \inf_{G\subset F}\left\{\sum_{U\in G}\xi(Z,U,\alpha):\Psi(U)\le\varepsilon\ ,\ \bigcup_{U\in G} U\supset Z\right\}\ ,$$

where G is a finite or countable subset in F . It is easy to see that $M(F,\alpha,Z,\varepsilon)$ does not decrease when ε tends to 0 . Therefore there exists a limit

$$m(F,\alpha,Z) = \lim_{\varepsilon\to 0} M(F,\alpha,Z,\varepsilon)\ .$$

We describe some properties of the function $m(F,\alpha,Z)$.

Theorem 1. The function $m(F,\alpha,\cdot)$ is a regular Borel outer measure on Y . That is:

1. $m(F,\alpha,\Phi) = 0$;
2. $m(F,\alpha,Z_1) \le m(F,\alpha,Z_2)$, if $Z_1\subset Z_2\subset Y$;
3. $m\left(F,\alpha,\bigcup_{i=1}^{\infty} Z_i\right) \le \sum_{i=1}^{\infty} m(F,\alpha,Z_i)$, $Z_i\subset Y$;
4. every Borel set is measurable, i.e.

$$m(F,\alpha,Z) = m(F,\alpha,Z\cap B) + m(F,\alpha,Z\setminus B)$$

for any set Z and any Borel set $B\subset Y$; in addition $m(F,\alpha,\cdot)$ is a σ -additive Borel measure on the σ -algebra of Borel sets;

5. for any $Z\subset Y$ there exists a Borel set B , $Z\subset B\subset Y$ such that $m(F,\alpha,Z) = m(F,\alpha,B)$.

The proof of this theorem is an easy modification of the arguments given in [1]. The following assertion is the direct consequence of our conditions A1 and A2 .

Theorem 2. The function $m(F,\cdot,Z)$ has the following property: there exists an α_0 such that

$$M(F,\alpha,Z) = \begin{cases} \infty , & \text{if } \alpha<\alpha_0 , \\ 0 , & \text{if } \alpha>\alpha_0 . \end{cases}$$

Let

$$\dim_C Z = \alpha_0 = \sup\{\alpha : m(F,\alpha,Z)=\infty\} = \inf\{\alpha : m(F,\alpha,Z)=0\} .$$

The value $\dim_C Z$ is called the Carateodory's dimension of Z. It depends of course on the choice of the family F and on the functions ξ, η, ψ. Therefore, we will sometimes use the notation $\dim_{C,*} Z$, where one or several parameters, which we would like to take into consideration, will be written instead of "*" .

2°. Now we formulate the basic properties of Carateodory's dimension. Their proofs follow from the definitions given above and Theorems 1 and 2.

Theorem 3.

1) $\dim_C \Phi = 0$;

2) $\dim_C Z_1 \le \dim_C Z_2$, if $Z_1 \subset Z_2 \subset Y$;

3) $\dim_C \left(\bigcup_{i=1}^{\infty} Z_i \right) = \sup_i \dim_C Z_i$, $Z_i \subset Y$.

Let X' be a compact topological space, $Y' \subset X'$, F' be a collection of open subsets in Y' , ξ', η', ψ' be three functions, satisfying conditions A1 and A2.

Theorem 4. Suppose that there exists a continuous map $\chi: X \to X'$ such that

1) $\chi(Y) = Y'$; 2) for any $Z' \subset Y'$, $U' \in F'$, $\alpha \in \mathbb{R}$ we have

$$\chi^{-1}(U') \in F , \quad \xi'(Z',U',\alpha) = \xi(\chi^{-1}(Z'),\chi^{-1}(U'),\alpha) ,$$

$$\eta'(Z',U',\alpha) = \eta(\chi^{-1}(Z'), \chi^{-1}(U'),\alpha), \psi'(U') = \psi(\chi^{-1}(U'))\} .$$

Then

$$\dim_{C,F',\xi',\eta',\psi'} Z' \ge \dim_{C,F,\xi,\eta,\psi} \chi^{-1}(Z') .$$

Moreover, if χ is a bijective map and $\chi(U) \in F'$ for any $U \in F$, then

$$\dim_{C,F',\xi',\eta',\psi'} Z' = \dim_{C,F,\xi,\eta,\psi} \chi^{-1}(Z') .$$

3°. We introduce another general class of characteristics of dimension type which are also used in the applications as well as the Carateodory's dimension. Let again X be a compact metric space, $Y \subset X$, F be some collection of open sets in Y, ξ, η, ψ be three functions, satisfying conditions A1 and A2. We put

$$R(F,\alpha,Z,\varepsilon) = \inf_{G \subset F} \left\{ \sum_{U \in G} \xi(Z,U,\alpha) : \Psi(U) = \varepsilon , \bigcup_{U \in G} U \supset Z \right\}$$

where G is a finite or countable subset of F. We denote

$$\bar{r}(F,\alpha,Z) = \overline{\lim_{\varepsilon \to 0}} R(F,\alpha,Z,\varepsilon) ,$$

$$\underline{r}(F,\alpha,Z) = \underline{\lim_{\varepsilon \to 0}} R(F,\alpha,Z,\varepsilon) .$$

One can show that the functions $\bar{r}(F,\cdot,Z)$ and $\underline{r}(F,\cdot,Z)$ have the property described in Theorem 2 and define the values which we call respectively upper and lower Carateodory's capacity of Z and denote by

$$\overline{\mathrm{Ca}}_{P_C} Z = \inf\{\alpha : \bar{r}(F,\alpha,Z) = 0\} ,$$

$$\underline{\mathrm{Ca}}_{P_C} Z = \inf\{\alpha : \underline{r}(F,\alpha,Z) = 0\} .$$

It is obvious that $\dim_C Z \le \underline{\mathrm{Cap}}_C Z \le \overline{\mathrm{Cap}}_C Z$.

4°. We consider some examples. Let X be a compact metric space, $Y=X$, F be the collection of all open sets in X. For $Z \subset X$, $U \in F$, $\alpha \in \mathbb{R}$ we put

$$\xi(Z,U,\alpha) = \eta(Z,U,\alpha) = (\mathrm{diam}\ U)^{\alpha} , \quad \Psi(U) = \mathrm{diam}\ U .$$

The dimension $\dim_{C,F,\xi,\eta,\Psi} Z$ is the Hausdorff dimension of Z (cf. [1]). We denote it by $\dim_H Z$.

5°. Let X be a compact metric space, $Y \subset X$, $f: Y \to Y$ be a continuous map, $\phi: X \to \mathbb{R}$ be a continuous function. Let also $\mathcal{U}$ be a finite open cover of X. Denote by

$\underline{U} = U_{i_0} \ldots U_{i_m}$ a collection consisting of the elements of the cover $\mathcal{U}$ ($m=m(\mathcal{U})$ is the length of the collection); $\mathcal{W}(\mathcal{U})$ is the set of all such collections. We put

$$Y^+(\underline{U}) = \{x \in Z: f^k(x) \in U_{i_k}, \ k=0,\ldots,m(U)\}, \qquad (1)$$

$$F^+ = \{Y^+(\underline{U}): U \in \mathcal{W}(\mathcal{U})\},$$

$$\xi(Z, Y^+(\underline{U}), \alpha) = \exp\left\{-\alpha m(\underline{U}) + \sup_{x \in Z \cap Y^+(\underline{U})} \sum_{k=0}^{m-1} \phi(f^k(x))\right\} \qquad (2)$$

$$\eta(Z, Y^+(\underline{U}), \alpha) = \exp(-\alpha m(\underline{U})),$$

$$\psi(Y^+(\underline{U})) = [m(\underline{U})]^{-1}.$$

It is easy to see that the functions ξ, η, ψ satisfy conditions A1, A2. The dimension $\dim_C Z$ depends also on the map f, the function ϕ and the cover $\mathcal{U}$. We denote it by

$$P^+_{f,Z}(\mathcal{U}, \phi).$$

One can show that there exists the limit

$$P^+_{f,Z}(\phi) = \lim_{\operatorname{diam} \mathcal{U} \to 0} P^+_{f,Z}(\mathcal{U}, \phi).$$

It is called the topological pressure of the function ϕ on the set Z (with respect to f). Our definition of the topological pressure includes the case of discontinuous maps and also the case of a noncompact and noninvariant set Z. The value $h^+_Z(f) = P^+_{f,Z}(0)$ is called the topological entropy of Z (with respect to f).

6°. The definition of the topological pressure for non-compact subsets of compact metric spaces were introduced by B. S. Pizkel and the author in [4]. In that paper the problems of variational principle for the topological pressure and the existence of equilibrium states were also considered. We formulate some of the results obtained there.

Theorem 5. Let f be a continuous map of a compact metric space X. Then the definition of the topological pressure given above is equivalent to the "usual" one (cf., for example [2]), and the definition of the topological entropy is equivalent to Bowen's definition (cf. [3]).

Denote by $M_f(X)$ the set of probability Borel f-invariant measures in X; by $M_f(Y)$ the set of $\mu \in M_f(X)$ satisfying $\mu(Y) = 1$ and similarly $M_f(Z)$ for an f-invariant subset $Z \subset Y$.

Theorem 6. For $\mu \in M_f(Y)$

$$h_\mu(f|Y) + \int_Y \phi d\mu \le P^+_{f,Y}(\phi) .$$

Let $x \in Y$. Consider the sequence of measures $\mu_{x,n} = \frac{1}{n} \sum_{k=0}^{n-1} \delta_{f^k(x)}$, where δ_y is measure concentrated on y. Denote by $V(x)$ the set of limit points (in the weak topology) of the sequence $\mu_{x,n}$. It is easy to see that $V(x) \subset M_f(X)$.

Theorem 7. Let $Z \subset Y$ be an f-invariant subset, $Z_1 = \{x \in Z : V(x) \cap M_f(Z) \neq \emptyset\}$. Then for any continuous function ϕ on X

$$\sup_{\mu \in M_f(Z)} (h_\mu(f|Z) + \int_Z \phi d\mu) = P^+_{f,Z_1}(\phi) .$$

Consequence 1. Suppose that for any $x \in Y$ the intersection $V(x) \cap M_f(Y) \neq \emptyset$. Then

$$\sup_{\mu \in M_f(Y)} (h_\mu(f|Y) + \int_Y \phi d\mu) = P^+_{f,Y}(\phi) .$$

Let $\mu \in M_f(Y)$ be an ergodic measure. Denote by G_μ the union of all forward generic points for μ, $G_\mu = \{x \in Y : \mu_{x,n}$ weakly tend to $\mu\}$.

Consequence 2. For any ergodic $\mu \in M_f(Y)$

$$h_\mu(f|Y) + \int_X \phi d\mu = P^+_{f,G_\mu}(\phi) .$$

The measure $\mu = \mu_\phi$ is called the equilibrium state for the function ϕ if $\mu_\phi \in M_f(Y)$ and

$$h_{\mu_\phi}(f|Y) + \int_X \phi d\mu_\phi = \sup_{\mu \in M_f(Y)} (h_\mu(f|Y) + \int_X \phi d\mu) .$$

The next assertion is the generalization of Bowen's

criterion [2].

Theorem 8. Suppose that f satisfies the following conditions:

1) f is a homeomorphism of Y ;

2) f separates the points of Y ;

3) the set $M_f(Y)$ is closed in the weak topology in $M_f(X)$.

Then for any continuous function ϕ on X there exists the equilibrium state μ_ϕ .

7°. We put

$$Y^-(\underline{U}) = \{x \in Y : f^{-k}(x) \in U_{i_k} , \ k=0,\dots,m(\underline{U})\} ,$$

$$F^- = \{Y^-(\underline{U}) : \underline{U} \in \mathcal{W}(\mathcal{U})\} .$$

It is easy to see that the topological pressure $P^-_{f,Z}(\phi)$ constructed using the collection F^- and the functions ξ, η, ψ (defined by (2) with $-k$ instead of k) coincides with $P^+_{f^{-1},Z}(\phi)$.

We put for $\underline{U}^1, \underline{U}^2 \in \mathcal{W}(\mathcal{U})$

$$Y(\underline{U}^1,\underline{U}^2) = \{x \in Y : \ f^{-k}(x) \in U^1_{i_k} , \ k=1,\dots,m(U^1), \quad f^k(x) \in U^2_{i_k} , \ k=0,\dots,m(\underline{U}^2)-1\} .$$

Let also

$$F = \{Y(\underline{U}^1,\underline{U}^2) : \ \underline{U}^1,\underline{U}^2 \in \mathcal{W}(\mathcal{U})\} .$$

Denote by $P_{f,Z}(\phi)$ the topological pressure, defined using the collection F as pointed out above. Recently B. S. Pizkel has shown that

$$P_{f,Z}(\phi) \le \min\{P^+_{f,Z}(\phi), \ P^-_{f,Z}(\phi)\}$$

and constructed an example of a noncompact f -invariant set (where f is a subshift of finite type) for which these three values are different.

8°. We consider some more examples of Carateodory's dimension. Let X be a compact metric space with the metric ρ , $Y \subset X$ and let $f : Y \to Y$ be a continuous map, ν a probability Borel measure in X, for which $\nu(Y)=1$. Let $W\subset Y$ be some f-invariant subset. Fix $\delta>0$ and put for $x\in W$

$$Y_\delta(n,x) = \{y\in Y: \rho(f^k(x), f^k(y)) \le \delta, k=0,\dots,n\}.$$

Suppose that the map f satisfies the following condition A3) for every small $\gamma>0$ there exist the sets $W_k^{(\gamma)}$, $k=0,1,2,\dots$ having the following properties:
a) $W_k^{(\gamma)} \subset W_{k+1}^{(\gamma)}$, b) $\bigcup_{k\ge 0} W_k^{(\gamma)} = W$, c) for any $k\ge 0$, $\varepsilon>0$ there is $N=N(k,\varepsilon)$ such that for any $x\in W_k$ and $n>N$

$$\nu(Y_\delta(n,x)) \le \varepsilon. \tag{3}$$

We introduce condition A3 because inequality (3) usually does not hold uniformly in $x\in W$ for diffeomorphisms of general type. We put

$$F_{\delta,k}^{(\gamma)} = \{Y_\delta(n,x): x\in W_k^{(\gamma)}, n\in \mathbb{Z}^+\},$$

$$\xi_{\delta,k}^{(\gamma)}(Z,Y_\delta(n,x),\alpha) = \eta_{\delta,k}^{(\gamma)}(Z,Y_\delta(n,x),\alpha) = \nu(Y_\delta(n,x))^\alpha,$$

$$\psi_{\delta,k}^{(\gamma)}(Y_\delta(n,x)) = n^{-1}.$$

It is easy to see that the functions $\xi_{\delta,k}^{(\gamma)}$, $\eta_{\delta,k}^{(\gamma)}$, $\psi_{\delta,k}^{(\gamma)}$ satisfy conditions A1, A2 and define Crateodory's dimension which we denote by $\dim_{C,\delta,k,\gamma} Z$, $Z\subset W_k^{(\gamma)}$. For $Z\subset W$ we call the value

$$\dim_f Z = \overline{\lim_{\gamma\to 0}} \sup_{k\ge 0} \overline{\lim_{\delta\to 0}} \dim_{C,\delta,k,\gamma}(W_k^{(\gamma)} \cap Z)$$

the dimension of Z with respect to f. It has the properties formulated in Theorem 3.

In a similar way one can define upper and lower Carateodory's capacities of the set Z with respect to the map f. We denote them by $\bar{C}_f(Z)$ and $\underline{C}_f(Z)$ respectively.

Now let μ be a Borel measure on X for which $\mu(W)=1$ (we do not suppose here that μ,ν are f-invariant, but later on we will consider only the case where μ is f-invariant). The values

$$\dim_f \mu = \inf\{\dim_f Z : Z\subset W,\ \mu(Z)=1\} ,$$

$$\bar{C}_f(\mu) = \overline{\lim_{\delta\to 0}} \inf_Z\{\bar{C}_f(Z) : Z\subset W,\ \mu(Z)\geq 1-\delta\} ,$$

$$\underline{C}_f(\mu) = \overline{\lim_{\delta\to 0}} \inf\{\underline{C}_f(Z) : Z\subset W,\ \mu(Z)\geq 1-\delta\}$$

are called respectively measure dimension with respect to f , upper and lower capacity with respect to f .

Theorem 9. Let μ, ν be Borel measures in X , for which $\mu(W)=1$, $\nu(Y)=1$. Suppose that the topological dimension of X is finite. Assume also that there is $Z\subset W$, $\mu(Z)=1$ having the following properties: for any $\gamma>0$, $k\geq 0$ there exist a $\delta_{k,\gamma} > 0$ such that for any $\delta\leq\delta_{k,\gamma}$ and $x \in W_k^{(\gamma)} \cap Z$

$$\underline{\alpha}(\delta,\gamma) \leq \underline{\lim}_{n\to\infty} \frac{\log \mu(V_\delta^{(\gamma)}(n,x))}{\log \nu(V_\delta^{(\gamma)}(n,x))} \leq$$

$$\leq \overline{\lim_{n\to\infty}} \frac{\log \mu(V_\delta^{(\gamma)}(n,x))}{\log \mu(V_\delta^{(\gamma)}(n,x))} \leq \bar{\alpha}(\delta,\gamma) .$$

Let also $\underline{\alpha} \leq \alpha(\delta,\gamma) \leq \bar{\alpha}(\delta,\gamma) \leq \bar{\alpha}$ for any δ, γ . Then

$$\underline{\alpha} \leq \dim_f\mu \leq \bar{\alpha} ,\ \underline{\alpha} \leq \underline{C}(\mu) \leq \bar{C}(\mu) \leq \bar{\alpha} .$$

Consequence 3. If the hypotheses of Theorem 9 hold and there exists the limit

$$\overline{\lim_{\gamma\to 0}}\ \overline{\lim_{\delta\to 0}}\ \underline{\alpha}(\delta,\gamma) = \overline{\lim_{\gamma\to 0}}\ \overline{\lim_{\delta\to 0}}\ \bar{\alpha}(\delta,\gamma) = \alpha ,$$

then $\dim_f \mu = \underline{C}(\mu) = \bar{C}(\mu) = \alpha$.

10°. Let f be a diffeomorphism of class C^2 of a smooth compact Riemannian p -dimensional manifold M ; let μ be an f -invariant ergodic Borel measure with nonzero Lyapunov exponents (cf. [8]) $\chi_\mu^1 \geq \ldots \geq \chi_\mu^k > 0 > \chi_\mu^{k+1} \geq \geq\ldots\geq \chi_\mu^p$.

We put $X=M$, $Y=X$. Let W be a set of regular (in the sense of Lyapunov) points in X whose exponents are equal to χ_μ^i , $i=1,\ldots,p$, $W_k^{(\gamma)}$ be a set of points with uniform estimates (cf. [8]); ν be a Riemannian volume in X .

Theorem 10. There is a set $Z \subset W$, satisfying the hypotheses of Theorem 9 and Consequence 3, and

$$\alpha = h_\mu(f) \Big/ \sum_{i=1}^{k} \chi_\mu^i , \tag{4}$$

where $h_\mu(f)$ is the metric entropy of f.

So, the measure dimension, upper and lower capacities with respect to f are equal to the expression in the right side of (4). If we consider the map f^{-1} instead of f then one can prove that

$$\dim_{f^{-1}} \mu = h_\mu(f) \Big/ \Big| \sum_{i=k+1}^{p} \chi_\mu^i \Big| .$$

It is easy to show that $\dim_{f^{-1}} \mu$ coincides with the measure dimension constructed by means of the collection $\tilde{F}_\delta$, formed by the sets

$$\tilde{Y}_\delta(n,x) = \{y\ Y : \rho(f^{-k}(x), f^{-k}(y)) \le \delta,\ k=0,\dots,n\} .$$

The differences between the collections F_δ and $\tilde{F}_\delta$ involve the differences between $\dim_f \mu$ and $\dim_{f^{-1}} \mu$. Therefore, it may be interesting to study the measure dimension with respect to f constructed with the help of a symmetric (with respect to forward and backward iterations) collection of sets

$$\{Y_\delta(n,x) \cap \tilde{Y}_\delta(m,x) : x \in W,\ n,m \in \mathbb{Z}^+\} .$$

However, I do not know any results for such a dimension.

11°. In some cases it is interesting to know not the dimensions of the set itself, but of its cross-section in some direction. For example, in the case of a hyperbolic attractor one calculates the dimension of its intersection with the stable layer because it has Cantor structure in this direction. The dimensions $\dim_f \mu$ and $\dim_{f^{-1}} \mu$ measure just the dimensions of the invariant sets in the direction of unstable or respectively stable layers (it is connected with the choice of F_δ and $\tilde{F}_\delta$). Another approach to the definition of the dimension with respect to a map in the direction of unstable or stable layers is given in [5].

In that paper there are formulae similar to (4) (and also other results) for the calculation of this dimension for locally maximal hyperbolic sets.

Results similar to (4) were obtained recently in [9] for attractors of Lorenz type.

References

[1] Federer H. Geometric Measure Theory. Berlin, Springer-Verlag, 1969.

[2] Bowen R. Equilibrium States and the Ergodic Theory of Anosov Diffeomorphisms. Lect. Notes Math. V. 470, p. 108, 1975.

[3] Bowen R. Topological Entropy for Noncompact Sets. Trans. Amer. Math. Soc. 184, 125-136 (1973).

[4] Pesin Ya. B., Pizkel B. S. Topological Pressure and Variational Principle for Noncompact Sets. Funct. Anal. and Its Appl. (in Russian), 1984 (to appear).

[5] Pesin Ya. B. On the Notion of the Dimension with Respect to a Dynamical System. Ergod. Theory and Dyn. Syst. 1983 (to appear).

[6] Afraimovich V. A., Pesin Ya. B. Ergodic Properties and Dimension of Lorenz Type Attractors. Proc. of Intern. Symp. of Non-linear and Turbulence Processes in Phys. Kiev, 1983.

[7] Young L.-S. Dimension, Entropy and Lyapunov Exponents. Ergod. Theory and Dyn. Syst. N 2, 109-124 (1982).

[8] Pesin Ya. B. Characteristic Lyapunov Exponents and Smooth Ergodic Theory. Russ. Math. Surveys N 32, 55-114 (1977).

[9] Afraimovich V. A., Pesin Ya. B. The Dimension of the Attractors of Lorenz Type (to appear).

CONVERGENCE OF IMAGES OF CERTAIN MEASURES

M. Misiurewicz and A. Zdunik

Abstract. Let X be a compact metric space, let $f: X \to X$ be an expansive homeomorphism satisfying the specification property and let μ be an invariant probabilistic totally ergodic measure with the whole X as a support. We prove that then there exists a probabilistic measure κ with $\operatorname{supp} \kappa = X$ such that $f_*\kappa$ is absolutely continuous with respect to κ and the averages $\frac{1}{n}\sum_{i=0}^{n-1} f_*{}^i\kappa$ converge to μ, but the averages $\frac{1}{n}\sum_{i=0}^{n-1} f_*{}^i\delta_x$ are dense in the set of all invariant probabilistic measures for κ-almost every x.

0. Introduction and statement of the result

We start with some motivations. For a diffeomorphism Φ of a compact manifold M (with or without a boundary) into itself there are usually many invariant probabilistic measures. Then one would like to know whether any of them are of special interest from some point of view. Certainly, the Lebesgue measure or any measure equivalent to it (at least this equivalence class is well defined on M; we take one of its representants and denote it by λ) would be such measure. If none of these measures is invariant or even if it is not obvious that one of them is invariant, we can use the following procedure, motivated by the ideas from physics. We choose a point $x \in M$ "at random" and look at the sequence of averages

$$\frac{1}{n}\sum_{j=0}^{n-1} \Phi_*{}^i\delta_x \tag{1}$$

where δ_x is the probabilistic measure concentrated at x and Φ_* is the transformation on measures induced by Φ, i.e. for a measure ξ and a Borel set A, $(\Phi_*\xi)(A) = \xi(\Phi^{-1}A)$. If for λ-almost every x (this explains what "at random" means) this sequence converges in the weak $-*$ topology to some measure μ, then μ is our special measure. Existence of such measures is known in some cases (see e.g. [4]).

By integrating the averages with respect to λ, we get

$$\frac{1}{n}\sum_{i=0}^{n-1} \Phi_*^{\,i}\lambda \to \mu \quad \text{as} \quad n \to \infty . \tag{2}$$

This fact alone is also interesting. Such convergence is often used, for instance for finding absolutely continuous invariant measures for interval maps (see e.g. [3]). In such a way, a problem arises: whether (under some reasonable conditions on the system (M,Φ,μ)) it can happen that (2) holds but there is no convergence of (1) to μ for λ-almost every x.

The conditions imposed on the system should eliminate trivial examples, like the measure μ concentrated at two points y and z and the averages (1) approaching δ_y and δ_z on subsequences. From this point of view, the assumption (iii) below seems reasonable.

In this paper, we consider the problem similar to the above one, but in a more general case. Instead of the manifold M, a diffeomorphism Φ and a Lebesgue measure λ, we take a compact metric space X, a homeomorphism f and any measure κ. Since we have less assumptions, it is easier to find an example. However, we obtain the result much stronger than the example. Now we shall state this result precisely.

Let (X,d) be a compact metric space, μ a probabilistic measure on X, $f: X \to X$ a homeomorphism such that:

(i) f is expansive: there exists $\varepsilon>0$ such that if $x,y \in X$, $x \neq y$, then $d(f^n x, f^n y) > \varepsilon$ for some integer n.

(ii) f satisfies the specification property (cf. [1] Def. (21.1)): for every $\delta>0$ there exists a positive integer $\alpha(\delta)$ such that for any sequence of trajectories

$$\begin{aligned}&\left\{x_1, fx_1, \ldots, f^{t_1-1}x_1\right\}\\&\left\{x_2, fx_2, \ldots, f^{t_2-1}x_2\right\}\\&\ldots\ldots\ldots\ldots\ldots\ldots\\&\left\{x_k, fx_k, \ldots, f^{t_k-1}x_k\right\}\end{aligned}$$

and integers $\alpha_1, \alpha_2, \ldots, \alpha_k \geq \alpha(\delta)$ there exists a periodic point y of period $\alpha_1+\ldots+\alpha_k+t_1+\ldots+t_k$ such that $d(f^i x_p, f^m y) < \delta$ where $m = i + \sum_{j=1}^{p-1} (t_j+\alpha_j)$ for $1 \leq p \leq k$, $0 \leq i < t_p$ (we say that the orbit of y δ-specifies the trajectories $\left\{(f^i x_p)_{i=0}^{t_p-1}\right\}_{p=1}^{k}$.

(iii) μ is f-invariant, totally ergodic (i.e. ergodic for all f^n , $n \geq 1$) and supp $\mu = X$ (i.e. $\mu(A) > 0$ for every non-empty open set A).

We consider the space $\mathfrak{M}(X)$ of all probabilistic measures on X and its subspace $\mathfrak{M}(X,f)$ of all probabilistic f-invariant measures on X , both with the weak -* topology.

Theorem. If X, f, μ satisfy conditions (i)-(iii), then there exists a measure $\kappa \in \mathfrak{M}(X)$ such that

(iv) supp $\kappa = X$,

(v) $f_*\kappa$ is absolutely continuous with respect to κ ,

(vi) $\lim_{n\to\infty} \frac{1}{n} \sum_{i=0}^{n-1} f_*^1 \kappa = \mu$,

(vii) for κ-almost every point $x \in X$, the set of all points of condensation of a sequence $\left(\frac{1}{n} \sum_{i=0}^{n-1} f_*^i \delta_x\right)_{n=1}^{\infty}$ is equal to $\mathfrak{M}(X,f)$.

To prove Theorem, we construct (using coding technique) a Borel map $g\colon X \to X$ such that the measure $\nu = g_*\mu$ has properties (vi) and (vii) (with κ replaced by ν). Then we show that the measure $\kappa = \sum_{k=0}^{\infty} 2^{-k-1} f_*^k \nu$ has properties (iv)-(vii).

We denote by N the set of all non-negative integers.

1. Decomposition of N

Suppose we have four sequences of positive integers $(c_n)_{n=1}^{\infty}$, $(i_n)_{n=1}^{\infty}$, $(p(k))_{k=1}^{\infty}$ and $(r(k))_{k=1}^{\infty}$ and that the sequence $(i_n)_{n=1}^{\infty}$ is increasing.

We shall call a finite set of consecutive integers a <u>block</u> and denote by $|\cdot|$ its length (i.e. the number of elements). We divide N into consecutive blocks $T_1, T_2, \ldots$ (which we write $N = T_1 T_2 \ldots$), then we subdivide each T_n: $T_1 = R_1 R_2 \ldots R_{i_1}$, $T_n = C_n^o R_{i_{n-1}+1} R_{i_{n-1}+2} \ldots R_{i_n}$ for $n>1$; and at last we subdivide each R_k :

$$R_k = P_k^1 C_n^{k,1} P_k^2 C_n^{k,2} \ldots P^{r(k)} C_n^{k,r(k)} \quad \text{where} \quad i_{n-1}+1 \le k \le i_n .$$

In such a way we obtain

$$N = \underbrace{\underbrace{P_1^1 C_1^{1,1} \ldots P_1^{r(1)} C_1^{1,r(1)}}_{R_1} \ldots \underbrace{P_{i_1}^1 C_1^{i_1,1} \ldots P_{i_1}^{r(i_1)} C_1^{i_1,r(i_1)}}_{R_{i_1}}}_{T_1}$$

$$\underbrace{\ldots C_n^o \underbrace{P_{i_{n-1}+1}^1 C_n^{i_{n-1}+1,1} \ldots P_{i_{n-1}+1}^{r(i_{n-1}+1)} C_n^{i_{n-1}+1,r(i_{n-1}+1)}}_{R_{i_{n-1}+1}}}_{T_n}$$

$$\underbrace{\ldots \underbrace{P_{i_n}^1 C_n^{i_n,1} \ldots P_{i_n}^{r(i_n)} C_n^{i_n,r(i_n)}}_{R_{i_n}}}_{T_n} \ldots$$

We advise the reader to consult this scheme often while reading the paper.

We require that $|C_n^o| = |C_n^{k,j}| = c_n$ and $|P_k^j| = p(k)$.

Hence, the sequences $(c_n)_{n=1}^{\infty}$, $(i_n)_{n=1}^{\infty}$, $(p(k))_{k=1}^{\infty}$ and $(r(k))_{k=1}^{\infty}$ determine the whole division described above.

2. Sequences

Our four sequences have to satisfy certain conditions. In order to state them, we set

$$u(k) = \sum_{j=1}^{k} p(j) \quad \text{for } k=1,2,3,\ldots ,$$

$$v(k) = \begin{cases} |R_1|+\ldots+|R_{k-1}| & \text{if } R_k \subset T_1 \\ |T_1|+\ldots+|T_{n-1}|+|C_n^o|+|R_{i_n+1}|+\ldots+|R_{k-1}| & \text{if } R_k \subset T_n \text{ and } n \geq 2 \end{cases}$$

for $k=1,2,3,\ldots$;

$$A(x,n) = \{z \in X : d(f^j x, f^j z) < \varepsilon \text{ for } j=1,\ldots,n\}$$

for $x \in X$ and $n>0$ (ε is the constant of expansiveness);

$$B(x,y) = \{k>0 : f^{u(k)} y \in A(x,p(k+1))\} \quad \text{for } x,y \in X .$$

The conditions are the following:

(S1) $c_n \geq \alpha(2^{-n-1}\varepsilon)$ for $n=1,2,\ldots$,

(S2) $c_{n+1} \geq c_n$ for $n=1,2,\ldots$,

(S3) $i_{n+1} \geq i_n$ for $n=1,2,\ldots$,

(S4) $\lim_{n\to\infty} \dfrac{c_n}{p(i_{n-1})} = 0$,

(S5) $p(k+1) \geq p(k)$ for $k=1,2,\ldots$,

(S6) $\lim_{k\to\infty} p(k) = \infty$,

(S7) for μ-almost all $y \in X$ and all periodic points $x \in X$ the set $B(x,y)$ is infinite,

(S8) $\lim_{k\to\infty} \dfrac{v(k)}{p(k)r(k)} = 0$.

We have to show that such sequences exist. The existence of a sequence $(c_n)_{n=1}^{\infty}$ satisfying (S1) and (S2) is obvious. Once sequences $(c_n)_{n=1}^{\infty}$ and $(p(k))_{k=1}^{\infty}$ are chosen and (S6) is satisfied, the existence of a sequence $(i_n)_{n=1}^{\infty}$ satisfying (S3) and (S4) is also obvious. Then a sequence $(r(k))_{k=1}^{\infty}$ satisfying (S8) can be defined easily by induction (notice that $v(k)$ depends only on $r(j)$ with $j<k$). Thus, the only problem is to show that a sequence $(p(k))_{k=1}^{\infty}$ satisfying (S5)-(S7) exists.

Since f is expansive, so is f^n , and thus the set of fixed points of f^n is finite. For positive integers m, n set $D(n,m) = \{y\in X\colon$ if $f^n x = x$ then there exists $j \in \{0,1,\dots,m-1\}$ such that $f^{jn}y \in A(x,n)\}$.

Since $\operatorname{supp}\mu = X$, we have $\mu(A(x,n)) > 0$. Since μ is totally ergodic, by the ergodic theorem for f^n there exists $m_n>0$ such that $\mu(D(n,m_n)) \geq 1-2^{-n}$. Now we set

$$p(k) = n \quad \text{if} \quad \sum_{j=1}^{n-1} m_j < k \leq \sum_{j=1}^{n} m_j .$$

Clearly, (S5) and (S6) are satisfied. In order to prove (S7), fix n and x such that $f^n x = x$. If $y \in \bigcap_{i=1}^{\infty} f^{-t_i} D(in,m_{in})$, where $t_i = u\left(\sum_{j=1}^{in-1} m_j\right)$, then for each $i\geq q$, by the definition of $D(in,m_{in})$, we have $f^{t_i+k_i in} y \in A(x,in)$ for some $k_i \in \{0,1,\dots,m_{in}-1\}$ (if $f^n x = x$ then $f^{in}x = x$ for all i). By the definition of $p(k)$ and $u(k)$ we have $t_i+k_i in = u\left(\sum_{j=1}^{in-1} m_j\right)+k_i in = u(q_i)$ and $p(q_i+1) = in$ for some $q_i \geq \sum_{j=1}^{in-1} m_j$. By the definition of $B(x,y)$, this q_i belongs to $B(x,y)$. Since $\lim_{i\to\infty} \sum_{j=1}^{in-1} m_j = \infty$, this proves that for such y the set $B(x,y)$ is infinite.

Since the measure μ is invariant, we have for each n

$$\liminf_{q\to\infty} \mu\left(\bigcap_{i=q}^{\infty} f^{-t_i} D(in,m_{in})\right) \geq$$

$$\geq \liminf_{q\to\infty} \left(1-\sum_{i=q}^{\infty} (1-\mu(D(in,m_{in})))\right) \geq$$

$$\geq \liminf_{q\to\infty} \left(1-\sum_{i=1}^{\infty} 2^{-in}\right) = 1 ,$$

and consequently (S7) is satisfied.

3. The map g

We shall define the map g as a limit of Borel maps, attaining only finite number of values each. We proceed by induction.

To define g_1, for $x \in X$ consider a sequence of trajectories

$$\left.\begin{matrix} (x, fx, \ldots, f^{u(1)}x) \\ \cdots\cdots\cdots\cdots \\ (x, fx, \ldots, f^{u(1)}x) \end{matrix}\right] \quad r(1) \text{ times}$$

$$\left.\begin{matrix} (f^{u(1)+1}x, \ldots, f^{u(2)}x) \\ \cdots\cdots\cdots\cdots \\ (f^{u(1)+1}x, \ldots, f^{u(2)}x) \end{matrix}\right] \quad r(2) \text{ times}$$

$$\cdots\cdots\cdots\cdots$$

$$\left.\begin{matrix} \cdots\cdots\cdots\cdots \\ (f^{u(i_1-1)+1}x, \ldots, f^{u(i_1)}x) \\ \cdots\cdots\cdots\cdots \\ (f^{u(i_1-1)+1}x, \ldots, f^{u(i_1)}x) \end{matrix}\right] \quad r(i_1) \text{ times}$$

Denote this sequence by $G_1(x)$. Set $U_x^1 = \{y \in X : d(f^k x, f^k y) < 2^{-3}\varepsilon$ for all k with $0 \leq k \leq u(i_1)\}$. The family $\{U_x^1\}_{x \in X}$ is an open cover of a compact space X. There exists a finite subcover $\{U_j^1\}_{j=1}^{k_1}$. It is easy to show (see e.g. [2], proof of Theorem 4.2) that there exists a Borel partition $\{A_j^1\}_{j=1}^{k_1}$ of X such that $A_j^1 \subset U_j^1$ and $\mu(\partial A_j^1) = 0$ for $j = 1, 2, \ldots, k_1$, where ∂A denotes the boundary of A. We may assume that all sets A_j^1 are non-empty. Then we can fix a point $x_j^1 \in A_j^1$ for each j. From (S1) it follows that there exists a periodic point $g_1(x_j^1)$ of period $|T_1|$ which orbit $2^{-2}\varepsilon$-specifies the sequence $G_1(x_j^1)$. We set $g_1 y = g_1(x_j^1)$ for all $y \in A_j^1$. By the definition of $U_{x_j^1}^1$, for each $y \in X$ the orbit of $g_1 y$ $2^{-1}\varepsilon$-specifies $G_1(y)$.

Note the connection between the above construction and the structure of T_1.

Now we assume that the partition $\{A_j^{n-1}\}_{j=1}^{k_{n-1}}$ and the map g_{n-1} are already defined, and that g_{n-1} is constant on the sets A_j^{n-1}. We consider the open cover of the closure of A_j^{n-1} by the sets $U_x^n = \{y \in X : d(f^k x, f^k y) < 2^{-n-2}\varepsilon$ for all k with $0 \le k \le u(i_n)\}$. Then we choose its finite subcover and take a finer partition of A_j^{n-1} into Borel sets with boundaries of measure zero. By taking the union of these partitions over j from 1 to k_{n-1}, we obtain a partition $\{A_i^n\}_{i=1}^{k_n}$ finer than $\{A_j^{n-1}\}_{j=1}^{k_{n-1}}$ such that for each i, $A_i^n \subset U_x^n$ for some x and $\mu(\partial A_i^n) = 0$. Again we can assume that each A_i^n is non-empty and fix a point $x_i^n \in A_i^n$.

To define g_n, for $x \in X$ consider a sequence of trajectories $G_n(x)$:

$$
\begin{array}{ll}
\left\{g_{n-1}x, f(g_{n-1}x), \ldots, f^{|T_1|+\ldots+|T_{n-1}|}(g_{n-1}x)\right\} & \\
\left.\begin{array}{l}
\left\{f^{u(i_{n-1})+1}x, \ldots, f^{u(i_{n-1}+1)}x\right\} \\
\cdots\cdots\cdots\cdots\cdots\cdots \\
\left\{f^{u(i_{n-1})+1}x, \ldots, f^{u(i_{n-1}+1)}x\right\}
\end{array}\right] & r(i_{n-1}+1) \text{ times} \\
\cdots\cdots\cdots\cdots\cdots\cdots & \\
\cdots\cdots\cdots\cdots\cdots\cdots & \\
\left.\begin{array}{l}
\left\{f^{u(i_n-1)+1}x, \ldots, f^{u(i_n)}x\right\} \\
\cdots\cdots\cdots\cdots\cdots\cdots \\
\left\{f^{u(i_n-1)+1}x, \ldots, f^{u(i_n)}x\right\}
\end{array}\right] & r(i_n) \text{ times}
\end{array}
$$

From (S1) it follows that for each i there exists a periodic point $g_n(x_i^n)$ of period $|T_1|+\ldots+|T_n|$ which orbit $2^{-n-1}\varepsilon$-specifies the sequence $G_n(x_i^n)$. We set $g_n y = g_n(x_i^n)$ for all $y \in A_i^n$.

If $y \in A_i^n$ then by the definition of U_x^n we have $d(f^k y, f^k x_i^n) < 2^{-n-1}\varepsilon$ for all k with $0 \le k \le u(i_n)$. Since $\{A_i^n\}_{i=1}^{k_n}$ is finer than $\{A_j^{n-1}\}_{j=1}^{k_{n-1}}$, we have $g_{n-1}y = g_{n-1}x_i^n$.

Therefore for all $y \in X$ the orbit of $g_n y$ $2^{-n}\varepsilon$ -specifies $G_n(y)$.

Note again the connection between the above construction and the structure of the blocks $T_1, T_2, \ldots, T_n$.

Set $P = \bigcup_{k,j} P_k^j$, $C = N \setminus P$. For $m \in P$ define $s(m) = u(k-1)+i$ if m is the i -th element of P_k^j . Let $y \in X$. Since the orbit of $g_n y$ $2^{-n}\varepsilon$ -specifies $G_n(y)$, we obtain:

if $m \in T_1 \cup \ldots \cup T_{n-1}$ then $d(f^m(g_n y), f^m(g_{n-1} y)) < 2^{-n}\varepsilon$,

if $m \in T_n$ and $m \in P$ then $d(f^m(g_n y), f^{s(m)} y) < 2^{-n}\varepsilon$.

We can draw two conclusions from this.

The first conclusion is that for all n and y $d(g_n y, g_{n-1} y) < 2^{-n}\varepsilon$. Hence, the sequence $(g_n y)_{n=1}^{\infty}$ converges. We set $gy = \lim_{n\to\infty} g_n y$. The map g is Borel as a limit of Borel maps.

The second conclusion is that if $m \in P \cap T_j$ and $j \le n$ then $d(f^m(g_n y), f^{s(m)} y) \le (2^{-j} + 2^{-j-1} + \ldots + 2^{-n})\varepsilon$. Therefore

$$\text{if } m \in P \cap T_j \text{ then } d(f^m(gy), f^{s(m)} y) \le 2^{-j+1}\varepsilon \quad . \tag{3}$$

4. The measures $\frac{1}{n} \sum_{i=0}^{n-1} f_*^i (g_* \mu)$

Set $\nu = g_* \mu$. We are going to prove that

$$\mu = \lim_{n\to\infty} \frac{1}{n} \sum_{i=0}^{n-1} f_*^i \nu \; .$$

Fix a Borel set A with $\mu(\partial A) = 0$. For $\eta > 0$ denote

$A+\eta = \{x \in X: \operatorname{dist}(x, A) \le \eta\}$,

$A-\eta = X \setminus \{x \in X: \operatorname{dist}(x, X \setminus A) \le \eta\}$.

Fix $\gamma > 0$. Since $\mu(\partial A) = 0$, there exists $\delta > 0$ such that $\mu(A+\delta) < \mu(A) + \gamma$ and $\mu(A-\delta) > \mu(A) - \gamma$.

Now we choose $n_o \in N$ such that

$$2^{-n_o+1}\varepsilon < \delta \tag{4}$$

and

$$\frac{c_j}{p(i_{j-1})} < \gamma \quad \text{for all } j \ge k_o \tag{5}$$

where $n_o \in T_{k_o}$ (this can be done by (S4)).

It follows from (4) and (3) that if $n \geq n_o$, $n \in P$ and $x \in X$ then $d(f^n(gx), f^{s(n)}x) < \delta$. Hence if $n \geq n_o$ and $n \in P$ then $f^{-s(n)}(A-\delta) \subset g^{-1}(f^{-1}A) \subset f^{-s(n)}(A+\delta)$. Therefore for $n \geq n_o$ if $n \in P$ then

$$\mu(A)-\gamma < \mu(A-\delta) = \mu(f^{-s(n)}(A-\delta)) \leq \mu(g^{-1}(f^{-n}A)) \leq$$

$$\leq \mu(f^{-s(n)}(A+\delta)) = \mu(A+\delta) < \mu(A)+\gamma .$$

Since $\mu(g^{-1}(f^{-n}A)) = (f^n_*(g_*\mu))(A) = (f^n_*\nu)(A)$, we obtain

$$|(f^n_*\nu)(A)-\mu(A)| < \gamma \quad \text{for } n \geq n_o , \ n \in P . \tag{6}$$

We have to estimate the number of elements of C between n_o and $n-1$. The set $\{n_o, n_o+1, \ldots, n-1\}$ is divided into blocks of elements of C and P, as in Section 1 (we call them C-blocks and P-blocks respectively). If the first or/and the last block is shorter than in Section 1 (i.e. if n_o or/and $n-1$ is not the first (resp. last) element of the corresponding block from Section 1), we either take the whole Section 1 block - if it is a C-block, or omit it - if it is a P-block. After such modification, we shall have perhaps more elements of C and less elements of P. If n is sufficiently large, we have at least one P-block. Then to each of considered C-blocks we assign one of considered P-blocks (the closest one to the left or to the right). If the length of a C-block is c_j then the length of the assigned P-block is $p(m)$ for some $m \geq i_{j-1}$. By (5) and (S5) their ratio is smaller than γ. Since there are no more than 2 consecutive C-blocks, to each P-block we assigned at most 4 C-blocks.

Hence, we get for n sufficiently large

$$\frac{\text{Card}(\{n_o, n_o+1, \ldots, n-1\} \cap C)}{\text{Card}(\{n_o, n_o+1, \ldots, n-1\} \cap P)} < 4\gamma ,$$

and consequently

$$\frac{\text{Card}(\{n_o, n_o+1, \ldots, n-1\} \cap C)}{n-n_o} < 4\gamma .$$

Thus, in view of (6), we obtain for n sufficiently large $\left|\sum_{i=0}^{n-1} (f_*^i\nu)(A) - n\mu(A)\right| \leq \sum_{i=0}^{n-1} |(f_*^i\nu)(A) - \mu(A)| \leq$

$$\leq n_o + 4\gamma(n-n_o) + \sum_{\substack{i=n_o \\ i\in P}}^{n-1} |(f_*^i\nu)(A) - \mu(A)| \leq n_o + 5\gamma(n-n_o) .$$

Hence, $\limsup_{n\to\infty} \left|\frac{1}{4} \sum_{i=0}^{n-1} (f_*^i\nu)(A) - \mu(A)\right| \leq 5\gamma$. Since $\gamma>0$ was chosen arbitrarily, we obtain

$$\lim_{n\to\infty} \frac{1}{n} \sum_{i=0}^{n-1} (f_*^i\nu)(A) = \mu(A) .$$

Since A was an arbitrary Borel set with $\mu(\partial A) = 0$, we obtain $\lim_{n\to\infty} \frac{1}{n} \sum_{i=0}^{n-1} f_*^i\nu = \mu$ in the weak-* topology (see e.g. [1], Proposition (2.7)).

5. The measures $\frac{1}{n} \sum_{i=0}^{n-1} f_*^i\delta_y$

We start by proving that for every Lipschitz continuous function $\phi: X \to R$ and every $y\in X$

$$\lim_{k\to\infty} \left(\frac{1}{v(k+1)} \sum_{m=0}^{v(k+1)-1} (f_*^m(g_*\delta_y))(\phi) - \frac{1}{p(k)} \sum_{i=u(k-1)+1}^{u(k)} (f_*^i\delta_y)(\phi)\right) = 0 . \tag{7}$$

Let ϕ satisfy the Lipschitz condition with a constant L . We fix $\eta>0$ and then take k_o such that

$$\frac{c_n}{p(i_{n-1})} < \eta \quad \text{if} \quad R_k\subset T_n \quad \text{and} \quad k\geq k_o \tag{8}$$

$$\frac{v(k)}{p(k)r(k)} < \eta \quad \text{if} \quad k\geq k_o \tag{9}$$

$$\frac{1}{p(k)} < \eta \quad \text{if} \quad k\geq k_o . \tag{10}$$

Such k_o exists by (S4), (S8) and (S6). We have

$$(f_*^m(g_*\delta_y))(\phi) = \delta_y(\phi\circ f^m\circ g) = \phi(f^m(gy)) \tag{11}$$

and

$$(f^i\delta_y)(\phi) = \delta_y(\phi\circ f^i) = \phi(f^i y) . \qquad (12)$$

Consider a block $P_k^j \subset T_n$. For all $m \in P_k^j$ we have by (3) $|\phi(f^m(gy))-\phi(f^{s(m)}y)| \le 2^{-n+1}\varepsilon L$, and hence

$$\sum_{m\in P_k^j} |\phi(f^m(gy))-\phi(f^{s(m)}y)| \le 2^{-n+1}\varepsilon L\cdot p(k) . \qquad (13)$$

By arguments similar as in Section 4, we get in view of (8) if $k \ge k_o$

$$\frac{\text{Card } (\{v(k),v(k)+1,...,v(k+1)-1\} \cap C)}{v(k+1)-v(k)} < 2\eta$$

(the estimate is slightly better now because of the definitions of $v(k)$ and $v(k+1)$. Thus,

$$\sum_{\substack{m=v(k)\\ m\in C}}^{v(k+1)-1} |\phi(f^m(gy))| \le 2\eta v(k+1)\|\phi\| \qquad (14)$$

(where $\|\cdot\|$ is the sup norm).

If m runs over the set $\{v(k),v(k)+1,...,v(k+1)-1\}\cap P$, then $s(m)$ runs over the set $\bigcup_{j=1}^{r(k)} P_k^j$. Therefore from (11)-(14) we obtain for $k \ge k_o$

$$\left|\sum_{m=0}^{v(k+1)-1} (f_*^m(g_*\delta_y))(\phi) - r(k) \sum_{t=u(k-1)+1}^{u(k)} (f_*^t\delta_y)(\phi)\right| \le$$

$$\le \left|\sum_{\substack{m=v(k)\\ m\in P}}^{v(k+1)-1} \phi(f^m(gy)) - r(k) \sum_{t=u(k-1)+1}^{u(k)} \phi(f^t y)\right| +$$

$$+ \sum_{\substack{m=v(k)\\ m\notin P}}^{v(k+1)-1} |\phi(f^m(gy))| + \sum_{m=0}^{v(k)-1} |\phi(f^m(gy))| \le$$

$$\le 2^{-n+1}\varepsilon\cdot L\; p(k)r(k) + \|\phi\|(2\eta v(k+1)+v(k)) .$$

Since $p(k)r(k) < |R_k| \le v(k+1)$ and by (9), we obtain

$$\left|\frac{1}{v(k+1)} \sum_{m=0}^{v(k+1)-1} (f_*^m(g_*\delta_y))(\phi) - \frac{r(k)}{v(k+1)} \sum_{t=u(k-1)+1}^{u(k)} (f_*^t\delta_y)(\phi)\right| \le$$

$$\le 2^{-n+1}\varepsilon L + \|\phi\|\left(2\eta + \frac{v(k)}{p(k)r(k)}\right) \le 2^{-n+1}\varepsilon L + 3\eta\|\phi\| .$$

Notice that in view of (8) and (9)

$$1-\frac{p(k)r(k)}{v(k+1)}=\frac{(v(k+1)-v(k)-p(k)r(k)+v(k)}{v(k+1)}\le$$

$$\le\frac{c_n r(k)+c_{n+1}+v(k)}{p(k)r(k)}<\frac{2c_{n+1}}{p(k)}+\eta<3\eta \quad \text{. Hence,}$$

$$\left|\frac{1}{v(k+1)}\sum_{m=0}^{v(k+1)-1}(f_*^m(g_*\delta_y))(\phi)-\frac{1}{p(k)}\sum_{t=u(k-1)+1}^{u(k)}(f_*^t\delta_y)(\phi)\right|\le$$

$$\le 2^{-n+1}\varepsilon L+3\eta\|\phi\|+\frac{3\eta}{p(k)}\sum_{t=u(k-1)+1}^{u(k)}|\phi(f^t y)|\le 2^{-n+1}\varepsilon L+6\eta\|\phi\| \ .$$

Since $n\to\infty$ as $k\to\infty$ and $\eta>0$ was arbitrary, we obtain (7).

Now we fix a point $x\in X$ such that $f^n x=x$ for some n and set $\lambda_x=\frac{1}{n}(\delta_x+\delta_{fx}+\ldots+\delta_{f^{n-1}x})$. We shall prove that for μ-almost all $y\in X$ the sequence $\left(\frac{1}{m}\sum_{i=0}^{m-1}f_*^i(g_*\delta_y)\right)_{m=1}^{\infty}$ has a subsequence convergent to λ_x.

Take y for which the set $B(x,y)$ is infinite (see Section 2). Let ϕ be as before. Fix $\eta,\theta>0$. Since f is expansive with the constant ε, there exists a positive integer t such that for each $z_1,z_2\in X$ if $d(f^i z_1,f^i z_2)<\varepsilon$ for all i with $|i|<t$ then $d(x,y)<\eta$ (see [1], Corollary 1, p. 109).

Assume that $k\in B(x,y)$ and

$$p(k)\ge\frac{t}{\theta} \ . \tag{15}$$

For all m with $u(k-1)+t+1\le m\le u(k)-t$ we have (by the definitions of $B(x,y)$ and $A(x,n)$) $d(f^{m+j}y,f^{m-u(k-1)+j}x)<\varepsilon$ for all j with $|j|<t$ and consequently

$$d(f^m y,f^{m-u(k-1)}x)<\eta \ . \tag{16}$$

Thus, we have in view of (15) and (16) (if θ is sufficiently small)

$$\left|\frac{1}{p(k)}\sum_{i=1}^{p(k)}(f_*{}^i\delta_x)(\phi)-\frac{1}{p(k)}\sum_{m=u(k-1)+1}^{u(k)}(f_*^m\delta_y)(\phi)\right|=$$

$$=\frac{1}{p(k)}\left|\sum_{m=u(k-1)+1}^{u(k)}\phi(f^{m-u(k-1)}x)-\sum_{m=u(k-1)+1}^{u(k)}\phi(f^m y)\right|\le$$

$$\leq \frac{1}{p(k)} \left(L \sum_{m=u(k-1)+t+1}^{u(k)-t} d(f^{m-u(k-1)}x, f^m y) + 2t\|\phi\| \right) \leq$$

$$\leq \frac{1}{p(k)} (Lp(k)\eta + 2p(k)\theta\|\phi\|) = L\eta + 2\theta\|\phi\| .$$

Since $\theta>0$ is arbitrary, the set $B(x,y)$ is infinite and $\lim_{k\to\infty} p(k) = \infty$, we obtain

$$\lim_{\substack{k\to\infty \\ k\in B(x,y)}} \left(\frac{1}{p(k)} \sum_{i=1}^{p(k)} (f_*^i \delta_x)(\phi) - \frac{1}{p(k)} \sum_{m=u(k-1)+1}^{u(k)} (f_*^m \delta_y)(\phi) \right) = 0 .$$

Together with (7) this gives

$$\lim_{\substack{k\to\infty \\ k\in B(x,y)}} \left(\frac{1}{v(k+1)} \sum_{m=0}^{v(k+1)-1} (f_*^m (g_* \delta_y))(\phi) - \frac{1}{p(k)} \sum_{i=1}^{p(k)} (f_*^i \delta_x)(\phi) \right) = 0 .$$

Since $f^n x = x$, we have $\lim_{k\to\infty} \frac{1}{p(k)} \sum_{i=1}^{p(k)} (f_*^i \delta_x)(\phi) = \lambda_x(\phi)$ and therefore

$$\lim_{\substack{k\to\infty \\ k\in B(x,y)}} \frac{1}{v(k+1)} \sum_{m=0}^{v(k+1)-1} (f_*^m (g_* \delta_y))(\phi) = \lambda_x(\phi) .$$

This equality holds for every Lipschitz continuous function ϕ . But Lipschitz continuous functions are dense in the space of all continuous functions on X (this follows for instance from Stone-Weierstrass' theorem) and it follows easily that this equality holds also for every continuous function $\phi: X \to R$. Therefore

$$\lim_{\substack{k\to\infty \\ k\in B(x,y)}} \frac{1}{v(k+1)} \sum_{m=0}^{v(k+1)-1} f_*^m (g_* \delta_y) = \lambda_x . \tag{17}$$

The space $\mathfrak{M}(X,f)$ is metrizable. Since the maps $z \mapsto \delta_z$ and f are continuous, so is the map $z \mapsto \frac{1}{n} \sum_{i=0}^{n-1} f_*^i \delta_z$ for each n . Therefore the sets $E(x,n,m) = \left\{ z\in X: \operatorname{dist}\left(\frac{1}{n} \sum_{i=0}^{n-1} f_*^i \delta_z , \lambda_x \right) < \frac{1}{m} \right\}$ are open, and consequently the set $F(x) = \{z\in X: \lambda_x$ is a point of condensation of the sequence

$$\left\{\frac{1}{n}\sum_{i=0}^{n-1} f^{i}\delta_z\right\}_{n=1}^{\infty} = \bigcap_{m=1}^{\infty}\ \bigcap_{k=0}^{\infty}\ \bigcup_{n=k}^{\infty} E(x,n,m)$$

is a Borel set. By (17) and (S7) and since $g_*\delta_y = \delta_{gy}$, we have $\nu(F(x)) = 1$.

Denote $F = \left\{z \in X:\right.$ the set of all points of condensation of the sequence $\left\{\frac{1}{n}\sum_{i=0}^{n-1} f^{i}\delta_z\right\}_{n=1}^{\infty}$ is equal to $\left.\mathfrak{M}(X,f)\right\}$.

It is clear that every point of condensation of such sequence belongs to $\mathfrak{M}(X,f)$. On the other hand, if z belongs to $F(x)$ for all periodic points x then, since the set $\{\lambda_x : x \text{ is periodic}\}$ is dense in $\mathfrak{M}(X,f)$ (see [1], Proposition (21.8)), z belongs also to F. Consequently, $F = \bigcap\{F(x) : x \text{ is periodic}\}$. The map f is expansive and therefore the set of all periodic points is at most countable. Hence, F is a Borel set and $\nu(F) = 1$.

6. The measure κ

In previous sections we have constructed the measure $\nu \in \mathfrak{M}(X)$ satisfying the properties (vi) and (vii) (with κ replaced by ν).

We set now $\kappa = \sum_{k=0}^{\infty} 2^{-k-1} f_*^k\nu$. Clearly, $\kappa \in \mathfrak{M}(X)$. It is also clear that κ satisfies (v). We shall show that it satisfies (iv). If U X is an open non-empty set then there exists a continuous non-negative function $\phi: X \to R$ which is equal to 0 outside U, but is positive on some open non-empty set contained in U. Since $\operatorname{supp}\mu = X$, we have $\mu(\phi) > 0$. Since $\lim \frac{1}{n}\sum_{i=0}^{n-1} f_*^i\nu = \mu$, we have $\left(\frac{1}{n}\sum_{i=0}^{n-1} f_*^i\nu\right)(\phi) > 0$ for some n, and hence $(f_*^i\nu)(\phi) > 0$ for some i. Therefore $\kappa(\phi) > 0$ and consequently $\kappa(U) > 0$. Thus, $\operatorname{supp}\kappa = X$.

Now we have to show that κ satisfies the conditions (vi) and (vii). If $\lambda \in \mathfrak{M}(X)$ and $\phi: X \to R$ is a continuous function then we have

$$\left|\left(\frac{1}{n}\sum_{i=0}^{n-1} f_*^{i+k}\lambda\right)(\phi) - \left(\frac{1}{n+k}\sum_{j=0}^{n+k-1} f_*^{j}\lambda\right)(\phi)\right| =$$

$$= \left|\sum_{j=0}^{k-1} \frac{1}{n+k}(f_*^j\nu)(\phi) + \sum_{j=k}^{n+k-1}\left(\frac{1}{n}-\frac{1}{n+k}\right)(f_*^j\lambda)(\phi)\right| \le \qquad (18)$$

$$\le \left[k\,\frac{1}{n+k} + n\left(\frac{1}{n}-\frac{1}{n+k}\right)\right]\|\phi\| \le \frac{k+n+k-n}{k+n}\|\phi\| \le \frac{2k}{n}\|\phi\| ,$$

where $\|\phi\| = \sup_{x\in X} |\phi(x)|$.

Since $\lim_{n\to\infty} \frac{1}{n}\sum_{i=0}^{n-1} f^i\nu = \mu$, also $\lim_{n\to\infty} \sum_{k=0}^{\infty} 2^{-k-1} \frac{1}{n+k}\sum_{j=0}^{n+k-1} f_*^k\nu = \mu$. We have by (18), for all continuous functions $\phi: X \to R$,

$$\left|\left(\frac{1}{n}\sum_{i=0}^{n-1} f_*^i\left(\sum_{k=0}^{\infty} 2^{-k-1} f_*^k\nu\right)\right)(\phi) - \right.$$

$$\left. - \left(\sum_{k=0}^{\infty} 2^{-k-1}\frac{1}{n+k}\sum_{j=0}^{n+k-1} f_*^j\nu\right)(\phi)\right| \le$$

$$\le \sum_{k=0}^{\infty} 2^{-k-1}\left|\left(\frac{1}{n}\sum_{i=0}^{n-1} f_*^{i+k}\nu\right)(\phi) - \left(\frac{1}{n+k}\sum_{j=0}^{n+k-1} f_*^j\nu\right)(\phi)\right| \le$$

$$\le \sum_{k=0}^{\infty} 2^{-k-1}\frac{2k}{n}\|\phi\| = \frac{1}{n}\sum_{k=0}^{\infty} k\,2^{-k}\|\phi\| \xrightarrow[n\to\infty]{} 0 ,$$

and hence $\lim_{n\to\infty} \frac{1}{n}\sum_{i=0}^{n-1} f_*^i\kappa = \mu$. This proves (vi).

By (18) applied to $\lambda=\delta_x$, the sets of points of condensation of the sequences $\left(\frac{1}{n}\sum_{i=0}^{n-1} f_*^{i+k}\delta_x\right)_{n=1}^{\infty}$ and $\left(\frac{1}{n}\sum_{i=0}^{n-1} f_*^i\delta_x\right)_{n=1}^{\infty}$ are equal. Since $f_*^{i+k}\delta_x = f_*^i\delta_{f^k x}$, this means that the set F is f -invariant. Therefore $(f_*^k\nu)(F) = \nu(F) = 1$ for all k , and consequently $\kappa(F)=1$. This proves (vii).

References

[1] Denker M., Grillenberger C., Sigmund K. Ergodic theory on compact spaces. Lect. Notes in Math. 527, Springer, Berlin-Heidelberg-New York, 1976.

[2] Misiurewicz M. Topological conditional entropy. Studia Math. 55, 175-200 (1976).

[3] Misiurewicz M. Maps of an interval, in: Chaotic Behaviour of Deterministic Systems (Les Houches, Session XXXVI, 1981), North-Holland, 1983.

[4] Ruelle D. A measure associated with Axiom A attractors. Amer. J. Math. 98, 619-654 (1976).

CLUSTER EXPANSION FOR UNBOUNDED NON-FINITE POTENTIAL

R. R. Akhmitzjanov, V. A. Malyshev and E. N. Petrova

Let $\mathbb{Z}^\nu \subset R^\nu$, $\nu \geq 2$, be the ν-dimensional lattice, $r_{tt'}$ the distance between $t,t' \in \mathbb{Z}^\nu$.

Let $\Lambda \subset \mathbb{Z}^\nu$ be a finite volume. We consider the Gibbs measure μ_Λ on $\mathbb{R}^\Lambda$:

$$\mu_\Lambda(x) = Z_\Lambda^{-1} \exp\{-\lambda \cdot U_\Lambda(x) - \sum_{t\in\Lambda} |x_t|^m\} dx \tag{1}$$

where $x=\{x_t, t\in\Lambda\} \in R^\Lambda$, $dx = \prod_{t\in\Lambda} dx_t$, $\lambda>0$ is small, $m>0$, and

$$U_\Lambda(x) = \sum_{t,t'\in\Lambda} U_{tt'}(x_t,x_{t'}) = \sum_{t,t'\in\Lambda} x_t^{2\kappa} x_{t'}^{2\kappa} r_{tt'}^{-(\nu+\varepsilon)} \tag{2}$$

$\kappa>0$ is an integer, $\varepsilon>0$, the sum runs over all pairs t,t' from Λ .

The partition function is

$$Z_\Lambda = \int_{R^\Lambda} \exp\{-\lambda U_\Lambda(x) - \sum_{t\in\Lambda} |x_t|^m\} dx .$$

Our main result is the following

Theorem. Let the parameters ν, m, ε, κ satisfy the inequality

$$m\nu + m\varepsilon - 2\nu\kappa \geq 0 . \tag{3}$$

Then there exists a $\lambda_o>0$ such that for each $0<\lambda<\lambda_o$ the partition function Z_Λ has the cluster expansion

$$Z_\Lambda = \sum_{\Gamma_1,\ldots,\Gamma_n} C^{|\Lambda\setminus\cup\Gamma_i|} \kappa_{\Gamma_1}\cdot\ldots\cdot\kappa_{\Gamma_n} \tag{4}$$

where the sum runs over all collections $\{\Gamma_1,\ldots,\Gamma_n\}$ of pairwise nonintersecting subsets of Λ : $\Gamma_i \subset \Lambda$, $i=1,\ldots,n$, C depends on the parameters λ and m , but is bounded from below by an absolute constant:

$$C = C(\lambda,m) \geq 2e^{-1} \tag{5}$$

We denote by $|A|$ the cardinality of the set $A \subset \mathbb{Z}^\nu$.

Moreover, the values of κ_Γ satisfy the cluster estimate: for each N

$$\sum_{\substack{\Gamma:\Gamma\ni o \\ |\Gamma|=N}} |\kappa_\Gamma| \quad (\delta(\lambda))^N \tag{6}$$

The sum in (6) runs over all sets $\Gamma \subset \mathbb{Z}^\nu$ such that Γ contains the origin and has fixed cardinality N , and $\delta(\lambda) \to 0$ when $\lambda \to 0$.

The potential $U_{tt'}$ in consideration is non-finite and unbounded. In the case of finite unbounded potential the only condition for the existence of the cluster expansions is the boundedness of the potential from below. For non-finite, but bounded from above potential condition $\varepsilon>0$ is sufficient. Both these results were obtained in [2]. In both of these cases the initial independent measure is arbitrary and not necessarily has the density $\exp\{-\sum_{t\in\Lambda} |x_t|^m\}$ as we have in (1). Cluster expansion for non-finite unbounded potential is also established in [1]. In terms of our paper conditions in [1] are as follows: $m>4\kappa$, $\varepsilon>5\nu$. We improve these conditions.

The main idea of the expansion is the following. We choose some barrier B and in the case when the values of the random field in consideration are less than B we use the known techniques (see [2]) of expansion and estimation. We build some neighbourhoods of such $t\in\mathbb{Z}^\nu$, for which $|x_t| > B$, and unite them into clusters. In this case we get the cluster estimate because of the smallness of $\exp(-|x_t|^m)$.

Note that because of (4)

$$Z_\Lambda \cdot C^{-|\Lambda|} = \sum_{\Gamma_1,\ldots,\Gamma_n} \left(\kappa_{\Gamma_1} C^{-|\Gamma_1|}\right) \cdots \left(\kappa_{\Gamma_n} C^{-|\Gamma_n|}\right)$$

and hence the standard cluster techniques can be used only in the presence of an estimate:

$$\sum_{\substack{\Gamma:\Gamma\ni o \\ |\Gamma|=N}} |\kappa_\Gamma| C^{-|\Gamma|} \le (\delta(\lambda))^N$$

but since we have (5) it is sufficient to prove (6).

Proof of the theorem. Cluster expansion

We need to formulate some definitions. We call a set $A \subset \mathbb{Z}^\nu$ connected iff for each $t,t' \in A$ there exists a sequence $t_1,\ldots,t_n$ such that $t_i \in A$, $i=1,\ldots,n$ and $r_{tt_1} = r_{t_1t_2} = \ldots = r_{t_{n-1}t_n} = r_{t_nt'} = 1$.

We say that a collection of sets $T = \{A_1,\ldots,A_n\}$, $A_i \subset \mathbb{Z}^\nu$, $i=1,\ldots,n$ is connected iff for each $\ell,m \in \{1,\ldots,n\}$ there exists a sequence $i_1,\ldots,i_\kappa$; $i_j \in \{1,\ldots,n\}$, $j=1,\ldots,\kappa$, such that $A_\ell \cap A_{i_1} \neq 0$, $A_{i_j} \cap A_{i_{j-1}} \neq \emptyset$ for all j and $A_{i_\kappa} \cap A_m \neq \emptyset$. We call $\Gamma = \bigcup_{i=1}^n A_i$ the support of the collection $T = \{A_1,\ldots,A_n\}$.

Now we shall describe the construction of the expansion (4). First we fix an arbitrary configuration $x = \{x_t, t\in\Lambda\}$ and construct clusters $\Gamma_1,\ldots,\Gamma_n$ corresponding to the fixed configuration.

Let us put

$$B = B(\lambda) = \lambda^{-1/8\kappa} . \tag{7}$$

For each $t\in\Lambda$ with $|x_t|>B$ we construct the ν-dimensional neighbourhood 0_t having the center t and radius R_t:

$$R_t = (|x_t| \cdot B^{-1})^{2\kappa/(\nu+\varepsilon)} . \tag{8}$$

Denote $M = \{t\in\Lambda: |x_t|>B\}$. Let $\mathcal{D}_1,\ldots,\mathcal{D}_p$ be the maximal connected components of the set $\bigcup_{t\in M} 0_t$. We shall refer to a $\mathcal{D}_i$ as a *drop*.

Let G be a graph with vertices $1,\dots,p$ (note that p is the number of constructed drops), a line connecting i and j, $i\neq j$, exists iff there exist $t\in\mathcal{D}_i\cap M$ and $t'\in\mathcal{D}_j\cap M$ such that

$$\lambda^{1/2}x_t^{2\kappa}\,x_{t'}^{2\kappa}\,r_{tt'}^{-(\nu+\varepsilon)}\geq 1\ . \tag{9}$$

In general G is not a connected graph. For each maximal connected component $\tilde{G}$ of G consider the union $\bigcup_{i\in\tilde{G}}\mathcal{D}_i$ with i running over all vertices of $\tilde{G}$. Changing the components $\tilde{G}$ we get the sets $A_1,\dots,A_\ell$, $\bigcup_{i=1}^{\ell}A_i=\bigcup_{i=1}^{p}\mathcal{D}_i$. We will refer to $A_1,\dots,A_\ell$ as *fragments*. So, the number of constructed fragments is equal to the number of connected components of G.

Let us denote by $T=T(x)$ the collection of such pairs (t,t'), that t and t' do not belong simultaneously to one and the same fragment. Note that for each $(t,t')\in T$

$$\lambda\cdot\mathcal{U}_{tt'}(x_t,x_{t'})=\lambda\,x_t^{2\kappa}\,x_{t'}^{2\kappa}\,r_{tt'}^{-(\nu+\varepsilon)}\leq\sqrt{\lambda}\ . \tag{10}$$

In fact, if $|x_t|<B$ and $|x_{t'}|<B$ (10) follows from (7). If $|x_t|>B$ and $|x_{t'}|>B$, then since t and t' belong to different fragments, (9) is not fulfilled and hence (10) is true. If $|x_t|>B$ and $|x_{t'}|<B$ then since $t'\notin 0_t$ $r_{tt'}>R_t$, so

$$\lambda x_t^{2\kappa}\,x_{t'}^{2\kappa}\,r_{tt'}^{-(\nu+\varepsilon)}\leq\lambda\cdot B^{2\kappa}\,x_t^{2\kappa}\,R_t^{-(\nu+\varepsilon)}\leq\lambda\cdot B^{4\kappa}\leq\sqrt{\lambda}\ .$$

The following identity will be useful for us:

$$\exp\left\{-\lambda\sum_{(t,t')\in T}\mathcal{U}_{tt'}\right\}=\sum_{Q\subset T}\ \prod_{(t,t')\in Q}a_{tt'} \tag{11}$$

where the sum runs over all subsets $Q\subset T$ (including the empty set) and

$$a_{tt'}=\exp\{-\lambda\mathcal{U}_{tt'}\}-1 \tag{12}$$

If $Q=\emptyset$ we put the corresponding term equal to 1.

We call each pair (t,t') a *link*. Let us fix an arbitrary $Q\subset T$. Let $\mathcal{T}$ be the collection of sets, consisting of all constructed fragments $A_1,\dots,A_\ell$ and all links belonging to Q. Let $T_1,\dots,T_n$ be the maximal connected subcollections of T, and respectively $\Gamma_1,\dots,\Gamma_n$ be their supports. We call each of $\Gamma_1,\dots,\Gamma_n$ a cluster, corresponding to fixed configuration x and $Q\subset T(x)$ and define

$$f_{T_i}(x) = \prod_{A\in T_i} \exp\{-\lambda U_A(x)\} \prod_{(t,t')\in T_i} a_{tt'} \prod_{t\in\Gamma_i} \exp\{-|x_t|^m\} \tag{13}$$

where the product $\prod_{A\in T_i}$ runs over all fragments belonging to T_i, $\prod_{(t,t')\in T_i}$ is meant over all links $(t,t')\in Q$ belonging to T_i, and

$$U_A(x) = \sum_{t,t'\in A} U_{tt'}(x_t,x_{t'}) = \sum_{t,t'\in A} x_t^{2\kappa}\, x_{t'}^{2\kappa}\, r_{tt'}^{-(\nu+\varepsilon)}$$

So, we have constructed a collection of clusters $\Gamma_1,\dots,\Gamma_n$ which corresponds to the fixed configuration x and fixed $Q\subset T(x)$, and defined the "weights" f_{T_i} of these clusters.

Consider an arbitrary collection $\Gamma_1,\dots,\Gamma_n$ of pairwise disjoint subsets $\Gamma_i\subset\Lambda$, $i=1,\dots,n$ (Γ_i is not necessarily connected). Let $X(\Gamma_1,\dots,\Gamma_n)\subset R^\Lambda$ be the set, consisting of configurations x with the following property: there exists $Q\subset T(x)$ such that $\Gamma_1,\dots,\Gamma_n$ is just the collection of clusters corresponding to (x,Q). Note that restriction of any $x\in X(\Gamma_1,\dots,\Gamma_n)$ on $\Lambda\setminus\bigcup_{i=1}^n\Gamma_i$ belongs to $[-B,B]^{\Lambda\setminus\bigcup_i\Gamma_i}$, hence, the set $X(\Gamma_1,\dots,\Gamma_n)$ can be represented as a direct product

$$X(\Gamma_1,\dots,\Gamma_n) = \left(\mathop{\times}_{i=1}^{n} X_{\Gamma_i}\right) \times [-B,B]^{\Lambda\setminus\bigcup_i\Gamma_i}$$

where X_{Γ_i} is exactly the set of configurations $x\in R^{\Gamma_i}$ such that for any $x\in X_{\Gamma_i}$ there exists $Q_i\subset T(x)$ such that the pair (x,Q_i) generates Γ_i.

For any finite $\Gamma\subset\mathbf{Z}^\nu$ denote

$$\kappa_\Gamma = \int_{X_\Gamma} \sum_Q f_T(x_\Gamma)\,dx_\Gamma \tag{14}$$

where $X_\Gamma \subset R^\Gamma$ is the set of such configurations x_Γ that there exists $Q \subset T(x_\Gamma)$ such that the pair (x_Γ, Q) generates just the cluster Γ, the sum $\sum\limits_Q$ runs over all such Q, and $f_T(x_\Gamma)$ is defined in (13), $dx_\Gamma = \prod\limits_{t\in\Gamma} dx_t$.

We assume $\kappa_\Gamma = 0$ if $X_\Gamma = \emptyset$. Taking into account (14) we obtain the expansion (4) with

$$c = \int_{-B}^{B} \exp(-|y|^m)\,dy \geq \int_{-1}^{1} \exp(-|y|^m)\,dy \geq 2e^{-1} .$$

Indeed,

$$Z_\Lambda = \sum_{A_1,\ldots,A_\ell} \int_{\{x:A_1,\ldots,A_\ell\}} \prod_{i=1}^{\ell} \exp(-\lambda U_{A_i}) \exp\left(-\lambda \sum_{(t,t')\in T} U_{tt'}\right) \cdot$$

$$\cdot \exp\left(-\sum_{t\in\Lambda} |x_t|^m\right) dx .$$

The sum is meant over all collections of pairwise disjoint (not necessarily connected) sets $A_1,\ldots,A_\ell$, $A_i \subset \Lambda$, $i = 1,\ldots,\ell$ and integration is over the set of configurations $x \in R^\Lambda$ such that $A_1,\ldots,A_\ell$ are exactly all fragments, generated by x, $T = (\Lambda\times\Lambda) \setminus \bigcup\limits_{i=1}^{\ell} (A_i \times A_i)$.

Using now (11) for $\exp\left(-\lambda \sum\limits_{(t,t')\ T} U_{tt'}\right)$ we obtain

$$Z_\Lambda = \sum_{A_1,\ldots,A_\ell} \sum_{Q\subset T} \int_{\{x:A_1,\ldots,A_\ell\}} \prod_{i=1}^{\ell} \exp(-\lambda U_{A_i}) \prod_{(t,t')\in Q} a_{tt'} \cdot$$

$$\cdot \exp\left(-\lambda \sum_{t\in\Lambda} |x_t|^m\right) dx .$$

Since a collection $A_1,\ldots,A_\ell$ together with Q determines uniquely the clusters $\{\Gamma_1,\ldots,\Gamma_n\}$, we may represent a summation as follows:

$$Z_\Lambda = \sum_{\Gamma_1,\ldots,\Gamma_n} \sum_{\{A_1,\ldots,A_\ell,Q\}} \int_{\{x:A_1,\ldots,A_\ell\}} \prod_{i=1}^{\ell} \exp(-\lambda U_{A_i}) \cdot$$

$$\cdot \prod_{(t,t')\in Q} a_{tt'} \exp\left(-\lambda \sum_{t\in\Lambda} |x_t|^m\, dx\right) .$$

Here the summation runs over all collections $\{\Gamma_1,\ldots,\Gamma_n\}$

of pairwise disjoint clusters $\Gamma_1,\dots,\Gamma_n$, then over all fragments $A_1,\dots,A_\ell$ and over all collections $Q\subset(\Lambda\times\Lambda)\setminus\bigcup_{i=1}^{\ell}(A_i\times A_i)$ of links, such that the collection of supports of maximal connected subcollections of the collection $\{A_1,\dots,A_\ell, Q\}$ coincides with $\{\Gamma_1,\dots,\Gamma_n\}$.

Performing now the integration over x_t , $t\in\Lambda\setminus\bigcup_{i=1}^{\ell}\Gamma_i$ and taking into account the definition of X_Γ , we get:

$$Z_\Lambda=\sum_{\Gamma_1,\dots,\Gamma_n} C^{|\Lambda\setminus\bigcup_i\Gamma_i|}\prod_{i=1}^{n}\int_{X_{\Gamma_i}}\sum_{Q_i} f_{T_i}(x_{\Gamma_i})\,dx_{\Gamma_i} =$$

$$=\sum_{\Gamma_1,\dots,\Gamma_n} C^{|\Lambda\setminus\bigcup_i\Gamma_i|}\kappa_{\Gamma_1}\cdots\kappa_{\Gamma_n} .$$

Proof of the theorem. Cluster estimate

First of all we obtain the cluster estimate for clusters, consisting only of fragments but not of links.

Let us fix the cardinality of Γ : $|\Gamma|=N$. We regard only the clusters Γ containing the origin: $\Gamma\ni 0$.

Moreover, let us assume at first that Γ is a ν-dimensional spherical neighbourhood of some t . Then, obviously, $|x_t|>B$. We will show now that

$$\exp(-\tfrac{1}{3}|x_t|^m)\le(\sqrt{\lambda})^N . \tag{15}$$

Let 0 be a ν-dimensional sphere having radius R , and N - the number of points in 0 : $N=|\mathbb{Z}^\nu\cap 0|$. There exist some constants c_1 and c_2 (depending on ν) such that

$$c_2R^\nu\le N\le c_1R^\nu \tag{16}$$

and therefore it is sufficient to show that

$$\exp(-\tfrac{1}{3}|x_t|^m)\le(\sqrt{\lambda})^{c_1R_t^\nu}$$

or, taking the logarithm and using (8)

$$|x_t|^m\ge\frac{3}{2}c_1(x_t\cdot B^{-1})^{2\kappa\nu/(\nu+\varepsilon)}\cdot(-\ln\lambda) .$$

Together with (7) it leads to

$$|x_t|^{\frac{m(\nu+\varepsilon)-2\kappa\nu}{(\nu+\varepsilon)}} \geq \frac{3}{2} c_1 \lambda^{\nu/4(\nu+\varepsilon)} \cdot (-\ln \lambda) \ .$$

If (3) holds and λ is sufficiently small, the last inequality holds, too. Consequently, if Γ is a sphere and $|\Gamma|=N$, then

$$|\kappa_\Gamma| \leq (\sqrt{\lambda})^N \int \exp\left\{-\frac{1}{3}\sum_{t\in\Gamma} |x_t|^m\right\} \prod_t dx_t \leq (C\sqrt{\lambda})^N \qquad (17)$$

where the constant C depends on m.

Now, let Γ be a drop, i.e. $\Gamma = \bigcup_t 0_t$ where 0_t is a spherical neighbourhood of t. Note that Γ is connected in this case. Let us choose $M\subset\Gamma$ a subset of the centers of spherical neighbourhoods for which $\Gamma = \bigcup_{t\in M} 0_t$. (It can be done in at most 2^N different ways.) For any $t\in M$ similarly to (15) we have

$$\exp(-\frac{1}{3}|x_t|^m) \leq (\sqrt{\lambda})^{|0_t|}$$

and since $\sum_{t\in M} |0_t| \geq |\Gamma| = N$ we have (17) with another constant C. (C denotes different constants not depending on λ.)

Note that Γ is connected, $|\Gamma|=N$ is fixed and $\Gamma\ni 0$. The number of different sets $\Gamma\subset\mathbb{Z}^\nu$ with these properties does not exceed C^N for some constant C, depending on ν. Therefore, taking into account (17), we get (6) for connected Γ.

Now we are going to prove the cluster estimate in the case when Γ is not connected. Then, according to our construction, if Γ contains no links it consists only of one fragment. We regard at first fragments containing only two drops, i.e. Γ is the union of two connected sets. Let us denote these drogs by $\mathcal{D}_1$ and $\mathcal{D}_2$ and let $\mathcal{D}_1\ni 0$. Let us fix $\mathcal{D}_1$ and let $|\mathcal{D}_1|=N_1$. We fix also $t_1\in\mathcal{D}_1$ such that

$$|x_{t_1}| = \max_{t\in\mathcal{D}_1} |x_t| = b_1 \ .$$

Since the drops $\mathcal{D}_1$ and $\mathcal{D}_2$ form one fragment, there exists $t_2 \in \mathcal{D}_2$ such that

$$\sqrt{\lambda}\, b_1^{2\kappa} b_2^{2\kappa} r_{t_1 t_2}^{-(\nu+\varepsilon)} \geq 1$$

where $b_2 = |x_{t_2}|$. That is, if t_1, b_1 and b_2 are fixed, t_2 belongs to the spherical neighbourhood of t_1 having radius $(b_1 b_2)^{2\kappa/(\nu+\varepsilon)}$, so that the number of different ways of fixing t_2 does not exceed $c_1 (b_1 b_2)^{2\kappa\nu/(\nu+\varepsilon)}$. Denoting $N_2 = N - N_1$ and summing through all connected $\mathcal{D}_2$, having cardinality N_2 and containing fixed t_2, similarly to (17) we have:

$$\sum_{\substack{\mathcal{D}_2 : \mathcal{D}_2 \ni t_2 \\ |\mathcal{D}_2| = N_2}} \int \exp(-\lambda \mathcal{U}_{\mathcal{D}_2}) \exp\left(-\sum_{t \in \mathcal{D}_2} |x_t|^m\right) \prod_{t \in \mathcal{D}_2 \setminus t_2} dx_t \leq$$

$$\leq \exp(-\tfrac{2}{3} b_2^m) \cdot (C\sqrt{\lambda})^{N_2} . \tag{18}$$

The left hand side integral is over the set of such configurations on $\mathcal{D}_2$, that $\mathcal{D}_2$ is a drop and $x_{t_2} \equiv b_2$ is fixed.

Similarly,

$$\sum_{\substack{\mathcal{D}_1 : \mathcal{D}_1 \ni 0 \\ |\mathcal{D}_1| = N_1}} \int \exp(-\lambda \mathcal{U}_{\mathcal{D}_1}) \exp\left(-\sum_{t \in \mathcal{D}_1} |x_t|^m\right) \prod_{t \in \mathcal{D}_1 \setminus t_1} dx_t \leq$$

$$\leq \exp(-\tfrac{2}{3} b_1^m) \cdot (C\sqrt{\lambda})^{N_1}$$

where the sum runs over all connected $\mathcal{D}_1$ having cardinality N_1 and containing the origin, integration is over the set of such configurations on $\mathcal{D}_1$ that $\mathcal{D}_1$ is a drop and $x_{t_1} \equiv b_1$ is fixed.

Consequently,

$$\sum_{\substack{\Gamma:\Gamma\ni 0\\ |\Gamma|=N}} |\kappa_\Gamma| \le \int db_1 db_2 \sum_{\mathcal{D}_1\ni 0} \sum_{t_1\in\mathcal{D}_1} \Big\{\int \exp(-\lambda U_{\mathcal{D}_1}) \cdot$$

$$\cdot \exp\Big(-\sum_{t\in\mathcal{D}_1} |x_t|^m\Big) \prod_{t\in\mathcal{D}_1\setminus t_1} dx_t\Big\} \sum_{t_2} \sum_{\substack{\mathcal{D}_2\ni t_2\\ |\mathcal{D}_2|=N-|\mathcal{D}_1|}} \Big\{\int \exp(-\lambda U_{\mathcal{D}_2})\cdot$$

$$\cdot \exp\Big(-\sum_{t\in\mathcal{D}_2} |x_t|^m\Big) \prod_{t\in\mathcal{D}_2\setminus t_2} dx_t\Big\} .$$

Substituting (18) and (19) into the last inequality we obtain

$$\sum_{\substack{\Gamma:\Gamma\ni 0\\ |\Gamma|=N}} |\kappa_\Gamma| \le (C\sqrt{\lambda})^N \int_B^\infty b_1^{2\kappa\nu/(\nu+\varepsilon)} \exp(-\tfrac{2}{3}b_1^m) \cdot$$

$$\cdot \int_B^\infty b_2^{2\kappa\nu/(\nu+\varepsilon)} \exp(-\tfrac{2}{3}b_2^m)\, db_2 \le (C\sqrt{\lambda})^N .$$

Now we shall examine the case when a fragment contains an arbitrary number s of drops. Let us fix $\mathcal{D}_1\ni 0$ and $t_1\in\mathcal{D}_1$. For each fragment there exists a tree $\mathcal{Y}$ with vertices $t_1,t_2,\ldots,t_s$, $t_i\in\mathbb{Z}^\nu$ such that $t_i\in\mathcal{D}_i$ and a line between t_i and t_j meaning that

$$\sqrt{\lambda}\, x_{t_i}^{2\kappa}\, x_{t_j}^{2\kappa}\, r_{t_i t_j}^{-(\nu+\varepsilon)} \ge 1 . \tag{20}$$

Let us denote $b_i = x_{t_i}$, $i=1,\ldots,s$. Let us fix $t_2,\ldots,t_s$, $b_1,\ldots,b_s$, $b_i\ge B$ for each i, and a tree $\mathcal{Y}$, satisfying (20). Let us fix integers $N_1,\ldots,N_s$, such that $\sum N_i = N$ (i.e. we fix the cardinalities of the drops $\mathcal{D}_1,\ldots,\mathcal{D}_s$, of a fragment). Obviously,

$$\exp(-\lambda U_\Gamma)\exp\Big(-\sum_{t\in\Gamma}|x_t|^m\Big) \le \prod_{i=1}^{s} \Big\{\exp(-\lambda U_{\mathcal{D}_i})\cdot\exp\Big(-\sum_{t\in\mathcal{D}_i}|x_t|^m\Big)\Big\}. \tag{21}$$

Using (18) for each $i=2,\ldots,s$ we have

$$\sum_{\substack{\mathcal{D}_i:\mathcal{D}_i\ni t_i\\ |\mathcal{D}_i|=N_i}} \int \exp(-\lambda\mathcal{U}_{\mathcal{D}_i})\cdot\exp\left(-\sum_{t\in\mathcal{D}_i}|x_t|^m\right)\prod_{t\in\mathcal{D}_i\setminus t_i} dx_t \le \tag{22}$$

$$\le \exp(-\tfrac{2}{3}\, b_i^m)\,(C\sqrt{\lambda})^{N_i} .$$

Let us keep $b_1,\ldots,b_s$ fixed and estimate the number of trees $\mathcal{Y}$, containing a fixed vertex t_1 and satisfying (20). We shall describe an algorithm, which enumerates all such trees $\mathcal{Y}$.

An algorithm, enumerating the trees

1. step. Fix a vector of nonnegative integers $(n_1,\ldots,n_s)$ such that $n_1\neq 0$, $n_s=0$ and $\sum n_i=s-1$. Evidently, it can be done in at most 4^s different ways.

2. step. Choose n_1 vectors $v_1,\ldots,v_{n_1}$, $v_i\in\mathbb{Z}^\nu$, $i=1,\ldots,n_1$ satisfying the following conditions: v_i belongs to the spherical neighbourhood of the origin having radius $(b_1 b_{i+1})^{2\kappa/(\nu+\varepsilon)}$. Construct:

$$t_2 = t_1 + v_1$$
$$t_3 = t_1 + v_2$$
$$\cdots\cdots$$
$$t_{n_1+1} = t_1 + v_{n_1} .$$

Construct the lines between t_1 and each of the $t_2,\ldots,t_{n_1+1}$.

3. step. Choose the first (in lexicographic order) of the constructed vertices, excluding t_1. Let this vertex be t_j. If $n_2\neq 0$, choose n_2 vectors $v_{n_1+1},\ldots,v_{n_1+n_2}$, $v_i\in\mathbb{Z}^\nu$ for each i, such that v_i belongs to the spherical neighbourhood of the origin having radius $(b_j b_{i+1})^{2\kappa/(\nu+\varepsilon)}$. Construct

$$t_{n_1+2} = t_j + v_{n_1+1}$$
$$\cdots\cdots$$
$$t_{n_1+n_2+1} = t_j + v_{n_1+n_2} .$$

Construct the lines between t_j and each of

$t_{n_1+2},\dots,t_{n_1+n_2+1}$. If $n_1=0$, pass to the next step without any construction.

We proceed by induction. Let p steps be already performed and $n_1+n_2+\dots+n_{p-1}+1$ vertices be constructed.

(p+1). step. Choose the first (in lexicographic order) of all vertices having been constructed during the previous steps excluding those which have been already chosen earlier. Let this vertex be t_ℓ . If $n_p=0$, pass to the next step without any construction.

If $n_p\neq 0$, choose n_p vectors $v_{n_1+\dots+n_p+1},\dots,$ $v_{n_1+\dots+n_p}$, $v_i\in\mathbb{Z}^\nu$ for each i , such that v_i belongs to the spherical neighbourhood of the origin having radius $(b_\ell b_{i+1})^{2\kappa/(\nu+\varepsilon)}$.

Construct

$$t_{n_1+\dots+n_{p-1}+2} = t_\ell+v_{n_1+\dots+n_{p-1}+1}$$

$$\dots\dots\dots\dots\dots$$

$$t_{n_1+\dots+n_p+1} = t_\ell+v_{n_1+\dots+n_p} .$$

Construct the lines between t_ℓ and each of the vertices constructed in this step.

After s steps the construction is finished.

Thus, if $b_1,\dots,b_s$, $n_1,\dots,n_{s-1}$ and $v_1,\dots,v_{s-1}$ are chosen, the graph constructed by our algorithm is unique. Choosing all possible $b_1,\dots,b_s$, $n_1,\dots,n_{s-1}$, $v_1,\dots,v_{s-1}$ we can construct among other graphs all possible trees $\mathcal{Y}$. Moreover, we construct each tree more times. Indeed, let T be a tree having s vertices and a root t_1 . Let us denote by n_1 the number of lines in T incident with t_1 , (n_i+1) - the number of lines in T incident with t_i , $i=2,\dots,s$. Then the tree T is constructed by our algorithm at last $\prod_{i=1}^{s}(n_i!)$ times.

In fact, let $\tilde b_1,\dots,\tilde b_s$ and $\tilde v_1,\dots,\tilde v_{s-1}$ be the vectors, generating the tree T . Let us divide a collection $(\tilde b_1,\dots,\tilde b_s)$ into subcollections in the following way. The first subcollection consists of only one element, namely

$\tilde{b}_1$. The second subcollection consists of n_1 vectors: $\tilde{b}_2,\ldots,\tilde{b}_{n_1+1}$; the third one consists of the next n_2 vectors, and so on. Consider now a new collection $(\bar{b}_1,\ldots,\bar{b}_s)$, obtained from $(\tilde{b}_1,\ldots,\tilde{b}_s)$ by arbitrary permutations in each of the described subcollections (but without any permutations between the subcollections). Let $\bar{v}_1,\ldots,\bar{v}_{s-1}$ be a collection, obtained from $\tilde{v}_1,\ldots,\tilde{v}_{s-1}$ by the same permutations. Obviously, the tree generated by $\bar{b}_1,\ldots,\bar{b}_s, \bar{v}_1,\ldots,\bar{v}_{s-1}$ is T , and there are exactly $\Pi(n_i!)$ such permutations.

The number of different ways of choosing $v_1,\ldots,v_{s-1}$, having $b_1,\ldots,b_s$ and $n_1,\ldots,n_{s-1}$ fixed, is

$$C_1^s\, b_1^{2\kappa n_1 \nu/(\nu+\varepsilon)} \prod_{j=2}^{s} b_{i_j}^{2\kappa(n_j+1)/(\nu+\varepsilon)} \tag{23}$$

where we have denoted by i_j the number of the vertex, which is chosen as the first one in lexicographic order in the $(j+1)^{th}$ step of the algorithm, $j=2,\ldots,s-1$, and used (16).

Note that

$$\sup_{b_i} b_i^{2\kappa(n_i+1)/(\nu+\varepsilon)} \exp(-\tfrac{1}{3}b_i^m) \le (Cn_i)^{n_i} . \tag{24}$$

Taking into account (21), (22), (23) and (24) finally we have:

$$\sum_{\substack{\Gamma:\Gamma\ni 0\\ |\Gamma|=N}} |\kappa_\Gamma| \le (C\sqrt{\lambda})^N \sum_{s\le N} \sum_{(n_1,\ldots,n_{s-1})} \frac{1}{(\Pi n_i!)} \prod_{i=1}^{s} \Big[(Cn_i)^{n_i} \cdot \int_B^\infty \exp(-\tfrac{1}{3}b_i^m)\,db_i\Big] \le (C\sqrt{\lambda})^N .$$

Now we must consider the case when Γ contains links. Since

$$\sum_{t'\in\mathbb{Z}^\nu} |a_{tt'}| \le \sum_{t'\in\mathbb{Z}^\nu} \lambda B^{4\kappa} r_{tt'}^{-(\nu+\varepsilon)} \le C\sqrt{\lambda} \tag{26}$$

the case when Γ consists only of links is trivial. The general case follows from (26) and (25) by induction on the number of links and fragments in the cluster.

References

[1] Cammarota Camilla. Decay of correlations for infinite-range interactions in unbounded spin systems. Commun. Math. Phys. 85, no. 4, 517-528 (1982).

[2] Malyshev V. A. Semiinvariants of the unlocal functionals of Gibbs random fields. Matem. Zametky 34, no. 3, 443-452 (1983, September).

Thermal Layer Solutions of the Boltzmann Equation

Claude Bardos
Russel Caflisch
Basil Nicolaenko

Abstract

The steady Boltzmann equation in one space dimension with a small mean free path is solved for a gas contained between two plates at different temperatures. The solution consists of the Chapman-Enskog expansion together with boundary layer expansions near each plate. These expansions are truncated at a finite order, and an additional error term is added, so that the sum is an exact solution of the Boltzmann equation. The analysis of the boundary layer equations requires a solution of the Milne problem.

1. Introduction

In this paper we give a kinetic theory description of a gas contained between two plates at different temperatures. The state of the gas is described by the steady nonlinear Boltzmann equation in which the mean free path is assumed to be very small. The boundary conditions are diffuse reflection, i.e. a particle hitting the boundary is absorbed and reemitted with a Maxwellian distribution at the prescribed temperature.

In this problem gravity is excluded and uniformity in the lateral spatial directions is imposed. It follows that the average velocity of the gas is identically zero. The corresponding Navier-Stokes equations then reduce to the equations of constant pressure and Fourier's law for heat conduction. We show that the Boltzmann solution agrees with these two macroscopic equations, thus giving a basic derivation of Fourier's law. This continues a program [2] of relating solutions of the fluid equations with solutions of Boltzmann's equation.

As a future problem, we plan to include gravity and allow three dimensional spatial variation. A bifurcation in the steady solution, corresponding to the Rayleigh-Bénard instability, will then be sought. Note however that this instability is fully understood only for an incompressible fluid in the Boussinesq approximation [5]. A study of the Rayleigh-Bénard problem for a dense gas was made recently in [7].

The present solution for the steady thermal layer problem is found as a decomposition into several parts. The first part is the Chapman-Enskog expansion (modified to omit the Burnett and super-Burnett equations), which is valid in the interior of the region. The leading order term in this expansion is a Maxwellian distribution, which can be chosen to satisfy the boundary conditions. However the higher order terms in the expansion do not satisfy the boundary conditions, and so boundary layer terms must be added on. Since these terms are required only at high order, the resulting boundary layer equations are linear.

The Chapman-Enskog and boundary layer expansions may be only asymptotic, not convergent. In order to obtain an exact solution, these expansions are truncated at finite order and an error term is added. The error term is chosen so that the total sum gives an exact solution of Boltzmann's equation. The equation for the error term is simpler than the original Boltzmann equation because it is only weakly nonlinear.

The analysis of the boundary layer equations utilizes a solution of the linear Milne problem for the Boltzmann equation. This solution uses energy estimates similar to those in [1] and will be only briefly described here.

Some mathematical details in the analysis of the remainder equation and of the Milne problem are not yet completed; so the results here are not completely rigorous. The purpose of this paper is to set out the motivation and formulation of the thermal layer problem and its solution.

The plan of this paper is as follows: After the Boltzmann equation is summarized in section 2, the thermal layer problem and its solution are described in section 3. The Chapman-Enskog expansion is developed in section 4, the boundary layer expansion in section 5, and the error term in section 6. Finally in section 7 the Milne problem is discussed.

2. The Boltzmann Equation

The steady Boltzmann equation in one space dimension is

$$\xi_1 \frac{\partial}{\partial x} F = \frac{1}{\varepsilon} \underline{Q}(F,F). \tag{2.1}$$

in which the molecular distribution function is $F = F(x,\underset{\sim}{\xi})$ with $x \in R^1$ and $\underset{\sim}{\xi} = (\xi_1,\xi_2,\xi_3) \in R^3$. The mean free path ε is assumed to be small here.

For the collision operator $\underline{Q}$ we take the one corresponding to hard spheres orto the Krook model. In general $\underline{Q}$ has the following properties:

(i) $\underline{Q}(F,F) \equiv 0$ iff $F \equiv M$, in which the Maxwellian distribution M has the form

$$\begin{aligned} M &= M(\underset{\sim}{\xi};\rho,T,\underset{\sim}{u}) \\ &= \rho(2\pi T)^{-3/2} \exp(-(\underset{\sim}{\xi}-\underset{\sim}{u})^2/2T). \end{aligned} \tag{2.2}$$

(ii) $\int \psi\, \underline{Q}(F,G)\, d\underset{\sim}{\xi} = 0$ for all F,G iff ψ is a linear combination of the summational invariants ψ_i given by

$$\begin{aligned} \psi_0(\underset{\sim}{\xi}) &= 1 \\ \psi_i(\underset{\sim}{\xi}) &= \xi_i \\ \psi_4(\underset{\sim}{\xi}) &= \xi^2. \end{aligned} \tag{2.3}$$

Corresponding to these orthogonality relations, we decompose F as

$$F = \Phi_F + \Psi_F \tag{2.4}$$

with

$$\Psi_F(x,\underset{\sim}{\xi}) = \underset{\sim}{A}_F(x) \cdot \underset{\sim}{B}(\underset{\sim}{\xi}) \tag{2.5}$$

$$= \left\{ a(x) + \sum_{1}^{3} b_i(x)\xi_i + c(x)\xi^2 \right\} M(\underset{\sim}{\xi})$$

$$\underset{\sim}{A}_F(x) = (a, b_1, b_2, b_3, c) \tag{2.6}$$

and

$$\int \psi_i \Phi_F \, d\underset{\sim}{\xi} = 0 \tag{2.7}$$

for $i = 0, \ldots, 4$.

The linearized collision operator $\underline{L}$ is defined, relative to a Maxwellian M, by

$$\underline{L}F = -2\,\underline{Q}(M,F)\ . \tag{2.8}$$

It has the property that

$$\int M\, F\, \underline{L}F \, d\underset{\sim}{\xi} \geq \sigma \int M\, \Phi_F^2 \, d\underset{\sim}{\xi} \tag{2.9}$$

for some positive constant σ depending only on M.

3. The Thermal Layer Problem

The thermal layer problem is to solve the steady Boltzmann equation in a slab $0 < x < 1$ with specified temperature T_L at $x = 0$ and T_R at $x = 1$ and with specified total mass density $\bar{\rho} = \int_0^1 \int F \, d\underset{\sim}{\xi} \, dx$. There are thermal boundary conditions and the condition of no mass flux at $x = 0$ and $x = 1$. We seek a solution for which the macroscopic velocity is zero.

The Boltzmann equation for the thermal layer is

$$\xi_1 \frac{\partial F}{\partial x} = \frac{1}{\varepsilon} \underline{Q}(F,F)\ , \quad \text{for } 0 < x < 1 \tag{3.1}$$

with boundary conditions

$$F(x=0,\underset{\sim}{\xi}) = \rho_L M_L, \qquad \text{for } \xi_1 > 0 \tag{3.2}$$

$$F(x=1,\underset{\sim}{\xi}) = \rho_R M_R, \qquad \text{for } \xi_1 < 0 \tag{3.3}$$

$$\int \xi_1 F(0,\underset{\sim}{\xi})\, d\underset{\sim}{\xi} = \int \xi_1 F(1,\underset{\sim}{\xi})\, d\underset{\sim}{\xi} = 0 \tag{3.4}$$

and normalization

$$\int_0^1 \int F\, d\underset{\sim}{\xi}\, dx = \bar{\rho}. \tag{3.5}$$

The Maxwellian distributions at the boundaries are

$$M_L = (2\pi\, T_L)^{-3/2}\, e^{-(\xi^2/2)T_L} \tag{3.6}$$

$$M_R = (2\pi\, T_R)^{-3/2}\, e^{-(\xi^2/2)T_R}. \tag{3.7}$$

In these equations T_L, T_R and $\bar{\rho}$ are given positive constants, while ρ_L and ρ_R are unknown. In addition we ask that

$$\int \xi_i F(x,\underset{\sim}{\xi})\, d\underset{\sim}{\xi} = 0 \tag{3.8}$$

for all x and for i = 1,2,3. For i = 1, (3.8) is implied by (3.4).

We shall find the solution $F = F^\varepsilon$ for the thermal layer problem (3.1)-(3.5) for ε small by decomposing F^ε into three parts: an interior solution F^C given by the Chapman-Enskog expansion, a boundary layer expansion F^B at x = 0 and at x = 1, and an error term F^E. The form of F^ε is

$$F^\varepsilon = F^C + F^B + \varepsilon^2 F^E \quad . \tag{3.9}$$

These three components of F^ε are analyzed in sections 4-6.

The main character of the solution F^{ε} is its agreement with the Navier-Stokes equations, which reduce to Fourier's law for heat conduction for this simple problem. We shall show that

$$F^{\varepsilon}(x,\underset{\sim}{\xi}) = M(\underset{\sim}{\xi};\rho_0,T_0,u_0) + \underline{O}(\varepsilon) \tag{3.10}$$

in which ρ_0, T_0, u_0 satisfy the steady one-dimensional Navier-Stokes equations with $u_0 \equiv 0$.

The Navier-Stokes equations are

$$\frac{\partial}{\partial x} \rho u = 0 \tag{3.11}$$

$$\frac{\partial}{\partial x} (\rho\mu^2+p) = \varepsilon \frac{\partial}{\partial x} \left(\mu(T) \frac{\partial}{\partial x} \mu\right) \tag{3.12}$$

$$\frac{\partial}{\partial x} (\rho u(\frac{1}{2} u^2+e) + up) = \varepsilon \frac{\partial}{\partial x} (\mu u \frac{\partial}{\partial x} u) + \varepsilon \frac{\partial}{\partial x} (\lambda(T) \frac{\partial}{\partial x} T) \tag{3.13}$$

with the equation of state

$$p = \rho T , \quad e = \frac{3}{2} T . \tag{3.14}$$

The viscosity $\mu(T)$ and the heat conductivity $\lambda(T)$ are those which result from the Chapman-Enskog expansion. In the thermal layer problem with $u = 0$, these equations reduce to the hydrostatic equation

$$\frac{\partial}{\partial x} \rho T = 0 \tag{3.15}$$

and Fourier's law

$$\frac{\partial}{\partial x} \left(\lambda(T) \frac{\partial}{\partial x} T\right) = 0 . \tag{3.16}$$

In addition we specify the total density $\bar{\rho}$ and the temperature boundary values T_L and T_R , i.e.

$$\int_0^1 \rho \, dx = \bar{\rho} \tag{3.17}$$

$$T(0) = T_L , \quad T(1) = T_R . \tag{3.18}$$

The problem (3.15)-(3.18) has a unique solution for any $\bar{\rho}$, T_L , T_R which are positive.

Our main result for the thermal layer problem is the following theorem.

<u>Theorem</u>. If ε, T_L , T_R and $\bar{\rho}$ are positive and if ε is sufficiently small, there is a unique solution F^{ε} of the Boltzmann thermal layer equations (3.1)-(3.5). Moreover F^{ε} satisfies the stationarity condition (3.8) and

$$\| F^{\varepsilon} - M(\rho_0, T_0, u_0) \| \leq c\, \varepsilon \tag{3.19}$$

in which c is a constant (independent of ε), $u_0 = 0$, ρ_0 and T_0 solve the Navier-Stokes equations (3.15)-(3.18), and $\|\cdot\|$ is an appropriate norm.

4. <u>The Chapman-Enskog Expansion</u>

Away from the boundaries x = 0 and x = 1, the solution F^{ε} of the thermal layer problem is given primarily by the Chapman-Enskog solution F^C. The Chapman-Enskog expansion is discussed in detail in [3] and [4]. Here we summarize the results for the steady problem with zero velocity.

The solution F^C has an expansion

$$F^C = M_0 + \varepsilon F_1^C + \varepsilon^2 F_2^C \tag{4.1}$$

in which M_0 is the Maxwellian

$$M_0(\underset{\sim}{\xi}) = M(\underset{\sim}{\xi},\rho_0,T_0,u_0 = 0) \tag{4.2}$$

with ρ_0 , T_0 solving (3.15)-(3.18). The higher order terms F_1^C , F_2^C can be decomposed as

$$F_n^C = \Phi_n + \Psi_n \tag{4.3}$$

such that Ψ_n is the hydrodynamic part and Φ_n is the nonhydrodynamic part, i.e.

$$\Psi_n = \{\rho_n/\rho_0 + (T_n/2T_0)(\xi^2/T_0 - 3)\}M_0 \tag{4.4}$$

$$\int \psi_i \Phi_n \, d\xi = 0 \quad \text{for} \quad i = 0,\ldots,4. \tag{4.5}$$

The fluid dynamic variable $\rho_n(x)$, $T_n(x)$ satisfy (forced) linearized Navier-Stokes equations. For n = 1 these equations are

$$\frac{\partial}{\partial x}(\rho_0 T_1 + \rho_1 T_0) = 0 \tag{4.6}$$

$$\frac{\partial}{\partial x}\left(\lambda(T_0)\frac{\partial}{\partial x}T_1 + \lambda'(T_0)T_1\frac{\partial}{\partial x}T_0\right) = 0 \tag{4.7}$$

$$\int_0^1 \rho_1 \, dx = \bar{\rho}_1 \tag{4.8}$$

$$T_1(0) = T_{L1} , \quad T_1(1) = T_{R1} . \tag{4.9}$$

The constants $\bar{\rho}_1$, T_{L1} and T_{R1} determine ρ_1 , T_1 uniquely and will be found in section 5 using the boundary conditions. The equations for n = 2 are similar but contain forcing terms.

The nonhydrodynamic components Φ_n are given by

$$\Phi_1 = - \underline{L}^{-1}(I-P)\, \xi_1 \frac{\partial}{\partial x} M_0 \tag{4.10}$$

$$\Phi_2 = - \underline{L}^{-1}(I-P)\, \{\xi_1 \frac{\partial}{\partial x} F_1^C - \underline{Q}(F_1^C , F_1^C)\} \tag{4.11}$$

in which $\underline{L} = -2\underline{Q}(M_0,\cdot)$ and P is the projection onto the null space of $\underline{L}^{+}$. Note that Φ_1 is completely determined by M_0 , and Φ_2 is determined once Ψ_1 is known.

The truncated expansion (4.1) does not exactly solve the Boltzmann equation (3.1) or the boundary conditions (3.2)-(3.3), but the expansion does satisfy (3.4). The boundary conditions (3.2), (3.3) can be rewritten for $F^\varepsilon - M_0$ as

$$F^\varepsilon - M_0 = \Delta\rho_L M_L , \quad \text{for } x = 0 , \; \xi_1 > 0 \tag{4.12}$$

$$F^\varepsilon - M_0 = \Delta\rho_R M_R , \quad \text{for } x = 1 , \; \xi_1 < 0 \tag{4.13}$$

in which $\Delta\rho_L$ and $\Delta\rho_R$ are undetermined as yet. The right hand sides of (4.12), (4.13) are purely hydrodynamic. In order to accommodate the nonhydrodynamic components Φ_n , which enter in the Chapman-Enskog expansion, we must introduce boundary layer terms. However since the nonhydrodynamic part of F^C is size $\underline{O}(\varepsilon)$, we need boundary layer terms only of size $\underline{O}(\varepsilon)$. It is useful to expand $\Delta\rho_L$ and $\Delta\rho_r$ as

$$\Delta\rho_L = \varepsilon\Delta\rho_{L1} + \varepsilon^2\Delta\rho_{L2} \tag{4.14}$$

$$\Delta\rho_R = \varepsilon\, \Delta\rho_{R1} + \varepsilon^2\Delta\rho_{R2} \; . \tag{4.15}$$

5. The Boundary Layer Expansion

The boundary layer expansion F^B provides rapid transition (in distance ε) from the imposed boundary conditions at $x = 0$ and $x = 1$ to the Chapman-Enskog expansion in the interior. There are actually two boundary expansions: one at $x = 0$ denoted F^L and one at $x = 1$ denoted F^R. These are written in terms of scaled variables as follows

$$F^B(x,\xi) = F^L\left(y = \frac{x}{\varepsilon}, \underset{\sim}{\xi}\right) + F^R\left(z = \frac{1-x}{\varepsilon}, \underset{\sim}{\xi}\right) . \tag{5.1}$$

The Boltzmann equation for F^L is

$$\xi_1 \frac{\partial}{\partial y} F^L = 2\underline{Q}(\tilde{F}^C, F^L) + \varepsilon\, \underline{Q}(F^L, F^L) \tag{5.2}$$

for $0 < y < \infty$. In addition F^L is required to satisfy a boundary condition at $y = 0$ and to not extend into the interior, i.e.

$$F^L(y=0) + F^C(x=0) = \rho_L M_L , \quad \text{for} \quad \xi_1 > 0 \tag{5.3}$$

$$\int \xi_1 F^L \, d\underset{\sim}{\xi} = 0 \qquad \text{at } y = 0 \tag{5.4}$$

$$F^L \to 0 \qquad \text{as } y \to \infty . \tag{5.5}$$

In (5.2) we have replaced F^C by its truncated power serise $\tilde{F}^C$ around $y = 0$, given by

$$\tilde{F}^C(y) = M_L + \varepsilon\left\{F_1^C(x=0) + y \frac{\partial}{\partial x} M_0(x=0)\right\}$$
$$+ \varepsilon^2\left\{F_2^C(x=0) + y \frac{\partial}{\partial x} F_1^C(x=0) + \frac{1}{2} y^2 \frac{\partial^2}{\partial x^2} M_0(x=0)\right\} , \tag{5.6}$$

so that $\tilde{F}^C$ is defined for $0 < y < \infty$. Similar equations are found for F^R, and F^L is coupled with F^R by the normalization condition

$$\varepsilon \int_0^\infty \int F^L \, d\underset{\sim}{\xi} \, dy + \varepsilon \int_0^\infty \int F^R \, d\underset{\sim}{\xi} \, dz + \int_0^1 \int F^C \, d\underset{\sim}{\xi} \, dx = \bar{\rho} \quad . \tag{5.7}$$

We ask only that (5.7) hold with an error size ε^2.

The boundar layer equations (5.2)-(5.5) and (5.7) are solved by expanding F^L as

$$F^L = \varepsilon \, F_1^L + \varepsilon^2 F_2^L \quad . \tag{5.8}$$

The resulting equations for F_1^L are

$$\xi_1 \frac{\partial}{\partial y} F_1^L = - \underline{L} \, F_1^L \qquad \text{for } y > 0 \tag{5.10}$$

$$F_1^L(y=0) + F_1^C(x=0) = \Delta\rho_{L1} M_L \,, \quad \text{for} \quad \xi_1 > 0 \tag{5.11}$$

$$\int \xi_1 F_1^L \, d\underset{\sim}{\xi} = 0 \,, \qquad \text{at } y = 0 \tag{5.12}$$

$$F_1^L \to 0 \,, \qquad \text{as } y \to \infty \quad . \tag{5.13}$$

There is a similar expansion for F^R and a set of equations for F_1^R. In addition (5.7) is rewritten as

$$\int_0^1 \int F_1^C \, d\underset{\sim}{\xi} \, dx = 0$$

$$\int_0^\infty \int F_1^L + F_1^R \, d\underset{\sim}{\xi} \, dy + \int_0^1 \int F_2^C \, d\underset{\sim}{\xi} \, dx = 0 \quad . \tag{5.14}$$

Equations (5.10)-(5.14) will be solved in several steps.

First solve the following Milne problem:

$$\xi_1 \frac{\partial}{\partial y} G^L = - \underline{L}\, G^L , \qquad \text{for } \ y \geq 0 \tag{5.15}$$

$$G^L(0,\underset{\sim}{\xi}) = - \Phi_1(0,\underset{\sim}{\xi}) , \qquad \text{for } \ \xi_1 > 0 \tag{5.16}$$

$$\int \xi_1 \, G^L \, d\underset{\sim}{\xi} = 0 , \qquad \text{for } y \geq 0 \tag{5.17}$$

in which $\underline{L} = -2\underline{Q}(M_L ,\cdot)$. The solution G^L is found (formally) in section 7. In particular it is shown that

$$G^L \to (a_L + c_L \xi^2) M_L \tag{5.18}$$

as $y \to \infty$, with a_L and c_L constants which are determined once Φ_1 is known. Recall that Φ_1 is determined by knowing M_0.

Now for F_1^L we choose

$$F_1^L = G^L - (a_L + c_L \xi^2) M_L \tag{5.19}$$

which satisfies (5.10), (5.12), (5.13). Similarly we choose F_1^R to be

$$F_1^R = G^R - (a_R + c_R \xi^2) M_R \tag{5.20}$$

with G^R, a_R , c_R solving

$$\xi_1 \frac{\partial}{\partial z} G^R = - \underline{L}\, G^R \quad \text{for } z \geq 0 \tag{5.21}$$

$$G^R(0,\underset{\sim}{\xi}) = -\Phi(1,\underset{\sim}{\xi}) , \qquad \text{for } \xi_1 < 0 \tag{5.22}$$

$$\int \xi_1 \, G^R \, d\underset{\sim}{\xi} = 0 , \qquad \text{for } z \geq 0 \tag{5.23}$$

$$G^R \to (a_R + c_R \xi^2) M_R$$

in which $\underline{L} = -2\underline{Q}(\cdot, M_R)$ and a_R , c_R constants.

The conditions (5.11), (5.14), and a condition for F_1^R analogous to (5.11) will be satisfied by the proper choice of the constants T_{L1} , T_{R1} , $\bar{\rho}_1$, $\bar{\rho}_2$, $\Delta\rho_{L1}$, $\Delta\rho_{R1}$. Evaluate (5.11) using (4.3), (4.4), (5.19), (5.16) to get $-(a_L + c_L \xi^2) M_L + \{\rho_1(0)/\rho_0(0) + (T_{L1}/2T_L)(\xi^2/T_L - 3)\} M_L = \Delta\rho_{L1} M_L$ for $\xi_1 > 0$, or

$$T_{L1} = 2c_L T_L^2 \tag{5.24}$$

$$\Delta\rho_{L1} = \rho_1(0)/\rho_0(0) - a_L - 3c_L T_L \ . \tag{5.25}$$

Similarly we find that

$$T_{R1} = 2c_R^2 T_R^2 \tag{5.26}$$

$$\Delta\rho_{R1} = \rho_1(1)/\rho_0(1) - a_R - 3c_R T_R \ . \tag{5.27}$$

The condition (5.14) is evaluated to obtain

$$\bar{\rho}_1 = 0$$

$$\bar{\rho}_2 = -\int_0^\infty \int F_1^R + F_1^L \, d\underset{\sim}{\xi} \, dy \ . \tag{5.28}$$

This completes the determination of F_1^L and F_1^R and of the constants T_{L1} , T_{R1} , $\bar{\rho}_1$, $\bar{\rho}_2$, $\Delta\rho_{L1}$, $\Delta\rho_{R1}$.

As seen above, the determination of the Chapman-Enskog and

boundary layer expansions are coupled through the choice of the constants T_{Ln}, T_{Rn}, $\bar{\rho}_n$, $\Delta\rho_{Ln}$, $\Delta\rho_{Rn}$ and the boundary values of Φ_n. The logical order to determine these two expansions is as follows: First M_0 is found as in (4.2). Then Φ_1 is found from (4.10). Third F_1^L and F_1^R are given by (5.19), (5.20). Fourth, T_{L1}, T_{R1}, $\bar{\rho}_1$, $\bar{\rho}_2$, $\Delta\rho_{L1}$, $\Delta\rho_{R1}$ are found by (5.24)-(5.28), while simultaneously Ψ_1 is found through (4.4) and (4.6)-(4.9). Then the process is started again to determine Φ_2, etc. There are equations similar to (5.10)-(5.14) for F_2^L and F_2^R.

The combination $\tilde{F} = F^C + F^B$ is not an exact solution of the Boltzmann thermal layer problem (3.1)-(3.5). Although F_C satisfies (3.4), the errors in the other equations are

$$E_1 = \xi_1 \frac{\partial}{\partial x} \tilde{F} - \varepsilon^{-1} \underline{Q}(\tilde{F},\tilde{F}) \tag{5.29}$$

$$E_2 = \tilde{F}(x=0) - \tilde{\rho}_L M_L = F^R(z=\varepsilon^{-1}) \quad \text{for} \quad \xi_1 > 0 \tag{5.30}$$

$$E_3 = \tilde{F}(x=1) - \tilde{\rho}_R M_R = F^L(y=\varepsilon^{-1}) \quad \text{for} \quad \xi_1 < 0 \tag{5.31}$$

$$E_4 = \int_0^1 \int \tilde{F} \, d\underset{\sim}{\xi} \, dx - \bar{\rho} = \varepsilon \int_{\varepsilon^{-1}}^{\infty} \int (F_1^L + F_1^R)(y,\underset{\sim}{\xi}) \, d\underset{\sim}{\xi} \, dy$$
$$+ \varepsilon^2 \int_0^{\infty} \int (F_2^L + F_2^R) d\underset{\sim}{\xi} \, dy. \tag{5.32}$$

Since F^L and F^R are found to decay exponentially for large y (or z), E_2, E_3 and E_4 are exponentially small in ε^{-1}. The error E_1 is size ε^2. We have denoted $\tilde{\rho}_L = \rho_0(0) + \Delta\rho_L$, $\tilde{\rho}_R = \rho_0(1) + \Delta\rho_R$.

6. The Error Equation

The equation for F^E is found by asking that $F^\varepsilon = F^C + F^B + \varepsilon^2 F^E$ be an exact solution of the thermal layer Boltzmann equation (3.1)-(3.5). It must make up for the error terms E_1, E_2, E_3, E_4 in (5.29)-(5.32). It follows that F^E satisfies

$$\xi_1 \frac{\partial}{\partial x} F^E = \frac{2}{\varepsilon} \underline{Q}(F^C + F^B, F^E) + \varepsilon \underline{Q}(F^E,F^E) - \varepsilon^{-2} E_1 , \tag{6.1}$$

for $x \geq 0$ with boundary conditions

$$F^E(x=0,\underset{\sim}{\xi}) = \Delta\rho_L^E M_L - \varepsilon^{-2}E_2 \quad \text{for } \xi_1 > 0 \tag{6.2}$$

$$F^E(x=1,\underset{\sim}{\xi}) = \Delta\rho_R^E M_R - \varepsilon^{-2}E_3 \quad \text{for } \xi_1 < 0 \tag{6.3}$$

$$\int \xi_1 F^E d\underset{\sim}{\xi} = 0$$

and normalization condition

$$\int_0^1 \int F^E d\underset{\sim}{\xi}\, dx = -\varepsilon^{-2} E_4 . \tag{6.4}$$

in which $\Delta\rho_L^E$ and $\Delta\rho_R^E$ are constants to be determined.

The analysis of (6.1)-(6.4) proceeds as in the analysis of the similar error equation in [2].

7. The Milne Problem

The Milne problem consists of solving the linearized Boltzmann equation with a prescribed incoming velocity distribution and prescribed average normal velocity, along with a condition of asymptotic behavior at ∞. The resulting equations are

$$\xi_1 \frac{\partial}{\partial x} F + \underline{L}F = 0 , \qquad \text{for } x \geq 0 \tag{7.1}$$

$$F(0,\underset{\sim}{\xi}) = g(\underset{\sim}{\xi}) , \qquad \text{for } \xi_1 > 0 \tag{7.2}$$

$$\int \xi_1 F(0,\underset{\sim}{\xi})\, d\underset{\sim}{\xi} = u \tag{7.3}$$

$$\int_0^\infty \int (1 + |\underset{\sim}{\xi}|)\Phi_F^2\, d\underset{\sim}{\xi}\, dx < \infty \tag{7.4}$$

in which the constant average normal velocity u and the incoming distribution g are to be given. In this section we use the decomposition $F = \Phi_F + \Psi_F = \Phi_F + \underset{\sim}{A}_F \cdot \underset{\sim}{B}$ described in section 2.

The Milne problem for neutron transport was analyzed in [1] and that solution has motivated this study. For the Krook model of the Boltzmann equation, the Milne problem has been partly analyzed in [4] and [8]. Partial results for the full Boltzmann equation are contained in [6] and [9]. The results presented here give a complete solution of the Milne problem for the Krook model; for the full Boltzmann equation there are still a number of unfinished technical details.

<u>Theorem 1</u> (Existence). Suppose that $\int_{\xi_1>0} (1+|\xi|)^2 g(\underset{\sim}{\xi})^2 d\underset{\sim}{\xi} < \infty$. Then there exists a solution F of the Milne problem (7.1)-(7.4) with $\underset{\sim}{A}_F \in L^2_{loc}(R^+_x)$.

<u>Theorem 2</u> (Uniqueness). Let F_1 and F_2, both in $L^2_{loc}(R^+_x, L^2(R^3_\xi))$, be solutions of the Milne problem (7.1)-(7.4). If $F_1(0,\underset{\sim}{\xi}) = F_2(0,\underset{\sim}{\xi})$ for $\xi_1 > 0$ and if $\int \xi_1 F_1 d\underset{\sim}{\xi} = \int \xi_2 F_2 d\underset{\sim}{\xi}$ at $x = 0$, then $F_1 = F_2$ for all $x > 0$, $\underset{\sim}{\xi}$ R^3.

<u>Theorem 3</u> (Orthogonality and Asymptotic Properties). Let $F \in L^2_{loc}(R^+_x, L^2(R^3_\xi))$ solve (7.1)-(7.4). Then F has the following properties:

$$\int \xi_1 \xi_2 \Phi_F \, d\underset{\sim}{\xi} = 0 \tag{7.5}$$

$$\int \xi_1 \xi_3 \Phi_F \, d\underset{\sim}{\xi} = 0 \tag{7.6}$$

$$\int \xi_1 |\xi|^2 \Phi_F \, d\underset{\sim}{\xi} = 0 \,. \tag{7.7}$$

Moreover $b_1 \equiv \int \xi_1 F \, d\xi$ is constant in x and equal to u. As x goes to infinity, a(x), $b_2(x)$, $b_3(x)$, and c(x) converge exponentially to constants a^∞, b_2^∞, b_3^∞, c^∞ while Φ_F converges exponentially to zero. To be precise if $\delta < \sigma$, with σ defined by (2.9), there is a constant κ such that

$$|a(x)-a^\infty| + |b_1(x)-b_2^\infty| + |b_3(x)-b_3^\infty|$$

$$+ |c(x)-c^{\infty}| + \int (1+|\xi|) |w(x,\underset{\sim}{\xi})|^2 d\underset{\sim}{\xi} \leq \kappa e^{-\delta x}. \tag{7.8}$$

Finally there is the following symmetry property: If $g(\xi_1,\xi_2,\xi_3) = g(\xi_1,-\xi_2,\xi_3) = g(\xi_1,\xi_2,-\xi_3)$ for all $\underset{\sim}{\xi} = (\xi_1,\xi_2,\xi_3)$, then $b_2 = b_3 = 0$.

8. References

[1] Bardos, Santos and Sentis. Diffusion approximation and computation of the critical size. Preprint.

[2] Caflisch. The fluid dynamic limit of the nonlinear Boltzmann equation. Comm. Pure Appl. Math. 33 (1980) 651-666.

[3] Caflisch. Fluid dynamics and the Boltzmann equation, in Nonequilibrium Phenomena, eds. Lebowitz and Montroll, (1983) North-Holland, 193-223.

[4] Cercignani. Theory and Application of the Boltzmann Equation (1975) Elsevier.

[5] Drazin and Reid. Hydrodynamic Stability (1981) Cambridge.

[6] Guiraud. Equation de Boltzmann linéaire dans un demi-space. Comptes-Rendus 274 (1974) 417-419.

[7] Kirkpatrick and Cohen. Kinetic theory of fluctuations near a convective instability. J. Stat. Phys. 33 (1983) 639-694.

[8] Pao. Application of kinetic theory to the problem of evaporation and condensation. Phys. Fl. 14 (1971) 306-312.

[9] Rigolot-Turbat. Probleme de Kramers pour l'équation de Boltzmann en théorie cinétique des gaz. Comptes-Rendus 273 (1971) 58-61.

Claude Bardos
Ecole Normale Supérieure
75005 Paris

Russel Caflisch
Courant Institute of Mathematical Sciences
New York, NY 10012

Basil Nicolaenko
Los Alamos National Laboratories
Los Alamos, NM 87544

Research supported by the National Science Foundation and the Department of Energy.

HYPERBOLICITY AND MÖLLER-MORPHISM FOR A MODEL OF CLASSICAL STATISTICAL MECHANICS

E. Presutti, Ya. G. Sinai and M.R. Soloviechik

Abstract. We consider a gas of point particles in $\mathbb{R}_+$. The first particle has mass M, the others m and $M>m$. The particles interact by elastic collisions (among themselves and with the wall at the origin). Let $\mathcal{X}$ be the phase space and μ a Gibbs measure for the system, S_t denotes the time flow and $(\mathcal{X},\mu,S_t)$ is a dynamical system.

We identify the m-particles during their evolution so that they keep the same velocity until they collide with the M-particle. Hence the motion is free, asymptotically far from the origin: free particles come from $+\infty$, interact with the M-particle and then move back free to $+\infty$. We prove that the Möller wave operators $\Omega_\pm$ exist, asymptotic completeness holds and that $\Omega_-^{-1}\Omega_+$ defines a non-trivial scattering matrix for the system. $\Omega_\pm$ define isomorphisms between the dynamical system $(\mathcal{X}^o,\mu^o,S_t^o)$ and $(\mathcal{X},\mu,S_t)$, $(\mathcal{X}^o,\mu^o,S_t^o)$ refers to the case when all the particles have mass m and μ^o has the same thermodynamical parameters as μ.

An independent generating partition is explicitely known for the system $(\mathcal{X}^o,\mu^o,S_t^o)$ and $\Omega_\pm$ transform it in an independent generating partition for $(\mathcal{X},\mu,S_t)$, thereby proving that this is a Bernoulli flow.

The proof of the existence of the wave operator is based on the (almost everywhere) existence of contractive manifolds. Namely we prove that for almost all configurations $x\in\mathcal{X}$ the following holds. Fix any finite subset I of particles in x and consider all the configurations y obtained by changing the coordinates of the particles in I while leaving all the others fixed. Then if the change is small enough S_tx and S_ty become (locally) exponentially close.

0. Introduction

Very little is known of the ergodic properties of systems with infinitely many degrees of freedom which describe the behavior of a gas in classical statistical mechanics. Results have been obtained in particular cases: the infinite ideal gas [25], [1], [12], the one dimensional hard rods system [10], [22], [2], [5], the infinite chain of harmonic oscillators [8], [11]. The analysis in the above models is based on a more or less complete knowledge of the time flow. For more complex systems it seems hopeless to look for explicit solutions of the equations of motion. It is not even completely clear what one would like to find, i.e. which features are responsible for the good ergodic properties of the system. Here we find a difference with the finite dimensional case where it is generally believed that the main mechanism of chaos relies on the existence of foliations of the phase space into stable and unstable manifolds [3]. The construction of Markov partitions [26], [27] frames then the problem within the classical theory of stochastic processes. It is conceivable that some analogue of such hyperbolic structure extends to the infinite systems. Besides it another mechanism, peculiar of the unboundedness of the system, is expected to play an important role. This is very well described in [16]: "... A subsystem undergoes all kinds of interactions with other parts of the system [due to the infinite size of the system] ... [and] owing to the complexity of [such] interactions [it] will pass sufficiently often through all its possible states...". In the ideal gas local perturbations move freely away and in the other models mentioned above similar mechanisms come into play. In a model considered in [14] the analysis does not exploit knowledge of the time flow. The authors consider a semi-infinite point particle system. The first particle, h.p., has mass M, all the others m and $M>m$. The h.p. is confined between 0 and L by two elastic walls, the latter being transparent to any other particle. The only interaction consists in the elastic collisions between the h.p. and the light ones. The states of the system are determined by the corresponding Gibbs measure. The mechanism responsible for the ergodic properties of the model (which in [14] has been proven to be a Bernoulli flow) is the following: whenever the h.p. bounces off from the wall at L , then the information of the whole past is only the velocity of the h.p. at that time. The process of the velocity at such times is then proven to be a Markov Döblin chain and from this the ergodic properites of the system follow. An extension of the result to the "infinite" case when light particles can also arrive from the left has been obtained in [13] (the analysis covers the case when the gas on the right has different temperature from that on the left). An extension to the two dimensional case has been recently worked out [15].

The main feature of the semi-infinite case considered in [14] is that the interaction is strictly localized, the system behaves like free in the interval L, ∞ . If one takes off the wall at L then the h.p. is free to move further to the right and it goes eventually past any fixed point in the line, since the Gibbs equilibrium measure gives a non-zero probability to such event. For such reason the Markov property exploited in [14] is lost. The analysis requires then a deeper study of the dynamical structure of the system and, as we shall see, it singles out both mechanisms we have outlined before: the existence of local contractive "manifolds" (hyperbolic structure) and the occurrence of clusters of particles which "come from infinity" and play a role analogous to that predicted in [16].

This model, as well as some of its variants, has been recently studied in [7], [6], [19], [4]. It has been shown to be a Bernoulli flow. The intuition one has of the system is that the incoming particles exert a pressure on the h.p. which is consequently kept close to the wall at 0 . Only rare fluctuations of the pressure allow the h.p. to escape from its confinement and to travel far away. Hence one expects that the dynamics is asymptotically (far away from the origin) free: namely free particles come from $+\infty$, arrive, interact with the h.p., then flow away to gain back asymptotic freedom at $+\infty$. Such picture calls to mind the Möller wave operators in scattering theory [21]. In the present paper we will prove their existence. Asymptotic completeness is also shown: the wave operators provide then isomorphisms between the dynamical system $(\mathcal{X},\mu,S_t)$, i.e. the system having the h.p., and $(\mathcal{X}^o,\mu^o,S_t^o)$, the system with only the light particles. (Here and in the following μ denotes a Gibbs measure at some fixed temperature and density for the system having the h.p.. μ^o has the same thermodynamic parameters and refers to the system where all particles have mass m.) Independent generating partitions are explicitely known for the discrete-time dynamical systems $(\mathcal{X}^o,\mu^o,S_{(n)})$, $S_{(n)}=S_{\tau n}$, for any $\tau>0$. Hence our results provide an explicit construction of the Bernoulli isomorphism for the system $(\mathcal{X},\mu,S_t)$. In [6] its existence was only implicitely granted by the general theorem of [20] via the estimates obtained in [6]. [The scattering matrix in infinite systems of Statistical Mechanics has been recently considered in [17] for a quantum model. Its relation with the existence of quasi-particle excitations has been established in a very general setup in [18].]

The existence of the wave operator in our system exploits essentially three characteristics of the time flow: (1) the existence of a contractive manifold, similar to a "leaf" of the stable foliation in the hyperbolic systems (we shall see that there are several analogies with the dispersed billiard problems) (2) the cluster structure of the dynamics [23], i.e. the h.p. interacts with a finite set of light particles and when such interaction finishes, a new finite group of particles arrives, when also this is over the same phenomenon starts again, and so on (3) the occurrence of rare, very large fluctuations in the flow of the incoming particles: they cause a traumatic change in the behavior of the h.p. which in turns determines an almost complete loss of the memory of the past history of the h.p..

Points (2) and (3) provide an example of the mechanism proposed by [16] and described before. Point (1) offers an analogy with the hyperbolic structure assumed to be relevant in the analysis of chaotic behavior of finite-dimensional systems. We think such remarks might come useful in the understanding of the approach to stochasticity in the infinite systems.

Let us consider a configuration $x \in \mathcal{X}$. The first particle (q_o, v_o) in x is the h.p., the coordinates of the other particles are denoted by (q_i, v_i). We denote by μ a fixed Gibbs measure for the system and by S_t the time flow, hence $S_t x$ is the configuration at time t starting from x at time 0. Let us consider a light particle (q_i, v_i) in x and denote by $(q_i(t), v_i(t))$ its coordinates at time t in $S_t x$. It is convenient to assume that whenever two light particles collide they pass through each other with unchanged velocities. It is easy to see, cf. [7], that almost surely each particle interacts in a finite interval of time, say t_-, t_+. Then the

$$\lim_{t\to\infty} S^o_{+t} \; (q_i(-t) \, , \, v_i(-t))$$

exists, S^o_t being the free evolution with elastic collisions at 0. By repeating the argument for each light particle in x we construct a configuration $\tilde{\psi}_+ x$ in $\mathcal{X}^o$ which can be written as

$$\tilde{\psi}_+ x = \lim S^o_t \, E \, S_{-t} x \tag{0.1}$$

E is the map from $\mathcal{X}$ onto $\mathcal{X}^o$ obtained by "erasing" the h.p.. The problem is then to reconstruct x out of $\tilde{\psi}_+ x$. This looks a very difficult task since the coordinates of the h.p. in $\tilde{\psi}_+ x$ are hidden in the requirement of compatibility which states that the asymptotic velocities of the light particles should be those given in $\tilde{\psi}_+ x$. If one can solve this problem then he can invert $\tilde{\psi}_+$ to obtain the wave operator $\tilde{\Omega}_+$ which maps $\mathcal{X}^o$ into $\mathcal{X}$. It turns out that a direct construction of $\tilde{\Omega}_+$ is easier than trying to invert $\tilde{\psi}_+$. Namely we start considering a $x^o \in \mathcal{X}^o$ and then we "try" to define

$$\tilde{\Omega}_+ x^o = \lim S_t \, I \, S^o_{-t} \, x^o \tag{0.2}$$

where I is a map from $\mathcal{X}^o$ into $\mathcal{X}$ which consists in inserting, in some way still to be specified, the h.p. in the available space. Obviously the existence of the limit in eq. (0.2) is linked to proving that the indeterminacy in the insertion of the h.p. is irrelevant in the limit when t goes to infinity. Let $\mathcal{X}^o$ be the set where $\tilde{\Omega}_+$ is defined and assume that $\mathcal{X}^o$ is S_t^o invariant. Then as a consequence of this and of eq. (0.2)

$$S_t\tilde{\Omega}_+ = \tilde{\Omega}_+ S_{-t}^o \quad \text{on} \quad \mathcal{X}_+^o \ . \tag{0.3}$$

Namely $\tilde{\Omega}_+$ carries trajectories into trajectories. With the help of eq. (0.3) one can then try to prove that the inverse of $\tilde{\Omega}_+$ is $\tilde{\psi}_+$. Notice that if $\mathcal{X}_+^o$ is measurable and $\mu^o(\mathcal{X}_+^o) > 0$, then, by the ergodicity of $(\mathcal{X}^o,\mu^o,S_t^o)$, $\mu^o(\mathcal{X}_+^o) = 1$. If we can also prove that $\tilde{\Omega}_+$ is an isomorphism between the probability spaces $(\mathcal{X}^o,\mu^o)$ and $(\mathcal{X},\mu)$ then $\tilde{\Omega}_+$ defines an isomorphism also between the dynamical systems $(\mathcal{X},\mu,S_t)$ and $(\mathcal{X}^o,\mu^o,S_t^o)$.

Unfortunately we have not been able to proceed in strict agreement to the above scheme. The main problem arises from the request that the limit in eq. (0.2) should exist with μ^o probability one. The way out is to modify the r.h.s. of eq. (0.2): we define a family I_t of insertion maps from $\mathcal{X}^o$ into $\mathcal{X}$ and we pose

$$\Omega_+ x^o = \lim S_t\, I_t\, S_t^o\, x^o \ . \tag{0.4}$$

It is clear that greater freedom in the choice of the insertion map makes it easier the proof of the limit in eq. (0.4). At this point, however, it is not any longer evident that $S_t\Omega_+ = \Omega_+ S_t^o$. Such property will be regained by choosing the I_t's so that they are constant in long time intervals which become infinitely long when t goes to infinity. We in fact prove the following

Theorem 0.1. One can define I_t in such a way, see Def. 3.5, that there exists a set $\mathcal{X}_+^o \subset \mathcal{X}^o$, $\mu^o(\mathcal{X}_+^o) = 1$, $S_t^o\mathcal{X}_+^o = \mathcal{X}_+^o$, such that the limit in eq. (0.4) exists. Ω_+ is a measurable map from $\mathcal{X}_+^o$ onto $\Omega_+\mathcal{X}_+^o$ which establishes a modulo zero isomorphism between $(\mathcal{X}^o, \mu^o, S_t^o)$ and $(\mathcal{X}, \mu, S_t)$. Analogous construction holds for $t<0$. Denote by Ω_- the corresponding wave operator. Then $\Omega_-^{-1}\Omega_+$ is a non-trivial isomorphism of $(\mathcal{X}^o, \mu^o, S_t^o)$ into itself (modulo zero). $\Omega_-^{-1}\Omega_+$ is the scattering matrix of $(\mathcal{X}, \mu, S_t)$.

In Section 1 we prove the existence of contractive manifolds through each configuration in a set of full measure.

In Section 2 we discuss the large fluctuation - nice event - which causes an almost complete loss of the memory of the past. At this stage we recall results obtained in [7], [6], [4].

In Section 3 we prove the existence of the wave operators.

In Section 4 we conclude the proof of Theorem 0.1.

Detailed proofs of our results require more space than what is available in the present Proceedings. So we did not reproduce proofs "essentially" existent in the literature. Moreover we took the attitude to give for granted all those probability estimates which origin from properties of Poisson processes and which are extensively considered in the analysis of models describing the systems of equilibrium statistical mechanics.

1. Contractive manifolds

The contractive nature of the system $(\mathcal{X}, \mu, S_t)$ is stated in Thm. 1.2, first we pose the following:

1.1 Notation. Let $x \in \mathcal{X}$ and let L and r be positive numbers. $V(x,L,r)$ denotes the set of all configurations which can be obtained by shifting all coordinates (both position and velocity) of any particle of x which is in $0,L$ by less than r .

If y is in $V(x,L,r)$ there is a natural labelling of the particles of y which is inherited from that of x . We consider the collisions of the h.p. both with the light particles and with the wall at the origin. $i(n)$ is defined as the label of the light particle at the n -th collision. If the collision is against the wall at the origin we set $i(n) = 0$. We will then say that the h.p. has the same order of collisions in x and $y \in \mathcal{V}(x,L,r)$ if $i(n)$ is the same for each n both in x and y.

1.2 Theorem. There exists a set $\mathcal{X}_c, \mu(\mathcal{X}_c) = 1$, such that for every $x \in \mathcal{X}$ the following holds. Let $x \in \mathcal{X}_c$ and $L>0$, then there is $r>0$ such that for every $y \in \mathcal{V}(x,L,r)$

$$\lim |q_o(S_t x) - q_o(S_t y)| = 0 \tag{1.1}$$

exponentially fast. [$q_o(x)$ denotes the position of the h.p. in the configuration x.] Furthermore the h.p. has the same order of collisions in x and y.

Given ε and L (positive) there is r and a set $\mathcal{X}_c(\varepsilon,L)$ such that for every $x \in \mathcal{X}_c(\varepsilon,L)$ and $y \in \mathcal{V}(x,L,r)$ eq. (1.1) holds and

$$\mu(\mathcal{X}_c(\varepsilon,L)) > 1-\varepsilon \tag{1.2a}$$

$$|q_o(S_t x) - q_o(S_t y)| < \varepsilon \quad \text{for all} \quad t . \tag{1.2b}$$

1.3 Proof of the "infinitesimal version of Thm. 1.2. We assume that the coordinates (q_i, v_i) of x for $q_i<L$ are changed by infinitesimal amounts, dq_i dv_i, while all the other coordinates are kept fixed. We will see that for x in a set of full measure $dq_o(t)$ $dv_o(t)$ will tend exponentially to zero. Since we consider infinitesimal changes the collisions of the h.p. with the light particles and the origin occur at the same times in x and in the varied configuration, here we assume that x is chosen so that there is no triple collision. Due to the exponential decay of $dq_o(t)$ and $dv_o(t)$ we will then be able to show that for finite, small enough changes of the q_i, v_i analogous property holds: namely the h.p. has the same order of collisions in x and y and this will be enough to extend the result to the finite variation case.

The main point in the proof is the reduction of the estimate to that concerning a finite number of particles.

We call <u>cluster time</u> a time t when the h.p. collides with a light particle for the first time and after t the h.p. does not collide with it anymore and furthermore no other collision occurs with any of the particles involved in the previous interactions. Cluster times exist with positive probability [7], and since the system is ergodic, they have non-zero frequency. Hence we introduce the set

$$\{N(t) \geq c_1 t - c_2 \quad \forall t \geq 0\} \tag{1.3a}$$

where $N(t)$ denotes the (random) number of cluster times up to time t. Then given any $\varepsilon>0$ there are $c_1>0$ and c_2 such that

$$\mu(\{N(t) \geq c_1 t - c_2 \quad \forall t \geq 0\}) > 1-\varepsilon . \tag{1.3b}$$

In the following we consider $x \in \{N(t) \geq c_1 t - c_2 \ \forall t \geq 0\}$. We also take x so that no triple collision occurs. Let us assume that the particles which collide with the h.p. up to the first positive cluster time are $(q_1, v_1), \dots, (q_n, v_n)$ and that the coordinates of the h.p. are (q_o, v_o). We denote by $(dq_o, dv_o), \dots, (dq_n, dv_n)$ the infinitesimal changes in the coordinates of such particles, not necessarily being all different from zero. The evolution of such systems is reduced to the motion of a particle in a $n+1$-dimensional billiard with planar boundaries [24]. Since the kinetic energy is conserved it is natural to consider a metric for which the square length of $(dv_o, \dots, dv_n)$ is $M dv_o^2 + \dots + m dv_n^2$. We then change variables: $y_o = M^{1/2} q_o$, $w_o = M^{1/2} v_o$; $y_i = m^{1/2} q_i$, $w_i = m^{1/2} v_i$; $i=1\dots n$. In such variables we consider the usual Euclidean metric. We assume to have labelled the particles in such a way that $y_i \leq y_{i+1}$. The time evolution read in the new coordinates induces a motion in the $n+1$-dimensional billiard $0 \leq y_o \leq (M/m)^{1/2} y_1$, $y_i \leq y_{i+1}$ for $i=1,\dots,n-1$. The dynamics is such that when the point gets to the boundary of the billiard it is elastically reflected. From this it follows that

$$|\vec{dw}(t)| = |\vec{dw}| \quad |\vec{dy}(t)| \leq |\vec{dw}|t + |\vec{dy}| \tag{1.4a}$$

$$\vec{w} = (w_o, \dots, w_n \quad \vec{y} = (y_o, \dots, y_n)$$

$$|dv_o(t)| \leq M^{-1/2} |\vec{dw}| \quad |dq_o(t)| \leq M^{-1/2}(|\vec{dw}|t + |\vec{dy}|) . \tag{1.4b}$$

There will be a k such that after the k-th cluster and for the first time the only indeterminacy concerns the h.p.. The "error" will be (dq_o', dv_o') and in general will be larger than the initial one. Let us call t^- the interval of time between the k-th and the $(k+1)$-th cluster times. t^- means that we are considering what happens just before the last collision. Using eq. (1.4b) we have

$$|dv_o(t^-)| \le |dv_o'| \qquad |dq_o(t^-)| \le |dq_o'| + t^-|dv_o'| \; .$$

We can compute explicitely the effect of the last collision, which, by definition of cluster time, involves a "new" particle (which is colliding for the first time). Hence we get

$$|dv_o(t^+)| \le \alpha|dv_o'| \qquad |dq_o(t^+)| \le \alpha(|dq_o'| + t^+|dv_o'|)$$
$$\alpha = (M+m)^{-1}(M-m) < 1 \; . \tag{1.5}$$

Iterating the argument for each other cluster we have an exponential decrease of $dq_o(t)$, $dv_o(t)$ once we notice that $x \in \{N(t) \ge c_1 t - c_2 \;, \; \forall t \ge 0\}$.

1.4 Proof of Thm. 1.2. In subsection 1.3 we have seen that the system behaves like a billiard with planar faces. Concave billiard have hyperbolic structure in the finite dimensional case. Concavity is here replaced by the infinite size of the system and the cluster structure of the dynamics. Like for concave billiards troubles arise from discontinuities. In our case they appear as changes in the order of the collisions of the h.p. and were disregarded in 1.3 under the assumption of infinitesimal perturbations. Our argument goes like this. We assume that the collisions of the h.p. with the light particles and with the wall at the origin have the same order both in x and y. In such case the bounds of the infinitesimal case extend to the present case due to the locally linear character of the dynamics. We then need a consistency check to lift the above assumption. We can assume that with probability 1 there are $c_1 > 0$ and c_2 so that $x \in \{N(t) \ge c_1 t - c_2 \;\; \forall t \ge 0\}$ [7]. Supposing that in y the order of collisions is the same as in x, we get

$$|\delta q_o(t)| \le a_o\, e^{-\gamma t}(|\vec{\delta q}|+|\vec{\delta v}|)\ ;\ |\delta v_o(t)| \le a_o\, e^{-\gamma t}(|\vec{\delta q}|+|\vec{\delta v}|)\,. \quad (1.6)$$

Here γ and a_o are some suitable positive coefficients and $|\vec{\delta q}|\,|\vec{\delta v}|$ are the Euclidean norms of $\delta q_o,\dots,\delta q_n$; $\delta v_o,\dots,\delta v_n$ respectively, these being the only varied coordinates in going from x to y .

For the light particles we have the following estimate: let t_i denote the time of the last collision of the particle with label i then

$$|\delta q_i(t)| \le a_o e^{-\gamma t_i}(|\vec{\delta q}|+|\vec{\delta v}|t)\ ;\ |\delta v_i(t)| \le a_o e^{-\gamma t_i}(|\vec{\delta q}|+|\vec{\delta v}|)\,. \quad (1.7)$$

The consistency of eqs. (1.6) and (1.7) with the assumption that x and y have the same order of collisions is based on some restrictions on the choice of x and on making the initial error $\vec{\delta q}+\vec{\delta v}$ small enough. We first give the conditions on x , we then show that the consistency check holds and, in the next subsection, that the above conditions hold in a set of full measure.

Eq. (1.7) shows that $|\delta q_i(t)|$ increases in time, hence the only way out is to show that in $S_t x$ the i -th particle gets so far away from the h.p. that the "error" $\delta q_i(t)$ becomes unimportant. The first requirement for x is therefore that

$$q_o(S_t x) \le c_3 \log_+ t + c_4 \quad \forall t\ge 0\,. \quad (1.8)$$

Then we need a lower bound on the speed of the particles (the bounds given below are not optimal)

$$|v_i(t)| \ge (k_1 + t_i^{3/2})^{-1} \quad \forall t\ge 0 \quad (1.9)$$

t_i being the last time that particle i interacts with the h.p. From eqs. (1.7) and (1.9), assuming that $|\vec{\delta q}|+|\vec{\delta v}|$ is small enough, it follows that $2|\delta v_i(t)|\le|v_i(t)|$, therefore

$$q_i(t) - |\delta q_i(t)| \ge \tfrac{1}{2}(k_1 + t_i^{3/2})^{-1}(t-t_i) \quad \text{for} \quad t\ge t_i\,.$$

Then there exists k_2 so that

$$q_i(t) - |\delta q_i(t)| \geq c_3 \log_+ t + c_4 \geq q_o(t) \quad \forall t \geq k_2 + t_i^2 \ . \qquad (1.10)$$

This shows that for any i after $k_2+t_i^2$ the i-th particle stays strictly to the right of the h.p. both in x and y : we only have to control the collisions of each particle for a finite interval of time. Anyway the collisions of the h.p. need to be "well separated" in time. We therefore require that for some constant k_3

$$\Delta_t(x) \geq (t^2 + k_3)^{-1} \qquad (1.11)$$

$$\delta_t(x) \geq (t^2 + k_3)^{-1} \ . \qquad (1.12)$$

By $\Delta_t(x)$ and $\delta_t(x)$ we mean the following. Given t let $t_- \leq t$ be the time of the last collision before t of the h.p. (at 0 or with a light particle) and t_+ the next collision (strictly after t). Then $\Delta_t(x) = t_+ - t_-$. Let q be the position of the h.p. at t_- . If $q=0$ then $\delta_t(x)$ denotes the position of the next light particle. If $q=0$ $\delta_t(x)$ is the minimum between q and the distance from q of the next particle (of course disregarding the particle which is involved in the collision at q).

Another request on x is motivated by the following fact. Assume the h.p. has a collision with a light particle, the two velocities being almost the same: then if the coordinates are even slightly changed the resulting delay in the collision might be large. Hence we require that there is k_4 such that at any collision time of the i-th particle with the h.p.

$$|v_i(t) - v_o(t)| \geq (k_4 + t_i)^{-1} \ . \qquad (1.13)$$

Finally we need an upper bound on the velocities of the particles. x is chosen so that if t is a collision time the outgoing velocities $v_o(t)$ and $v_i(t)$ of the h.p. and the light particle, respectively, satisfy the inequality

$$|v_o(t)| \leq c_5 \log_+ t + c_6 \qquad |v_i(t)| \leq c_5 \log_+ t + c_6 \ . \qquad (1.14)$$

It is now easy to see that under conditions (1.6)...(1.14)

the order of the collisions in x and y is the same, provided that $|\delta\vec{q}|+|\delta\vec{v}|$ is small enough.

Assume in fact that at time t there is a collision in x with parameters $(q_o,v_o);(q_o,v)$. Assume that in y the parameters of the same particles are $(q_o+\delta q,\ v_o+\delta v)$; $(q_o+\delta\bar{q}\ ,\ v+\delta\bar{v})$. The delay time for such collision in y is

$$|\Delta T| \le (|\delta q|+|\delta\bar{q}|)(|v-v_o|-|\delta\bar{v}|-|\delta v_o|)^{-1}\ . \qquad (1.15)$$

Since $t\le t_i$, by definition, by using eqs. (1.6), (1.7) and (1.13) we get a bound which becomes exponentially small in time. By eq. (1.11) and taking $|\delta\vec{q}|+|\delta\vec{v}|$ small enough we prove the required consistency for any such collision.

Assume now that at some time t we have $(q_o+\delta q_o\ ,\ v_o+\delta v_o)$ and $(\tilde{q}+\delta\tilde{q}\ ,\ \tilde{v}+\delta\tilde{v})$ with $\tilde{q}+\delta\tilde{q}=q_o+\delta q_o$, namely a collision in the varied configuration. We then need to show that (1) also in x there is a collision between these particles, let ΔT be the corresponding delay; (2) the previous and the next collision in x occur with a delay larger than ΔT. The argument now goes as follows. ΔT is again given by eq. (1.15). By eq. (1.10) $t\le k_2+t_i^2$ (i being the particle involved in the collision). Hence ΔT is exponentially small with t. In principle there are two possibilities: (a) the collision corresponding to (q_o,v_o) and $(\tilde{q},\tilde{v})$ is one of those occurring in the history of x; (b) this is not the case and in the x history another light particle collides with the h.p. preventing the (q_o,v_o)-$(\tilde{q},\tilde{v})$ collision. In case (a) everything goes like before and the consistency check is verified. Case (b) contradicts our conditions and cannot occur. In fact using eq. (1.14) we bound the velocities v_o and $\tilde{v}$ so that the light particle for $|t'|\le\Delta T$ will be at a distance from the h.p. which is less than $2(c_5\log_+ t+c_6)\Delta T$. This is exponentially small for t large and it is incompatible with eq. (1.12) if in the meantime another collision (case (b)) occurs. By choosing $|\delta\vec{q}|+|\delta\vec{v}|$ small enough the same reasoning leads to contradiction whatever the time t of the collision is.

The collisions of the h.p. against the origin can be analysed in the same way, so we are left with the proof that the conditions on x we have stated in this subsection are satisfied in a set of full measure.

1.5 Probability estimates. We just outline the proof that the estimates used in subsection 1.4 are verified with probability 1 . The estimate relative to eq. (1.8) is proven in [7].

Eqs. (1.9)...(1.14). We describe the dynamical system $(\mathcal{X},\mu,S_t)$ by introducing the following special flow representation [9], [23]. The basis of the flow is the set of all configurations where the h.p. is colliding with a light particle. Hence the base is described by (1) a configuration $x\in\mathcal{X}$ where the colliding light particle is ignored and (2) a velocity v_1 which represents the outgoing velocity of the colliding light particle. (The position of the collision is of course $q_o(y)$.) The special flow representation is a dynamical system in the space $\{(y,v_1,t)\}$ where t ranges from 0 to $\tau(y,v_1)$, which denotes the next time the h.p. collides again. (y,v_1,t) is the configuration evolved for a time t starting from the configuration corresponding to (y,v_1) . The measure on the base is

$$d\mu(y,v_1,t) = \exp(-\tfrac{1}{2}\beta m v_1^2)(\beta^{-1}m)^{-1/2}(v_1-v_o)\chi(0\le t\le \tau(y,v_1)) \qquad (1.16)$$
$$\mu(dy)dv_1\,dt$$

and the flow is represented as an upward lift and a transformation S on the base.

From eq. (1.16) it is easy to see that the number of collisions of the h.p. with the light particles and the origin grows linearly in time with probability 1 . The estimates given in subsection 1.4 then easily follow from the analogous estimates for the transformation S on the base.

2. Large fluctuations (the nive event)

We report here some results obtained in [6] and [4] which prove the existence of the nice event described in the introduction.

2.1 Definitions. Assume L,N,V be given positive numbers.

The set $\mathcal{D}(L,N,V)$ is the set of all configurations x which have the following properties:

C.1 The particles of x are in R_L, namely if $(q,v) \in x$ and $v \leq 0$ then $q \leq L$;

C.2 The number of particles of x in $[0,2L]$ is bounded by N and $q_o(x) < L/2$;

C.3 The speed of any particle of x in $[0,2L]$ is less than V.

In our applications we will consider the limit L going to infinity and we consequently choose N and V so that $\{x: x \cap R_L \in \mathcal{D}(L,N,V)\}$ has large probability.

The aim is to show that there is a set of configurations of incoming particles in (L,∞) so that after some time T the new configuration in $[0,L]$ is very close to some prescribed configuration x_L independently of $x \in \mathcal{D}(L,N,V)$. The incoming particles are specified in a bounded region of the one particle phase space, the region as well as the time T depend on the target x_L and the accuracy with which it should be reached.

2.2 Theorem [6], [4]. Let L,N,V,ε,x_L be given. Here x_L is a configuration of particles in $[0,L]$, $q_o(x_L) < L$, and for notational simplicity we assume that if $y \in \mathcal{V}(x_L,L,\varepsilon)$, cf. Notation 1.1, then also y is a configuration in $[0,L]$. Then there exist T, y and $\delta > 0$ which depend on L,N,V,ε, x_L so that the following holds. Let

$$C_L(0,T) = \{(q,v): q \geq L,\ v < 0,\ q+vT \leq L\}\ . \qquad (2.1)$$

Then there exists a set of configurations $\mathcal{C}(L,T,\varepsilon,x_L)$ (for ease of reference we only explicit the dependence of these parameters) which is cylindrical with base in $C_L(0,T)$ such that:

P.1 $\mathcal{C}(L,T,\varepsilon,x_L)$ in $C_L(0,T)$ is of the form $\mathcal{V}(y,\delta)$ where y is a configuration in $C_L(0,T)$ and $\mathcal{V}(y,\delta)$ is the set of all configurations in $C_L(0,T)$ obtained by shifting the coordinates of any particle of y by less than δ.

P.2 Let $x \in \mathcal{D}(L,N,V)$ and let $z \in \mathcal{C}(L,T,\varepsilon,x_L)$ be a configuration of particles with negative velocities such that the only particle in $[L,2L]$ are those specified by the condition that $z \in C(L,T,\varepsilon,x_L)$. Then $S_T(x \cup z) \in V(x_L,L,\varepsilon)$.

P.3 Let x and z be as in P.2. Then the h.p. does not collide with the light particles of z in $C_L(T,\infty) = \{(q,v): q \geq L,\ v<0,\ q+vT > L\}$ in the time interval $[0,T]$.

P.4 Let x and z be as in P.2. Then all the particles of x which are in $[0,L]$ as well as those which have collided with the h.p. before time T have at time T positive velocity larger than 1 . The same holds for all the particles of z which at time T have positive velocity and are to the right of L .

It is possible and convenient for the sequel to arrange the parameters so that

P.5 $20^{-1} \geq \mu^{o}(\exists t \in (0,1): S_t^o x^o \in C(L,T,\varepsilon,x_L))T > 0$.

We shall refer to the set $C(L,T,\varepsilon,x_L)$ as "the nice event", the specification of the nice event being the set of parameters: $L,N,V,\varepsilon,x_L,T,y,\delta$.

3. The wave operator

In this section we prove the existence of the wave operator, i.e. of the limit in eq. (0.4). Given x^o we consider $S_{-t}^o x^o$ as a function of $t \in \mathbb{R}_+$. Special values of t are when $S_{-t}^o x^o$ has a set of incoming particles like those specified by the nice event described in Thm. 2.2: for such t 's we define I_t by inserting the h.p. so that, at such times, the conditions of Thm. 2.2 are all satisfied and the nice event actually occurs. The problem arises for other t 's. Assume t is a time for which the nice event occurs and take $t'>t$. Then we would like that $S_{t'-t} I_{t'} S_{-t'}^o x^o$ is still a configuration for which the conditions of Thm. 2.2 apply. We do not know if this is true uniformly in $t'>t$. So we proceeded as follows. We fix a sequence of different nice events: the parameters L,N,V being chosen so that the probability that the configuration is not in $D(L,N,V)$ becomes very small along the sequence. We denote by t_1^* the first time when $S_{-t}^o x^o$ has a set of incoming particles which satisfy the conditions for the first nice event to occur. $t_2^*>t_1^*$ is the first time after t_1^* when the second nice event's conditions are verified and so on. We define I_t for $t\in[t_n^*,t_{n+1}^*)$ so to force $I_t S_{-t}^o x^o$ to have the nice event occurring at $S_{t-t_n^*} I_t S_{-t}^o x^o$. Then we need to show that this holds also when $t\in[t_{n+p}^*,t_{n+p+1}^*)$, $p>0$. After each nice event the state will be in some contractive leaf so it will be sufficient to check such property only for $p=1$ and for all n larger than some N. We will first define the sequence of nice events, then the stopping times t_n^*, the map I_t, the set $\mathcal{X}_+^o$, for which the above can be proven, and finally the wave operator.

3.1 Notation. For ease of reference we report here some notation which will be frequently employed in the sequel. For $L>0$

$$\Gamma_L = \{(q,v):\ q\geq L,\ 0\geq v\geq -L^{1/2}\exp(-qL^{-1/2}+1)\}\ ,$$

$$\Gamma_L^+ = \{(q,v):\ q\geq L,\ 0\leq v\leq L^{1/2}\exp(-qL^{-1/2}+1)\}\ ,$$

$$R_L = \{(q,v):\ \text{either } v\geq 0 \text{ or } v<0 \text{ and } q\leq L\}\ .$$

For $0\leq s<t$: $C_L(s,t) = \{(q,v):\ q\geq L,\ v<0,\ q+vs\geq L\geq q+vt\}$

$$C_L = C_L(0,\infty)\ .$$

3.2 Definition. We choose an increasing sequence L_n, $L_n\nearrow\infty$, so that

$$\mu^o(\{x\cap\Gamma_{L_n}^+=\emptyset\}) > 1-2^{-n} \tag{3.1}$$

$$\mu(\{q_o(S_t x)\leq L_n^{1/2}\log_+|t|,\ \forall t\}) > 1-2^{-n}\ . \tag{3.2}$$

Given L_n we define N_n and V_n so that

$$\mu(\{x:\ x\cap R_L\in\mathcal{D}(L_n,N_n,V_n)\}) > 1-2^{-n} \tag{3.3}$$

cf. Def. 2.1 and Notation 3.1.

3.3 Remarks. We will prove the existence of the wave operator by means of an iterative procedure, the "allowed" error at the n-th step will be 2^{-n}. Notice that in $\{x \cap \Gamma_L^+ = \emptyset\} \cap \{q_o(S_t x) \le L^{1/2} \log_+ t,\ \forall t \ge 0\}$ the h.p. never collides with any of the particles which at time 0 have positive velocity and are to the right of L. Eqs. (3.1) and (3.3) are easily proven since μ and μ^o are Poisson measures. Eq. (3.2) is proven in [7].

3.4 Definition. Given L_n as in Def. 3.2 we define a configuration x_{L_n} of particles in $[0,L_n]$, a time $\tilde{T}_n > 0$, a time $\hat{T}_n > 2\tilde{T}_n$ so that

C.1 $\mu(\mathcal{X}_i(2^{-n},L_n) | x_{L_n}) > 1-2^{-1/2\,n}$ where $\mathcal{X}_i(\varepsilon,L)$ is defined in Thm. 1.2.

r_n is defined in such a way that if $x \in \mathcal{X}_i(2^{-n},L_n)$ and $y \in \mathcal{V}(x,L_n,r_n)$ then eq. (1.2) holds.

C.2 For any set A measurable w.r.t. the σ-algebra generated by $\{q_o(t),\ t \ge \hat{T}_n\}$

$$|\mu(A|x_{L_n}) - \mu(A)| < 2^{-n} . \tag{3.4}$$

C.3 $\mu(\{\text{there is a cluster time in } (\tilde{T}_n, 2T_n)\} | x_{L_n}) > 1-2^{-n}$.

The definition of cluster time is given in subsection 1.3.

C.4 Let $J(s,t)$, $0<s<t$, be the set of configurations where the h.p. does not interact with any of the particles in $C_L(t,\infty)$ during the time interval $[0,s]$. Then

$$\mu(J(2\tilde{T}_n,\hat{T}_n) | x_{L_n}) > 1-2^{-n} . \tag{3.5}$$

C.5 $$\mu(\{q_o(S_t x) \le L_n^{1/2} \log_+ |t|, \forall t\} | x_{L_n}) > 1-2^{-1/2\,n} . \tag{3.6}$$

Remarks. We need to prove that the above definition is well posed, namely that x_{L_n}, $\tilde{T}_n$, $\hat{T}_n$ actually exist. By eq. (1.2) C.1 holds in a set of measure larger than $1-2^{-1/2\,n}$. By choosing T_n large enough eq. (3.4) holds in a set of measure larger than $1-2^{-n}$, cf. [6]. The same holds for condition C.3. Once $\tilde{T}_n$ is fixed $J(2\tilde{T}_n,t)$ has a measure which goes to 1 as t diverges, as it can be easily seen.

Hence the x_{L_n} for which eq. (3.5) holds have measure larger than $1-2^{-n}$ for $\hat{T}_n$ large enough. Eq. (3.6) holds in a set of measure larger than $1-2^{-1/2\,n}$ as it can be seen using eq. (3.2). The intersection of all the above sets has positive measure hence it is not empty.

The motivation for the above conditions is the following: x_{L_n} is (essentially) the configuration left in $[0,L_n]$ after the n -th nice event. $\tilde{T}_n$ gives the rate to which equilibrium is reestablished after the nice event. The other conditions ensure that particles far away from the origin behave like free, so that the "next" nice cluster of particles responsible for the new nice event is not "perturbed" before its arrival.

3.5 Definition of I_t . We first decompose the positive real axis into intervals $[\tau_n,\tau_{n+1})$ whose length increases very fast with n . In fact the τ_n 's are chosen according to the following requirements:

Given n let $C(L_n,T_n,r_n,x_{L_n})$ be as in Thm. 2.2 (r_n is defined in C.1 of Def. 3.4 and L_n,N_n,V_n in Def. 3.2). For notational simplicity we assume to have chosen the parameters so that

$$\mu^o(\{x^o \cap \{(q,v): q\in [L_n,2L_n],\ v\le 0,\ q+vT_n > L_n\} = \emptyset\}) > 1-2^{-n} \qquad (3.7)$$

$$\mu^o(\{\exists t\in (0,1):\ S^o_t x^o \in C(L_n,T_n,r_n,x_{L_n})\}) < 2^{-n}\ . \qquad (3.8)$$

If this were not the case we would choose $\varepsilon < r_n$ and correspondingly T_ε as in Thm. 2.2. By making ε small we would then have the above inequality.

Let now $\tilde{t}_n$ be the infimum of the times $t\ge\tau_n$ such that

$$S^o_{-t}x^o \in C(L_n,T_n,r_n,x_{L_n})\ .$$

We define $t^*_n=\tilde{t}_n$ if $\tilde{t}_n < \tau_n + \frac{1}{2}(\tau_{n+1}-\tau_n)$; otherwise we put $t^*_n=\infty$. The τ_n 's are chosen so that

$$\mu^o(\{x^o:\ t^*_n<\infty\}) > 1-2^{-n}\ .$$

We require that $\tau_{n+1} - \tau_n$ is much larger than the other times introduced so far. For instance $\tau_{n+1} - \tau_n >$ $> 10(\tilde{T}_{n+1}+T_{n+1}+\hat{T}_{n+1})$. $\tilde{T}_n$ is defined in C.2 of Def. 3.4, T_n in C.1 above and $\hat{T}_n$ in C.4 of Def. 3.4. I_t acts on $S^o_{-t}x^o$ as follows. Let $t \in [\tau_n, \tau_{n+1})$. Let $m \leq n$ be the largest integer for which $t \geq t^*_m$. Then $I_t S^o_{-t} x^o$ is obtained from $S^o_{-t}x^o$ by (1) erasing all the particles of $S^o_{-t}x^o$ which are not in $C_{L_m}(t-t^*_m, \infty)$, (2) inserting the h.p. at 1 with zero speed. If there is no t^*_m which is less than t then $I_t S^o_{-t} x^o$ is the configuration consisting of the only h.p. at 1 with zero speed. (Such case will not occur in the sequel.)

Next we introduce the set $\mathcal{X}^o_+$ which will be the domain of definition of the wave operator Ω_+. Several conditions of rather technical origin specify $\mathcal{X}^o_+$. We first give some qualitative description of these conditions. The basic request is that $\{t^*_n < \infty\}$ from some N on. Denote by z_n the configuration $S^o_{-t^*_n+T_n} x^o \cap C_{L_n}(0,\infty)$ and $y_n = x_{L_n} \cup z_n$. We then require that $y_n \in \mathcal{X}_c(2^{-n}, L_n)$ so that a contractive leaf through y_n exists. Furthermore we ask that $S_t y_n$ satisfy the conditions for applying Thm. 2.2 at $t =$ $= t^*_n - T_n - t^*_{n-1}$. We also need that the particles of $S_{\bar{t}}\, y_n$, $\bar{t} = t^*_n - T_n - t^*_{n-1} + T_{n-1}$, which have positive velocity and are to the right of L_{n-1} will not interact with the h.p.. In this way $S_{\bar{t}}\, y_n$ is in the same contractive leaf of y_{n-1} and an iterative procedure can be applied.

3.6 Definition. The set $\mathcal{X}^o_+ \cdot \mathcal{X}^o_+$ is the set of all configurations x^o for which there are N and $c>0$ with the following properties. For each $n \geq N$

C.1 $\{t^*_n < \infty\}$

C.2 We denote by $\{\hat{t}_n \notin I\}$ the event $\{S^o_t x^o \notin C(L_n, T_n, r_n, x_{L_n}) \forall t \in I\}$ and by $\{\hat{t}_n \in I\}$ its complement. Then

$\{\hat{t}_n \notin I_n\}$, $I_n = \{t: |t-\tau_n| < cn\} \cup \{t: |t-\tau_n-\frac{1}{2}(\tau_{n+1}-\tau_n)| \leq cn\}$.

(Condition C.2 will be used when proving that $\mathcal{X}_+^o$ is S_t^o invariant.)

C.3 $S^o_{-t_n^*} x^o \cap \{(q,v): q \in [L_n, 2L_n],\ v \leq 0,\ q+vT_n \geq L_n\} = \emptyset$

$S^o_{-t_n^*} x^o \cap \{(q,v): q \in [0, L_n],\ v \leq 0,\ q+vT_n \geq 0\} = \emptyset$

$S^o_{-t_n^*} x^o \cap \Gamma_{L_n} = \emptyset$.

C.4 We write here and in the following $y_n = x_{L_n} \cup z_n$,

$$z_n = S^o_{-t_n^*+T_n} x^o \cap C_{L_n}(0,\infty) .$$

Then $y_n \in \mathcal{X}_c(2^{-n}, L_n)$.

(Notice that $z_n = S_{T_n} I_{t_n^*} S^o_{-t_n} x^o \cap C_{L_n}$, hence by C.4 after the n -th nice event the configuration is in the contractive leaf of y_n .)

C.5 $q_o(S_t y_n) \leq L_n^{1/2} \log_+ t,\ \forall t \geq 0$.

C.6 For y_n there is a cluster time in the interval $[\tilde{T}_n, 2\tilde{T}_n]$.

C.7 $y_n \in J(2\tilde{T}_n, \hat{T}_n)$, cf. C.4 of Def. 3.4.

(Condition C.7 will ensure that the h.p. in y_n does not interact with the light particles of the $(n-1)$ nice event in the initial time $[0, 2\tilde{T}_n]$.)

C.8 Let $y_{n+1} = z^{(1)}_{n+1} + z^{(2)}_{n+1}$ where $z^{(2)}_{n+1} =$
$= S^o_{-t^*_{n+1}+T_{n+1}+t^*_n-T_n} z_n$. Then the h.p. in y_{n+1} does not interact with the light particles of $z^{(2)}_{n+1}$ in the time interval $[0,\ t^*_{n+1}-T_{n+1}-t^*_n]$.

(By C.8 we have that $S_{t^*_{n+1}-T_n-t^*_n} y_{n+1}$ has the light particles required for the n -th nice event. We need however to impose that (1) the configuration is in $\mathcal{D}(L_n, N_n, V_n)$, (2) that the particles initially in $x_{L_{n+1}} \cup z^{(1)}_{n+1}$ and which at $t^*_{n+1} - T_n - t^*_n$ have negative velocity are in $[0, L_n]$. Notice that for (1) and (2) we need not to worry about the light particles of $x_{L_{n+1}} \cup z^{(1)}_{n+1}$ which have interacted with the h.p. before $\tilde{T}_{n+1}$. In fact we know that in $[\tilde{T}_{n+1}, 2\tilde{T}_{n+1}]$ there is a cluster time and after that they will not interact anymore with the h.p. Finally we want to prove that $S_{t^*_{n+1} - T_{n+1} - t^*_n + T_n} y_{n+1}$ is in the same contractive leaf of y_n. Since y_n does not have light particles with non-negative velocity to the right of L_n we need to require that the h.p. in $S_{t^*_{n+1} - T_{n+1} - t^*_n + T_n} y_{n+1}$ does not interact with the non-negative particles which are to the right of L_n.)

C.9 $S_{t^*_{n+1} - T_{n+1} - t^*_n}(x_{L_{n+1}} \cup z^{(1)}_{n+1}) \subset R_{L_n}$.

C.10 Let η_{n+1} be the configuration $S_{t^*_{n+1} - T_{n+1} - t^*_n} y_{n+1}$ without the light particles which have interacted with the h.p. before $\tilde{T}_{n+1}$. Then

$$\eta_{n+1} \cap R_{L_n} \in \mathcal{D}(L_n, N_n, V_n)$$

Furthermore the h.p. in η_{n+1} does not collide for all $t \geq 0$ with any of the particles of η_{n+1} which (in η_{n+1}) are to the right of L_n and have non-negative velocity.

3.7 Theorem. The set $\mathcal{X}^o_+$ defined in Def. 3.6 is μ^o measurable. $\mu^o(\mathcal{X}^o_+) = 1$. $S^o_t \mathcal{X}^o_+ = \mathcal{X}^o_+$ for any t. The limit

$$\lim S_t I_t S^o_{-t} x^o := \Omega_+ x^o \quad \text{for any} \quad x^o \in \mathcal{X}^o_+$$

exists. Ω_+ is a measurable map from $\mathcal{X}^o_+$ into $\mathcal{X}$ with range $\Omega_+ \mathcal{X}^o_+$ and

$$\Omega_+ S^o_t = S_t \Omega_+ \quad \text{on} \quad \mathcal{X}^o_+ .$$

Proof. $\mathcal{X}^o_+$ is a countable intersection of measurable sets, hence it is measurable. By C.2 of Def. 3.6 $t^*_n(S^o_t x^o) = t^*_n(x^o) + t$ for all $n \geq \bar{N}$, where $\bar{N}$ might depend on t. It is easy to see that C.2...C.10 of Def. 3.6 also hold for $S_t^o x^o$, hence $\mathcal{X}_+^o$ is S_t^o invariant.

Proof that $\mu^o(\mathcal{X}_+^o) = 1$. We use Borel-Cantelli, so we need to estimate the probability of each of the events which specify $\mathcal{X}_+^o$. We denote by $\chi_i(n)$ the characteristic function of the events C.i , i=1,...,10 of Def. 3.6.

C.1 $\mu^o(\chi_1(n)) > 1-2^{-n}$ by C.1 of Def. 3.5.

C.2 We will use the notation C_n for $C(L_n,T_n,r_n,x_{L_n})$. We have by eq. (3.8):

$$\mu^o(1-\chi_2(n)) \le \sum_x \mu^o(\{\hat{t}_n \in [k,k+1]\}) < 2^{-n}2(2(cn+2))$$

where k ranges over the integers such that $|k-\tau_n| < cn+1$ and $|k-(\tau_n+\frac{1}{2}(\tau_{n+1}-\tau_n))| < cn+1$.

C.3 We condition on the event $\{t_n^*=t\}$. y_n and z_n are defined in C.4 of Def. 3.6. We have

$$\mu^o(1-\chi_3(n)|\,t_n^*=t) \le \mu^o(\{t_n>t\})^{-1}\mu^o\left(\bigcap_{k=0}^{k^*}\{\hat{t}_n\notin[t-(2k+2)T_n ,\; t-(2k+1)T_n]\}\cap\{S_{-t}^o x^o\cap\{C.3\}=\emptyset\}\right) \tag{3.9}$$

where {C.3} denotes the event in C.3 with $t_n^*=t$ and k^* is 0 if $t-\tau_n<3T_n$ otherwise it is the largest integer k such that

$$2(k+1)T_n \le t-\tau_n .$$

We have just relaxed the condition $\{t_n^*>t\}$ so that the events in the r.h.s. of eq. (3.9) become mutually independent. We shorthand

$$\delta_n = \mu^o(\{\hat{t}_n\in[0,1]\}) \tag{3.10}$$

we use Cauchy-Schwarz in the last term on the r.h.s. of eq. (3.9), then by eqs. (3.1) and (3.7) we get

$$\mu^o(1-\chi_3(n)\,|t_n^*=t) \le \mu^o(\{t_n^*>t\})^{-1}(1-\delta_nT_n)^{1/2k^*}\cdot 3\cdot 2^{-1/2n} .$$

Notice that $\delta_nT_n<1/20$ by P.5 of Thm. 2.2. For $k>2T_n$,

$T_n > 1$ we have $(\bar{t}_n := t_n^* - \tau_n)$

$$\mu^o[(1-\chi_3(n)) |\, \chi_3(\{\bar{t}_n \in [k,k+1]\})] \le 3\cdot 2^{-1/2n}(1-\delta_n T_n)^{\frac{K}{4T_n} - \frac{3}{8}} \cdot$$

$$\cdot\ \mu^o(\{\bar{t}_n > k+1\})^{-1}\ \mu^o(\{\bar{t}_n \in [k,k+1]\}) \qquad (3.11)$$

$$\mu^o(\{\bar{t}_n \in [k,k+1]\}) \le \delta_n \mu^o(\{\bar{t}_n > k - T_n\})$$

$$\mu^o(\{\bar{t}_n > k+1\})\mu^o(\{\bar{t}_n > k-T_n\})^{-1} \ge \mu^o(\{\bar{t}_n > k+T_n\} | \bar{t}_n > k-T_n) =$$

$$= \mu^o(\{\bar{t}_n > 3T_n\} | \bar{t}_n > T_n) \ge$$

$$\ge 1-(1-\mu^o(\{\hat{t}_n \in [0,T_n]\}))^{-1}\mu^o(\{\hat{t}_n \in [0,2T_n]\}) \ge$$

$$\ge (1-\delta_n T_n)^{-1}(1-3\delta_n T_n) \ .$$

By summing eq. (3.11) over all the positive k 's and using the above estimates we get

$$\mu^o(1-\chi_3(n)) \le \text{const } 2^{-1/2n} \ .$$

The procedure for proving C.4,...,C.7 is similar so we only outline it for the condition C.4. We use the same approach as for C.3, the only difference arises from the fact that here we need the estimate for $\mu^o(\{y_n \notin \mathcal{K}_c(2^{-n},L_n)\})$. We write

$$\mu^o(\{y_n \notin \mathcal{K}_c(2^{-n},L_n)\}) \le \mu(\mathcal{K}_c(2^{-n},L_n)^{\text{compl.}} \cap A \cap B \,|x_{L_n}) +$$

$$+\ \mu(A^{\text{compl.}} \cup B^{\text{compl.}} |x_{L_n})$$

$$A = \{q_o(S_t x) \le L_n^{1/2} \log_+ t \ \forall t \ge 0\}$$

$$B = \{x \cap \Gamma^+_{L_n} = \emptyset\}$$

since in $A \cap B$ the h.p. does not interact with the light particles which have positive velocity and are to the right of L_n . By C.1 and C.5 of Def. 3.4 and eq. (3.1) we have a bound which goes like $2^{-1/2n}$.

C.8 We condition to the event $\{t^*_{n+1}=t,\ t^*_n=s,\ z^{(2)}_{n+1}\}$ where, using the same notation as in C.8 of Def. 3.6,

$$z^{(2)}_{n+1} = S^o_{-t-T_{n+1}-s} z_n \ .$$

Proceeding like in the proof of C.3 and writing $x = x_{L_{n+1}} \cup z^{(1)}_{n+1} \cup z^{(2)}_{n+1}$ we come to the estimate of

$$\mu(\{z^{(1)}_{n+1}:\ \sigma_o < t-s\} | x_{L_{n+1}}, z^{(2)}_{n+1}) \tag{3.12}$$

where σ_o is the first time that in $S_t x$ the h.p. collides with a particle of $z^{(2)}_{n+1}$; $\sigma_{t'}$ is the first time after t' when this happens. We can reduce the estimate of eq. (3.12) to the following

$$\mu(\{z^{(1)}_{n+1}:\ \sigma_{2\tilde{T}_{n+1}} < t-s\} | x_{L_n}, z^{(2)}_{n+1}) \tag{3.13}$$

the "error" being as in C.7. The event to estimate is now measurable w.r.t. the history of the h.p. after $\tilde{T}_{n+1}$ hence we can disregard, by C.1, the conditioning on $x_{L_{n+1}}$. As in [7] we have that

$$\mu(\{x:\sigma_o<t-s\} | z^{(2)}_{n+1}) \chi(\{z_n \cap \Gamma_{L_n} = \emptyset\}) \le$$

$$\le \mu(\{x:\ q_o(S_{t-s-t'}x) \le L_n^{1/2} \log_+|t'|, \forall t'<0\}) \chi(\{z_n \cap \Gamma_{L_n} = \emptyset\})$$

which is exponentially small, cf. eq. (3.2). We are then left with the estimate of $\chi(\{z_n \cap \Gamma_{L_n} = \emptyset\})$ which can be estimated as in C.3, we omit the details.

C.9 We proceed like for C.8. We notice that the event

$$\{S_{t-T_{n+1}-s}(x_{L_{n+1}} \cup z^{(1)}_{n+1}) \in R_{L_n}\} \text{ if } \{\text{there is a cluster in } \tilde{T}_{n+1}, 2\tilde{T}_{n+1}\}$$

becomes measurable w.r.t. the history of the h.p. after $\tilde{T}_{n+1}$. In fact particles which have interacted with the h.p. before $\tilde{T}_{n+1}$ have positive velocity after $2\tilde{T}_{n+1}$ (because of the existence of the cluster time), hence they are in R_{L_n} . After such remark the proof is reduced by

C.2 of Def. 3.4 to the one without the conditioning on $x_{L_{n+1}}$ and an exponential bound can be obtained, cf. [7].

Proof of C. 10. The first condition in C.10 concerns the behavior of the h.p. after $\tilde{T}_{n+1}$ and can be easily reduced to an equilibrium estimate, the proof goes like in [7] and we omit giving the details. The other condition requires more care because just after the nice event the state is "atypical" and we cannot use (directly) equilibrium estimates . However, by C.5 of Def. 3.6 $q_o(S_t y_n) \leq \sqrt{L_n} \log_+ t$, $\forall t \geq 0$. If η_{n+1} is in the same contractive leaf of y_n then $q_o(S_t \eta_{n+1}) \leq \sqrt{L_n} \log_+ t + 2^{-n}$ $\forall t \geq 0$. We require that $\eta_{n+1} \cap \tilde{\Gamma}^+_{L_n} = \emptyset$,

$$\tilde{\Gamma}^+_{L_n} = \left\{ (q,v):\ q \geq L_n\ ,\ 0 \leq v \leq \exp\left(-\frac{q-2^{-n}}{\sqrt{L_n}} - 1\right)\right\}$$

so that the last condition in C.10 holds. On the other hand the condition $\{\eta_{n+1} \cap \tilde{\Gamma}^+_{L_n} = \emptyset\}$ can be proven to hold with large probability when $n \nearrow \infty$, just by using equilibrium estimates.

We will now prove the existence of the limit which defines the wave operator Ω_+ . We shall actually prove that

$$\Omega_+ x^o = \lim S_{t^*_n - T_n} y_n \ . \tag{3.14}$$

For notational simplicity let us consider $\tau_{n+2} > t \geq \tau_{n+1} + \frac{1}{2}(\tau_{n+2} - \tau_{n+1})$ where $n \geq N$, cf. Def. 3.6. By definition of I_t we have that

$$S_t I_t S^o_{-t} x^o = S_{t^*_{n+1} - T_{n+1}} \tilde{y}_{n+1}$$

where $\tilde{y}_{n+1}$ is in the same contractive leaf of y_{n+1} . Therefore the h.p. in y_{n+1} will remain $2^{-(n+1)}$ close to the h.p. in y_{n+1} at all later times. We then consider the evolution of y_{n+1} . At time

$$\bar{t} = t^*_{n+1} - T_{n+1} - t^*_n$$

the configuration $S_{\bar{t}} y_{n+1}$ is such that:

(i) $S_{\bar{t}}y_{n+1} \cap C_{L_n}(0,T_n) \in \mathcal{C}(L_n,T_n,r_n,x_{L_n})$ by C.8 and the definition of t_n^* .

(ii) There is no particle in $\{(q,v): q \in [L_n,2L_n],$ $v<0,\ q+vT_n > L_n\}$ because of C.3 and C.8, as far as the particles of $z_{n+1}^{(2)}$ are concerned, and by C.9 with regard to the particles of $z_{n+1}^{(1)}$.

(iii) In R_{L_n} the configuration is in $\mathcal{D}(L_n,N_n,V_n)$ except for particles which will never interact in the future, by C.10 of Def. 3.6.

Therefore by Thm. 2.2 $S_{\bar{t}+T_n}y_{n+1} \cap \{(q,v): v\le 0$ or $v>0$ and $q\le L_n\}$ is in the contractive leaf of y_n . By C.10, on the other hand, we know that the h.p. in $S_{\bar{t}+T_n}y_{n+1}$ does not interact with the light particles which are in the region: $\{(q,v): v\ge 0,\ q>L_n\}$. So the h.p. remains 2^{-n} close at all times to the h.p. in y_n . As a consequence the h.p. in the configuration $S_{t-t_n^*+T_n}I_t x^o$ is at all times $2^{-(n+1)}+2^{-n}$ close to the h.p. in y_n . From this eq. (3.14) follows.

From the invariance of $\mathcal{X}_+^o$ and eq. (3.14) it easily follows that

$$S_t\Omega_+ = \Omega_+S_t^o \quad \text{on} \quad \mathcal{X}_+^o .$$

4. The isomorphism theorem

In this section we prove the following

4.1 Theorem. The map Ω_+ on $\mathcal{X}_+^o$ defined in Thm. 3.7 establishes an isomorphism between $(\mathcal{X}^o,\mu^o,S_t^o)$ and $(\mathcal{X},\mu,S_t)$.

Proof. We first prove that Ω_+ is invertible on $\mathcal{X}_+^o$. We will prove that for $x=\Omega_+x^o$ the following limit exists:

$$\lim_{t\to\infty} S_t^oE_tS_{-t}x \;:=\; \psi_+x \tag{4.1}$$

and that

$$\psi_+\Omega_+x^o = x^o . \tag{4.2}$$

E_t is defined as follows. Let n be the largest integer such that $t \geq \tau_n + \frac{1}{2}(\tau_n + \tau_{n+1})$ then E_t acts on x by taking away from x the h.p. and all the light particles which are not in $C_{L_n} \setminus \Gamma_{L_n}$. By Thm. 3.7 and eq. (3.15) we have

$$S_{-t}x = \lim S_{t_n^* - T_n - t} y_n \ .$$

Given t let k be the largest integer so that

$$t \geq \tau_k + \frac{1}{2}(\tau_k + \tau_{k+1}) \ .$$

We choose t so large that $k \geq N$ and we take $n > k+1$. We consider S_t, y_n and we have that the h.p. does not interact with the particles of z_k for $t' \leq t_n^* - T_n - t$. In fact $S_{t_n^* - T_n - t_{n-1} + T_{n-1}} y$ is in the contractive leaf of y_{n-1} hence the h.p. has the same order of collisions as in y_{n-1}. By iterating the argument the h.p. in $S_{t_n^* - T_n + t_{k+1}^* + T_{k+1}} y$ has the same order of collisions as y_{k+1}.

By C.8 of Def. 3.6 we know that the h.p. in y_{k+1} interacts with the particles of z_k only after time $t_{k+1}^* - T_k - t_k^*$. By C.3 of Def. 3.6 we can see that the particles of z_k are not affected by E_t, hence $S^o_{t - t_k^* + T_k} E_t S_{-t} y_n = z_k$ in $C_{L_k}(0,\infty)$. Therefore

$$z_k \cap C_o(T_k,\infty) = S^o_{-t_k^* + T_k} x^o \cap C_o(T_k,\infty)$$

because, by C.3 of Def. 3.6,

$$S^o_{-t_k^*} x^o \cap \{(q,v) : q \in [0,L_k],\ v \leq 0,\ q + vT_k > 0\} = \emptyset \ .$$

We then have

$$S^o_t E_t S_{-t} y_n = x^o \text{ in } \{(q,v) : \text{either } v<0 \text{ or } v \geq 0 \text{ and } q - v(t_k^* - T_k) < 0\} \ .$$

The same holds in the limit n going to infinity, hence in the limit when t diverges $\psi_+ \Omega_+ x^o = x^o$.

Ω_+ transforms the measure μ^o in a measure λ on

$\Omega_+ \mathcal{X}_+^o$, we want to show that $\lambda=\mu$. We will only sketch the proof. Let $\tilde{f}$ be a bounded, Lipschitz continuous, cylindrical function on $C(\mathbb{R}, \mathbb{R}_+)$, i.e. the space of paths of the h.p.. We denote by $f(x)$ the value of $\tilde{f}$ at $\{q_o(S_t x), t \in \mathbb{R}\}$. By eq. (3.14) we have that

$$\lim_{n\to\infty} \lambda(f_n) = \lambda(f) \tag{4.3}$$

where $\lambda(f)$ denotes the λ expectation of f and

$$\lambda(f_n) = \int \mu^o(dx^o) f(S_{t_n^*-T_n} y_n)\chi(t_n<\infty) \ . \tag{4.4}$$

We condition on the value of t_n^*, say $t_n^*=t$. This affects the distribution of z_n, cf. C.4 of Def. 3.6; since $S^o_{-t'}x^o \notin C_n$ for $t'<t$. If such condition were absent then we would be through. In such case the integral in eq. (4.4) would be

$$\mu(f(S_{t-T_n}x)\,|\,x_{L_n})$$

and it would be close to $\mu(f)$, by C.2 of Def. 3.4, for t large. The problem of the conditioning can be avoided by going to the "next" nice event. We define s_n^* to be the largest time smaller than $\tau_n + \frac{1}{2}(\tau_{n+1}-\tau_n)$ when the n-th nice event appears and the conditions of Thm. 2.2 apply. By the proof of Thm. 3.7 we know that

$$s_{n-1}^* > \tau_{n-1}$$

with large probability. After conditioning on $s_{n-1}^* = s$ we can reduce ourselves to estimate

$$\mu(f(S_s\ x)\,|\,x'_{L_{n-1}}) \ .$$

Here $s > \tau_{n-1}$ and $x'_{L_{n-1}}$ is in a r_{n-1} neighborhood of $x_{L_{n-1}}$. We introduce the set $\tilde{\mathcal{X}}_c$ of all z in $\{(q,v):\ q>L_{n-1}\}$ such that

$$\mathcal{X}_c = \{x_{L_{n-1}} \cup z \in \mathcal{X}_c(2^{-(n-1)}, L_{n-1})\} \ .$$

Then

$$|\mu(f(S_s x)|x'_{L_{n-1}})-\mu(f)| \leq \|f\|\mu^o(\tilde{\mathcal{X}}_c^{compl.}) +$$

$$+ |\mu(f(S_s x)\chi(\{z\in\tilde{\mathcal{X}}_c\})|x'_{L_{n-1}})-\mu(f)| \ .$$

By definition of $\mathcal{X}_c(2^{-n},L_n)$ and for n large enough, $(s > \tau_{n+1})$ we have

$$|f(S_s(x'_{L_{n-1}} \cup z))-f(S_s(x_{L_{n-1}} \cup z))| \leq k_f \ 2^{-(n-1)}$$

where k_f is the Lipschitz coefficient of f . Hence

$$|\mu(f(S_s x)\chi((z\in\tilde{\mathcal{X}}_c))|x'_{L_{n-1}})-\mu(f(S_s x)$$

$$\chi(\{z\in\mathcal{X}_c\}|x_{L_{n-1}})| \leq k_f \ 2^{-n+1} \ .$$

We can now eliminate the condition $\{z\in\mathcal{X}_c\}$ with an error $\leq\|f\|\mu^o(\tilde{\mathcal{X}}_c^{compl.})$. We have

$$|\mu(f(S_s x)|x'_{L_{n-1}})-\mu(f)| \leq 2\|f\|2^{-1/2n}+2^{-(n-1)}\|f\|+k_f \ 2^{-(n-1)}$$

and this shows that $\lambda=\mu$.

Acknowledgements

We acknowledge many useful discussions with Carlo Boldrighini and Sandro Pellegrinotti. One of us (E.P.) acknowledges very kind hospitality at the Institute of Problems of Information and Transmission, Moscow, and the State University of Moscow, where this work was started.

References

[1] Aizenmann M., Goldstein S., Lebowitz J. L. Ergodic properties of infinite systems. Springer Lect. Notes in Physics 38, 112 (1975).

[2] Aizenmann M., Goldstein S., Lebowitz J. L. Ergodic Properties of an infinite one dimensional hard rods system. Comm. Math. Phys.

[3] Arnold V. I., Avez A. Problèmes ergodiques de la mécanique classique. Paris, Gauthier-Villars, 1967.

[4] Boldrighini C., De Masi A. Ergodic properties of a class of one dimensional systems of statistical mechanics. In preparation.

[5] Boldrighini C., Dobrushin R. L., Sukhov Yu. One dimensional hard rod caricature of hydrodynamics. J. Stat. Phys. 31, 577 (1983).

[6] Boldrighini C., De Masi A., Nogueira A., Presutti E. The dynamics of a particle interacting with a semi-infinite ideal gas is a Bernoulli flow. Preprint, 1984.

[7] Boldrighini C., Pellegrinotti A., Presutti E., Sinai Ya. G., Solovietchic M. R. Ergodic properties of a one dimensional semi-infinite system of statistical mechanics. Preprint, 1984.

[8] Boldrighini C., Pellegrinotti A., Triolo L. Convergence to stationary states for infinite harmonic systems. J. Stat. Phys. 30 (1983).

[9] Cornfeld I. P., Fomin S. V., Sinai Ya. G. Ergodic theory. Springer-Verlag, 1982.

[10] De Pazzis O. Ergodic properties of a semi-infinite hard rods system. Commun. Math. Phys. 22, 121 (1971).

[11] Dobrushin R. L., Pellegrinotti A., Sukhov Yu., Triolo L. In preparation.

[12] Dobrushin R. L., Sukhov Yu. The asymptotics for some degenerate models of evolution of systems with an infinite number of particles. J. Soviet Math. 16, 1277 (1981).

[13] Farmer J., Goldstein S., Speer E. R., Invariant states of a thermally conducting barrier. Preprint, 1983.

[14] Goldstein S., Lebowitz J. L., Ravishankar K. Ergodic properties of a system in contact with a heat bath: a one dimensional model. Comm. Math. Phys. 85, 419 (1982).

[15] Goldstein S., Lebowitz J. L., Ravishankar K. Approach to equilibrium in models of a system in contact with a heat bath. Preprint.

[16] Landau L. D., Lifschitz E. M. Statistical physics. Pergamon Press, London-Paris, 1959.

[17] Botnic, Malishev
Commun. Math. Phys. ~(1983-84).

[18] Narnhofer T., Requardt M., Thirring W. Quasi particles at finite temperature. Commun. Math. Phys. 92, 247 (1983).

[19] Nogueira A. Ergodic properties of a one dimensional open system of statistical mechanics. Preprint, 1984.

[20] Ornstein D. S. Ergodic theory, randomness and dynamical systems. Yale University Press, New Haven and London 1974.

[21] Reed M., Simon B. Methods of modern mathematical physics: III scattering theory. Academic Press, 1979.

[22] Sinai Ya. G. Ergodic properties of a gas of one dimensional hard rods with an infinite number of degrees of

freedom. Funct. Anal. Appl. 6, 35 (1972).

[23] Sinai Ya. G. Construction of dynamics in one dimensional systems of statistical mechanics. Theor. Math. Phys. 11, 248 (1972).

[24] Sinai Ya. G. Introduction to ergodic theory. Princeton University Press, 1977.

[25] Sinai Ya. G., Volkovysski K. Ergodic Properties of an ideal gas with an infinite number of degrees of freedom. Funct. Anal. Appl. 5, 19 (1971).

[26] Bowen R. Equilibrium states of the ergodic theory of Anosov diffeomorphisms. Springer, Lect. Notes in Math. 470 (1975).

[27] Ruelle D. Thermodynamics formalism. Addison Wesley, Boston, 1978. Encyclopedia of mathematics and its applications.

QUANTUM STOCHASTIC PROCESSES

L. Accardi

Contents

1.) <u>Quantum stochastic processes.</u>

Let $\mathscr{B}$ be a $*$-algebra with identity (usually it wil be a C*- or a W*-algebra). A <u>quantum stochastic process</u> over $\mathscr{B}$ indexed by $\mathbb{R}$ is defined by a triple $\{\mathscr{A}, (j_t)_{t\in\mathbb{R}}, \varphi\}$ where

- $\mathscr{A}$ is a $*$-algebra with identity.
- $j_t : \mathscr{B} \hookrightarrow \mathscr{A}$ is an embedding ($t\in\mathbb{R}$).
- φ is a state on $\mathscr{A}$.

<u>Example 1.)</u> <u>Classical real valued stochastic processes.</u>

Let $(\Omega, \mathscr{F}, P)$ be a probability space and let $X_t : (\Omega, \mathscr{F}, P) \longrightarrow \mathbb{R}$ ($t\in\mathbb{R}$) be a real valued stochastic process. By choosing

- $\mathscr{B} = L^\infty(\mathbb{R})$ = algebra of all complex valued, Borel-measurable functions on $\mathbb{R}$.
- $\mathscr{A} = L^\infty(\Omega, \mathscr{F}, P)$.
- $j_t : f\in\mathscr{B} \hookrightarrow j_t(f) = f\circ X_t = f(X_t)$; $(t\in\mathbb{R})$ (1.1
- $\varphi(a) = \int_\Omega a dP$; $a\in\mathscr{A}$.

The triple $\{\mathscr{A}, (j_t)_{t\in\mathbb{R}}, \varphi\}$ is a quantum stochastic process in the sense defined above. Conversely, one easily sees that to a given a quantum stochastic process $\{\mathscr{A}, (j_t), \varphi\}$ such that $\mathscr{A}$ is an abelian C*-algebra, one can associate a classical stochastic process, characterized (up to stochastic equivalence) by the property of having the same finite dimensional correlation functions as the initial one. Thus, since the quantum stochastic processes include the classical ones, in the following we shal only speak of <u>stochastic processes</u>.

<u>Example 2.)</u> (<u>A "small" quantum system interacting with a "larger" one</u>).

Let H_0 and F be two Hilbert spaces. One might regard H_0 as the quantum state space of a "small system" interacting with an "extended system" with state space F (a typical situation is : $H_0 \cong \mathbb{C}^n$; F - some Fock space); in this case $H_0 \otimes F$ will be the state space of the "composite system". The evolution of the "composite system" is given by a 1-parameter group ($\mathcal{U}_t$) of unitary operators on $H_0 \otimes F$:

$$\mathcal{U}_t \in \mathcal{B}(H_0 \otimes F) \cong \mathcal{B}(H_0) \otimes \mathcal{B}(F) \tag{1.2}$$

and there is a natural embedding $j_0 : \mathcal{B}(H_0) \hookrightarrow \mathcal{B}(H_0) \otimes \mathcal{B}(F)$ of the algebra of the "small system" into the algebra of the composite system, given by :

$$j_0 : b \in \mathcal{B}(H_0) \hookrightarrow j_0(b) = b \otimes 1 \in \mathcal{B}(H_0) \otimes \mathcal{B}(F) \tag{1.3}$$

denoting, for each $t \in \mathbb{R}$ and $a \in \mathcal{B}(H_0) \otimes \mathcal{B}(F)$:

$$\mathbf{u}_t(a) = \mathcal{U}_t \cdot a \cdot \mathcal{U}_t^+ \tag{1.4}$$

one can define, for each $t \in \mathbb{R}$, the embedding :

$$j_t : b \in \mathcal{B}(H_0) \hookrightarrow j_t(b) = \mathbf{u}_t(j_0(b)) \in \mathcal{B}(H_0 \otimes F) \tag{1.5}$$

Usually a state $\overline{\varphi}$ on $\mathcal{B}(H_0 \otimes F)$ is given ($\overline{\varphi} \geqq 0$ and $\overline{\varphi}(1_{H_0} \otimes 1_F) = 1$) and, if we are interested only in the time evolution of the "small system", then all the interesting physical quantities can be expressed in terms of the <u>correlation functions</u> :

$$\overline{\varphi}(j_{t_1}(b_1) \cdot \ldots \cdot j_{t_n}(b_n)) \tag{1.6}$$

where $b_j \in \mathcal{B}(H_0)$ $(j = 1,\ldots,n)$ and $t_1,\ldots,t_n$ are real numbers which need not to be neither time-ordered nor mutually different. Choosing $\mathcal{B} = \mathcal{B}(H_0)$, $\mathcal{A} = \mathcal{B}(H_0 \otimes F)$, and $(j_t)_{t \in \mathbb{R}}$ as in (1.5), one obtains a quantum stochastic process in the sense defined above.

<u>Remark 1.)</u> Both in examples (1.) and (2.) one could have choosen a smaller algebra $\mathcal{A}$ - for example the norm (in $L^\infty(\Omega, \mathcal{F}, P)$ or in $\mathcal{B}(H \otimes F)$) closure of the $*$-algebra generated by the family $\{j_t(\mathcal{B}) : t \in \mathbb{R}\}$. In general, if $\mathcal{A}$ is generated, algebraically or topologically, by the family $\{j_t(\mathcal{B}) : t \in \mathbb{R}\}$, we say that the stochastic process $\{\mathcal{A}, (j_t), \overline{\varphi}\}$ is <u>minimal</u>. In the following, unless explicitaly stated, by "stochastic process" we will mean "minimal stochastic process".

Remark 2.) The occurrence of not necessarily time-ordered correlation functions in (1.6) arises naturally, for example in the computation of moments of observables of the form

$$\sum_{k=1}^{n} j_{s_k}(b_k) \ ; \ s_1 < \dots < s_n \ ; \ b_1, \dots, b_n \in \mathscr{B}(H_0)$$

Usually some commutation or anti-commutation relations (arising for example from Einstein causality) are available, and one is reduced to time-ordered correlations. Finally, by polarization and eventually choosing some b_j equal to 1, one verifies that the correlations (1.6) are uniquely determined by the so called correlation Kernels :

$$W^{\overline{\varphi}}_{t_1,\dots,t_n}(b_1,\dots,b_n) = \overline{\varphi}(|j_{t_1}(b_1)\cdot\ldots\cdot j_{t_n}(b_n)|^2) \qquad (1.7$$

($b_j \in \mathscr{B}$; $t_j \in \mathbb{R}$; $j = 1,\dots,n$). In [3] an abstract characterization of the correlation Kernels is given, and it is shown that any family of correlation Kernels defines (uniquely up to stochastic equivalence) a stochastic process.

2.) The local algebras associated to a stochastic process

Given a stochastic process $\{\mathscr{A}, (j_t)_{t\in\mathbb{R}}, \varphi\}$ over a $*$-algebra with identity $\mathscr{B}$, one can define, for each sub-set $I \subseteq \mathbb{R}$, the algebra

$$\mathscr{A}_I = \bigvee_{t\in I} j_t(\mathscr{B}) \qquad (2.1$$

where the right-hand side of (2.1) denotes the algebra generated by the set $\{j_t(\mathscr{B}) : t \in I\}$ (we leave unspecified the topology under which this algebra is closed : this will be clear, case by case, from the context). We will use the notations :

$$\mathscr{A}_{t]} = \bigvee_{s\leq t} j_s(\mathscr{B}) \qquad (2.2$$

$$\mathscr{A}_{[t} = \bigvee_{s\geq t} j_s(\mathscr{B}) \qquad (2.3$$

$$\mathscr{A}_t = j_t(\mathscr{B}) \qquad (2.4$$

Clearly

$$s \leq t \Longrightarrow \mathscr{A}_{s]} \leq \mathscr{A}_{t]} \qquad (2.5$$

A family $(\mathscr{A}_{s]})_{s\in\mathbb{R}}$ of sub-algebras of $\mathscr{A}$, satisfying (2.5), is called a <u>filtration</u>. Given a family $\mathscr{T}$ of sub-sets of $\mathbb{R}$ a family $(\mathscr{A}_I)$ of sub-algebras of $\mathscr{A}$ satisfying :

$$I \subseteq J \Longrightarrow \mathscr{A}_I \subseteq \mathscr{A}_J \tag{2.6}$$

is called a family of local sub-algebras of $\mathscr{A}$ or simply a <u>localization</u> on $\mathscr{A}$ based $\mathscr{T}$.

<u>Example.</u> In the case of a classical stochastic process $(X_t)_{t\in\mathbb{R}}$ cf. the Example (1.) in Section (1.), the local algebras $\mathscr{A}_I$ $(I\subseteq\mathbb{R})$ are sub-algebras of $L^\infty(\Omega, \mathscr{F}_I, P)$, where $\mathscr{F}_I$ is the σ-algebra generated by the random variables $(X_t)_{t\in I}$.

Given a family $(\mathscr{A}_I)_{I\subseteq\mathbb{R}}$ of local algebras $(\subseteq\mathscr{A})$ a 1-parameter group of automorphisms (sometimes endomorphisms) of $\mathscr{A}$ is called a <u>shift</u> (with respect to that localization) if :

$$u_t \mathscr{A}_I = \mathscr{A}_{I+t} \quad ; \forall t\in\mathbb{R} \ ; \ I\subseteq\mathbb{R} \ ; \ \text{(covariance)} \tag{2.7}$$

for any $I\subseteq\mathbb{R}$ and $t\in\mathbb{R}$. If the localization $(\mathscr{A}_I)$ is defined by a stochastic process through (2.1), then (2.7) is equivalent to :

$$u_t \cdot j_s = j_{s+t} \quad ; \quad \forall\ s,t\in\mathbb{R} \tag{2.8}$$

<u>Example.</u> For a classical stochastic process (X_t), one has

$$j_t \ : \ f\in L^\infty(\mathbb{R}) \hookrightarrow j_t(f) = f(X_t) \in L^\infty(\Omega, \mathscr{F}, P) \tag{2.9}$$

$$u_t(f(X_s)) = f(X_{s+t}) \quad ; \ s,t\in\mathbb{R} \tag{2.10}$$

A stochastic process $\{\mathscr{A}, (j_t), \varphi\}$ on $\mathscr{B}$ is called <u>stationary</u>, if there exists a shift (u_t) on $\mathscr{A}$ (i.e. a 1-parameter automorphisms group of $\mathscr{A}$ satisfying (2.8)) such that :

$$\varphi \cdot u_t = \varphi \quad ; \forall\ t\in\mathbb{R} \tag{2.11}$$

Recall that a conditional expectation from $\mathscr{A}$ onto a sub-algebra $\mathscr{C}$ is a linear map $E : \mathscr{A} \longrightarrow \mathscr{C}$ satisfying :

$$E(1) = 1 \ ; \ E(ca) = cE(a) \ ; \ \forall a\in\mathscr{A} \ ; \ \forall c\in\mathscr{C} \tag{2.12}$$

Sometimes (in classical stochastic processes - always for a natural choice of the local algebras $(\mathscr{A}_I)$) for any local algebra $\mathscr{A}_I$ $(I\subseteq\mathbb{R})$ there exists a conditional expectation $E_I : \mathscr{A} \longrightarrow \mathscr{A}_I$ such that

$$\varphi \cdot E_I = \varphi \tag{2.13}$$

i.e. compatible with the state φ. The family (E_I) satisfies

$$I \subseteq J \Longrightarrow E_I \cdot E_J = E_I \text{ (projectivity)} \tag{2.14}$$

and if the state φ is shift-invariant, then :

$$u_t \cdot E_I = E_{I+t} \cdot u_t \quad ; \forall\ t\ , \forall\ I \tag{2.15}$$

Any family (E_I) of surjective conditional expectations $E_I : \mathscr{A} \longrightarrow \mathscr{A}_I$ will be called <u>projective</u> if it satisfies (2.14) and <u>covariant</u> if it satisfies (2.15). In particular, in case of a filtration (a localization based on the "past half-lines" $\{(-\infty,t] : t \in \mathbb{R}\}$) conditions (2.14) and (2.15) become :

$$s \leq t \Longrightarrow E_{s]} \cdot E_{t]} = E_{s]} \tag{2.16}$$

$$u_t \cdot E_{s]} = E_{s+t]} \cdot u_t \tag{2.17}$$

3.) <u>Markov processes and dilations.</u>

A <u>markovian</u> <u>stucture</u> on a $*$-algebra $\mathscr{A}$ is defined by the assignment of :

- a "past-filtration" $(\mathscr{A}_{t]})_{t \in \mathbb{R}}$ on $\mathscr{A}$.
- a "future-filtration" $(\mathscr{A}_{[s})_{s \in \mathbb{R}}$ on $\mathscr{A}$.
- for each $t \in \mathbb{R}$ – an "algebra at time t", $\mathscr{A}_t$ such that :

$$\mathscr{A}_t \subseteq \mathscr{A}_{t]} \cap \mathscr{A}_{[t} \tag{3.1}$$

- A projective system of conditional expectations $E_{t]} : \mathscr{A} \longrightarrow \mathscr{A}_{t]}$ i.e. :

$$s \leq t \Longrightarrow E_{s]} \cdot E_{t]} = E_{s]} \tag{3.2}$$

  enjoying the Markov property :

$$E_{t]}(\mathscr{A}_{[t}) \subseteq \mathscr{A}_t \quad ; \forall\ t \in \mathbb{R} \tag{3.3}$$

If the localization $\{(\mathscr{A}_{t]})\ ,\ (\mathscr{A}_{[t})\ ,\ (\mathscr{A}_t)\}$ admits a shift (u_t), i.e.

$$u_t \cdot \mathscr{A}_{s]} = \mathscr{A}_{s+t]}\ ;\ u_t \cdot \mathscr{A}_{[s} = \mathscr{A}_{[s+t}\ ;\ u_t \cdot \mathscr{A}_s = \mathscr{A}_{s+t} \tag{3.4}$$

and if the family $(E_{t]})$ of conditional expectations is covariant, i.e.

$$u_t \cdot E_{s]} = E_{s+t]} \cdot u_t \tag{3.5}$$

then we speak of a <u>covariant</u> <u>markovian</u> <u>structure</u> .

<u>Example</u> Let $X_t : (\Omega, \mathscr{F}, P) \longrightarrow \mathbb{R}$ $(t \in \mathbb{R})$ be a classical Markov process;

let $\mathscr{F}_{t]}$, $\mathscr{F}_{t}$, $\mathscr{F}_{[t}$ be respectively the past, present and the future σ-algebras; denote

$\mathscr{A} = L^{\infty}(\Omega, \mathscr{F}, P)$; $\mathscr{A}_{t]} = L^{\infty}(\Omega, \mathscr{F}_{t]}, P)$;...

and let $E_{t]} = E^{P}(\cdot \mid \mathscr{F}_{t]})$ be the P-conditional expectation on $\mathscr{F}_{t]}$. Clearly these objects define a markovian structure on $\mathscr{A}$. – a covariant markovian structure if the process (X_t) is stationary.

The connection between markovian structures and semi-groups is made precise by the following

<u>Theorem (3.1)</u> Let $(E_{s]})$ be a family of maps (not necessarily conditional expectations) satisfying conditions (3.2) – projectivity – and (3.5) – covariance. Denote $\tilde{\mathscr{A}}_o$ the vector space generated by the family:

$$\{E_{o]}\cdot \mathbf{u}_t \cdot E_{o]}(\mathscr{A}_o) : t \geq 0\} \tag{3.6}$$

$(\tilde{\mathscr{A}}_o = \mathscr{A}_o$ in the markovian case), and define

$$P^t = E_{o]}\cdot \;{}_{t} \mid \tilde{\mathscr{A}}_o \quad ; \quad t \in \mathbb{R} \tag{3.7}$$

then the family (P^t) is a semi-group of $\tilde{\mathscr{A}}_o$ into itself.

<u>Proof.</u> For $a_o \in \tilde{\mathscr{A}}_o$ and $s,t \in \mathbb{R}$ one has :

$$P^sP^t(a_o) = E_{o]}\cdot \mathbf{u}_s \cdot E_{o]}\cdot \mathbf{u}_t(a_o) = E_{o]}E_{s]}\,\mathbf{u}_{s+t}(a_o) =$$

$$= E_{o]}\,\mathbf{u}_{s+t}(a_o) = P^{s+t}(a_o)$$

If the maps $(E_{t]})$ are completely positive, identity preserving, (e.g. conditional expectations) then the semi-group (P^t) is completely positive identity preserving. Any such a semi-group will be called a <u>markovian semi-group</u> on $\tilde{\mathscr{A}}_o$. If $\tilde{\mathscr{A}}_o$ is a non-commutative algebra, one also speaks of a quantum dynamical semi-group.

In the following we shall only consider the markovian case, i.e. $\tilde{\mathscr{A}}_o = \mathscr{A}_o$. Thus, denoting $\mathscr{B} = \mathscr{A}_o$ and j_0 = the identity embedding $\mathscr{A}_o \hookrightarrow \mathscr{A}$, one obtains the commutative diagramme :

$$\begin{array}{ccccc} \mathscr{A} & \xrightarrow{\;t\;} & \mathscr{A} & \xrightarrow{\;E_{o]}\;} & \mathscr{A} \\ {\scriptstyle j_0}\uparrow & & & & \downarrow{\scriptstyle j_0^{-1}} \\ \mathscr{B} & & \xrightarrow{\quad P^t \quad} & & \mathscr{B} \end{array} \tag{3.8}$$

where j_o^{-1} denotes the left inverse of j_o.

<u>Definition (3.2)</u> Let $\mathscr{B}$ be a $\mathscr{C}^*$-algebra (with identity) and (P^t) - a markovian semi-group on $\mathscr{B}$. A $\mathscr{C}^*$-<u>dilation</u> of the pair $\{\mathscr{B}, (P^t)\}$ is a quadruple $\{\mathscr{A}, j_o, E_{o]}, (\mathbf{u}_t)\}$ making commutative the diagramme (3.8) and such that $j_o : \mathscr{B} \hookrightarrow \mathscr{A}$ is a $\mathscr{C}^*$-embedding; $(\mathbf{u}_t)$ is a 1-parameter automorphisms group of $\mathscr{A}$; $E_{o]} : \mathscr{A} \longrightarrow \mathscr{A}$ is a norm-one projection satisfying :

$$E_{o]} \cdot \mathbf{u}_t(j_o(\mathscr{B})) \subseteq j_o(\mathscr{B}) \ ; \forall \ t \geq 0 \tag{3.9}$$

If, moreover, denoting $\mathscr{A}_{[t}$ - the algebra generated by $\{ \mathbf{u}_s \cdot j_o(\mathscr{B}) : s \geq t\}$ one has :

$$E_{o]} \cdot \mathbf{u}_t \cdot E_{o]} \cdot \mathbf{u}_t^{-1} \big| \mathscr{A}_{[t} = E_{o]} \big| \mathscr{A}_{[t} \ ; \ t \geq 0 \tag{3.10}$$

then we speak of a (covariant) markovian <u>dilation</u> of $\{\mathscr{B}, (P^t)\}$. Finally if there exists a state (weight) φ on $\mathscr{A}$ satisfying :

$$\varphi \cdot E_{t]} = \varphi \quad ; \quad \varphi \cdot \mathbf{u}_t = \varphi \quad ; \ t \geq 0 \tag{3.11}$$

then we speak of a <u>stationary</u> <u>markovian</u> dilation of $\{\mathscr{B}, (P^t)\}$.

<u>Remark</u> One easily sees that there is a one-to-one correspondence between covariant markovian dilations of $\{\mathscr{B}, (P^t)\}$ and covariant markovian structures (as defined at the beginning of this section) with $\mathscr{A}_o \cong \mathscr{B}$ and $E_{o]} \cdot \mathbf{u}_t \cdot j_o = P^t$.

A beautiful classification theory of dilations of completely positive semi-groups has been developed by B. Kummerer and W. Schroder. In the classical case, i.e. when $\mathscr{B}$ is abelian we know that :

i) any markovian semi-group (P^t) on $\mathscr{B}$ has a covariant markovian dilation (obtained through the well known Daniell-Kolmogorov construction).

ii) (P^t) has a stationary markovian dilation if and only if there exists a state (weight) φ_o on $\mathscr{B}$ such that

$$\varphi_o = \varphi_o \cdot P^t \tag{3.12}$$

In the quantum case the situation is more complicated and only recently R. Hudson and K.R. Parthasarathy [6] have shown that the statement (i) holds; while A. Frigerio [5] (cf. also the paper by A. Frigerio and V. Gorini [4]) has found the correct quantum analogue of the statement (ii).

In the following sections we will describe the main technical tools through which the solution of the above mentioned problems has been achieved.

4.) Perturbations of semi-group : the Feynman-Kac formula.

Let $\{\mathcal{A}, (\mathcal{A}_{t]}), (\mathcal{A}_t), (\mathcal{A}_{[t}), (u_t^o), (E_{t]})\}$ be a given covariant markovian structure, and let be given a covariant family of local algebras $(\mathcal{A}_{[s,t]})$ $(s \leq t \;;\; s,t \in \mathbb{R}_+)$ such that

$$\mathcal{A}_{[s,t]} \subseteq \mathcal{A}_{[s} \cap \mathcal{A}_{t]} \tag{4.1}$$

A markovian cocycle (with respect to the structure defined above) is a 1-parameter family $(M_{o,s})_{s \geq o}$ of elements of $\mathcal{A}$ such that :

$$M_{o,t} \in \mathcal{A}_{[o,t]} \;;\; \forall\, t \geq 0 \;;\; \text{(markovianity)} \tag{4.2}$$

$$M_{o,t+s} = M_{o,s} \cdot u_s^o(M_{o,t}) \;;\; \text{(cocycle property)} \tag{4.3}$$

Denoting, for $s < t$, $M_{s,t} = u_s(M_{o,t-s})$, then the two parameter family $(M_{s,t})s<t$ is such that :

$$M_{s,t} \in \mathcal{A}_{[s,t]} \;;\; \forall\, s \leq t \tag{4.4}$$

$$M_{r,s} \cdot M_{s,t} = M_{r,t} \;;\; r < s < t \tag{4.5}$$

$$u_t(M_{r,s}) = M_{r+t,s+t} \tag{4.6}$$

and the three conditions above are those which, in classical probability theory, define the so-called multiplicative functionals associated to a given family $\{\mathcal{F}_{[s,t]}\}$ of σ-algebras. Typical examples are given by : $\mathcal{A} = L^\infty(\Omega, \mathcal{F}, P)$; $(\Omega, \mathcal{F}, P)$ – a Wiener probability space; $(W_t)_{t \geq 0}$ – a real valued Wiener process;

$$M_{s,t} = \exp \tfrac{1}{2}\{-\int_s^t V(W_r)dr + \int_s^t a(W_r)dW_r\} \tag{4.7}$$

with $V, a : \mathbb{R} \longrightarrow \mathbb{R}$ – sufficiently regular functions.

Theorem (4.1) Let $(M_{o,s})_{s>o}$ be a markovian cocycle and define, for $t > 0$

$$P^t(a_o) = E_{o]}(M_{o,t} \cdot u_t^o(a_o) \cdot M_{o,t}^+) \;;\; a_o \in \mathcal{A}_o \tag{4.8}$$

It follows that (P^t) is a semi-group $\mathcal{A}_o \longrightarrow \mathcal{A}_o$.

Proof. For $a_o \in \mathscr{A}_o$ and $s,t \in \mathbb{R}_+$, one has :

$$P^tP^s(a_o) = E_{o]}(M_{o,t}[\, u^o_t \cdot \{E_{o]}(M_{o,s} \cdot u^o_s(a_o) \cdot M^+_{o,s})\}] \cdot M^+_{o,t}) =$$

$$= E_{o]}(M_{o,t} \cdot E_{t]}[\, u^o_t(M_{o,s}) \cdot u^o_{s+t}(a_o) \cdot u^o_t(M_{o,s})^+] \cdot M^+_{o,t}) =$$

$$= E_{o]} \cdot E_{t]}(M_{o,t} \cdot u^o_t(M_{o,s}) \cdot u^o_{s+t}(a_o) \cdot u^o_t(M_{o,s})^+ \cdot M^+_{o,t}) =$$

$$= E_{o]}(M_{o,t+s} \cdot u_{t+s}(a_o) \cdot M^+_{o,t+s}) = P^{t+s}(a_o)$$

Any semi-group (P^t) defined as above, will be called a Feynman-Kac perturbation of the semi-group $P^t_o = E_{o]} \cdot u^o_t$, $(t > 0)$.

Formula (4.8) will be referred to as the Feynman-Kac formula.This formula generalizes several known constructions :

1.) The classical Feynman-Kac formula. This is obtained by choosing, in the notations of formula (4.7) :

$$M_{o,t} = \exp - {}^1\!/_2 \int_o^t V(W_s)ds \tag{4.9}$$

where V is a suitably regular function (e.g. measurable bounded below).

2.) The interaction representation . This is obtained by choosing the markovian structure to be trivial (i.e. all the local algebras are equal to $\mathscr{A}$ and $E_{o]}$ is the identity map on $\mathscr{A}$), and the cocycle $M_{o,t} = U_{o,t}$ to be unitary. In this case, writing u_t instead of P^t the Feynman-Kac formula becomes :

$$u_t(a) = U_{o,t} \cdot u^o_t(a) \cdot U^+_{o,t} \quad ; \quad a \in \mathscr{A} \tag{4.10}$$

The cocycle property then assures that (u_t) is a 1-parameter automorphisms group of $\mathscr{A}$ (cf. the proof of Theorem (4.1), with all the conditional expectations equal to the identity).

The pair $\{(u^o_t), (U_{o,t})\}$ where (u^o_t) is a 1-parameter automorphisms group and $(U_{o,t})$ is a unitary (markovian) (u^o_t)-cocycle is called an interaction representation for the 1-parameter automorphisms group (u_t) defined by (4.10). The connection with the notion of interaction representation usually met in physics is given by the following formal considerations : let (u^o_t) be of the form :

$$u^o_t(a) = \mathscr{U}^o_t \cdot a \cdot \mathscr{U}^{o\,+}_t \quad ; \quad a \in \mathscr{A} \tag{4.11}$$

with $\mathcal{U}_t^o = \exp itH_0$ - a unitary in $\mathcal{A}$, and let $H_I \in \mathcal{A}$ be a self-adjoint operator. Define

$$H_I(t) = \mathbf{u}_t^o(H_I) = \mathcal{U}_t^o \cdot H_I \cdot \mathcal{U}_t^{o+} \quad ; \ t\in\mathbb{R} \tag{4.12}$$

and let $(U_{o,t})$ be defined by :

$$\frac{d}{dt}U_{o,t} = iU_{o,t}\cdot H_I(t) \quad ; \ U_{o,o} = 1 \tag{4.13}$$

then $(U_{o,t})$ is a unitary $(\mathbf{u}_t^o)$-cocycle (markovian in an appropriate localization) and

$$\mathcal{U}_t = U_{o,t} \cdot \mathcal{U}_t^o \quad ;$$

is a 1-parameter unitary group in $\mathcal{A}$ satisfying the formal equation

$$\frac{d}{dt}\mathcal{U}_t = i\,\mathcal{U}_t\cdot[H_0 + H_I] \tag{4.14}$$

In many concrete examples either $H_0 + H_I$ or $H_I(t)$ are not well defined as operators so that equation (4.13) or (4.14) has no rigorous meaning. But we will see that in many cases it is still possible to define, using quantum stochastic calculus, a markovian cocycle $(U_{o,t})$ and a 1-parameter unitary group $(\mathcal{U}_t)$ having all the properties of the formal solutions of the equations (4.13) and (4.14) (cf. Section (6.) in the following).

3.) <u>Perturbations of the identity semi-group.</u> Consider a markovian structure as in the beginning of this section, and let $\mathcal{A}$ be of the form :

$$\mathcal{A} \cong \mathcal{B}(H_0)\otimes\mathcal{B}(F) \cong \mathcal{B}(H_0\otimes F) \tag{4.15}$$

where H_0 and F are complex separable Hilbert spaces. Assume that the shift $(\mathbf{u}_t^o)$ as the form :

$$\mathbf{u}_t^o = \ell_0 \times V_t^o \tag{4.16}$$

where ℓ_0 is the identity map on $\mathcal{B}(H_0)$ and (V_t^o) is a 1-parameter automorphisms group of $\mathcal{B}(F)$. In this case the semi-group $P_o^t = E_{o]}\cdot \mathbf{u}_t^o$ is the identity semi-group on $\mathcal{A}_o \cong \mathcal{B}(H_0)\otimes 1$, and its Feynman-Kac perturbation with respect to a unitary markovian cocycle $(U_{o,t})$ has the form:

$$P^t(a_o) = E_{o]}(U_{o,t}\cdot a_o\cdot U^+_{o,t}) \tag{4.17}$$

A semi-group of this form will be called a <u>Feynman-Kac perturbation</u> <u>of the identity semi-group</u>.

<u>Theorem (4.2)</u> (cf. R. Hudson - K.R. Parthasarathy [6] , A. Frigerio, V.Gorini [4] ,A. Frigerio [5]). Let H_0 be a complex separable

Hilbert space. Any markovian semi-group on $\mathscr{B}(H_0)$ admitting a Lindblad generator has a covariant markovian dilation which is a Feynman-Kac perturbation of the identity semi-group.

5.) Perturbation of stochastic process .

In the preceeding section we have shown that any markovian cocycle gives rise to a perturbation of a markovian semi-group. In this section we show that any unitary markovian cocycle gives rise to a perturbation of a covariant markovian structure which is still a covariant markovian structure. This is a purely quantum-probabilistic phenomenon, since in the abelian case unitary markovian cocycles give rise only to trivial (i.e. identity) perturbations.

Let $\mathscr{A},(\mathscr{A}_{t]})$, $(\mathscr{A}_t)$, $(\mathscr{A}_{[t})$, $(\mathscr{A}_{[s,t})$, (u_t^0) , $(E_{t]})$ be as in Section (5.); let $(U_{0,t})$ be a unitary markovian cocycle, and define

$$u_t(a) = U_{0,t}\cdot u_t^0(a)\cdot U_{0,t}^+ \quad ; \quad a\in\mathscr{A} \tag{5.1}$$

Then (u_t) is a 1-parameter antomorphisms group of $\mathscr{A}$ and defining :

$$\mathscr{B}_0 = \mathscr{A}_0 \quad ; \quad \mathscr{B}_t = u_t(\mathscr{A}_0) \subseteq \mathscr{A}_{[0,t]} \tag{5.2}$$

one easily verifies that for each $a\in\mathscr{A}$:

$$u_t\cdot E_{s]}(a) = E_{s+t]}\cdot u_t(a) \ ; \tag{5.3}$$

thus the family $(E_{t]})$ is also covariant for the evolution (u_t) defined by (5.1).

Define now, for $t \geqq 0$:

$$\mathscr{B}_{[t} = \bigvee_{s\geqq t} u_s(\mathscr{B}_0) = \bigvee_{s\geqq t} u_s(\mathscr{A}_0) \tag{5.4}$$

and similarly for $\mathscr{B}_{t]}$. Remark that :

$$\mathscr{B}_{[t} \subseteq U_{0,t}\cdot\mathscr{A}_{[t}\cdot U_{0,t}^+ \tag{5.5}$$

whence, due to the Markov property of (E_t) :

$$E_{t]}(\mathscr{B}_{[t})\subseteq U_{0,t}\cdot E_{t]}(\mathscr{A}_{[t})\cdot U_{0,t}^+ \subseteq U_{0,t}\cdot\mathscr{A}_t\cdot U_{0,t}^+ = u_t(\mathscr{A}_0) = \mathscr{B}_t \tag{5.6}$$

Thus : $(E_{t]})$ is markovian also with respect to the localization $(\mathscr{B}_{t]})$, $(\mathscr{B}_t)$,$(\mathscr{B}_{[t})$ or equivalently, defining :

$$\mathscr{B} = \bigvee_{t\in\mathbb{R}_+} u_t(\mathscr{A}_0) \tag{5.7}$$

the family $\{\mathscr{B}, (\mathscr{B}_{t]}), (\mathscr{B}_t), (\mathscr{B}_{[t}), (u_t), (E_{t]})\}$ is still a covariant markovian structure. In particular, for any state φ_o on $\mathscr{B}_o = \mathscr{A}_o$, defining $\varphi = \varphi_o \cdot E_{o]}$ (state on $\mathscr{B}$) ; j_0 = the identity embedding $\mathscr{B}_o \hookrightarrow \mathscr{B}$; $j_t = u_t \cdot j_0$ $(t \geq 0)$, the triple $\{\mathscr{A}, (j_t)_{t \geq o}, \varphi\}$ is a (markovian) stochastic process over $\mathscr{A}_o$, in the sense defined at the beginning of Section (1.). As shown by A. Frigerio and V. Gorini [4], [5], (in the case of boson dilations) the process will be stationary if and only if the associated semi-group satisfies a detailed balance conditions. More generally, in the framework of local algebras, it can be shown that the stationarity of the process is related to the behaviour of the semi-group under appropriate "time-reflections" (cf. [1], [2]).

6.) The Wigner-Weisskopf atom .

In this section I will outline some results obtained in collaboration with D. Applebaum and which will be published elsewhere. For the description of the Wigner-Weisskopf model we follow the exposition given by W. von Waldenfels in [9] and we also refer to this paper for a more complete discussion of the physical limits of this approximation. In its simplest version the model describes a 2-levels atom in interaction with an electro-magnetic field. In the "rotating wave approximation" the system is described on the Hilbert space

$$\mathscr{H} = \mathbb{C}^2 \otimes \mathscr{F}(\mathbb{C}^{|\Lambda|}) \cong \mathbb{C}^2 \otimes [\otimes_{\lambda \in \Lambda} \mathscr{F}_\lambda] \tag{6.1}$$

where Λ is a finite set (indexing the frequencies of the EM field), $|\boldsymbol{\lambda}|$ denotes the cardinality of Λ and, for each $\lambda \in \Lambda$, $\mathscr{F}_\lambda \cong \Gamma(\mathbb{C})$ is the Fock space over the Hilbert space $\mathbb{C}$ (with scalar product $\langle u,v \rangle = \bar{u}v$; $u,v \in \mathbb{C}$). On each space $\mathscr{F}_\lambda$ the creation and annihilation operators B^+_λ, B_λ are defined in the usual way and they satisfy the commutation relations :

$$[B_\lambda , B^+_{\lambda'}] = \delta_{\lambda\lambda'} \ ; \ [B_\lambda , B_{\lambda'}] = [B^+_\lambda , B^+_{\lambda'}] = 0 \tag{6.2}$$

Introducing the spin matrices :

$$\sigma_+ = \begin{pmatrix} 0 & 0 \\ 1 & 0 \end{pmatrix} \ ; \ \sigma_- = \begin{pmatrix} 0 & 1 \\ 0 & 0 \end{pmatrix} \ ; \ \sigma_3 = \tfrac{1}{2}\begin{pmatrix} -1 & 0 \\ 0 & 1 \end{pmatrix} \tag{6.3}$$

The hamiltonian of the system in the rotating wave approximation is :

$$H_{tot.} = H_{at.} + H_{EM} + H_I =$$
$$= (\omega_o \sigma_3 \otimes 1) + [\textstyle\sum_{\lambda\in\Lambda} (\omega_o + \omega_\lambda) 1 \otimes B^+_\lambda B_\lambda] +$$
$$+ [\textstyle\sum_{\lambda\in\Lambda} (g_\lambda \sigma_+ \otimes B_\lambda + \bar{g}_\lambda \sigma_- \otimes B^+_\lambda)]$$

where $\omega_o + \omega_\lambda$ is the frequency of the λ-th oscillator and g_λ is the coupling constant of the atom with the λ-th oscillator. Rewriting the hamiltonian as

$$H_{tot.} = H_o + H_1 = \omega_o[\sigma_3 \otimes 1 + \textstyle\sum_{\lambda\in\Lambda} 1 \otimes B^+_\lambda B_\lambda] +$$
$$+ [\textstyle\sum_{\lambda\in\Lambda} (\omega_\lambda 1 \otimes B^+_\lambda B_\lambda + g_\lambda \sigma_+ \otimes B_\lambda + \bar{g}_\lambda \sigma_- \otimes B^+_\lambda)]$$

and remarking that H_o and H_1 commute, we we reduce ourselves to the consideration of the single term

$$H_1 = H_{1o} + H = [\textstyle\sum_{\lambda\in\Lambda} \omega_\lambda 1 \otimes B^+_\lambda B_\lambda] + [\textstyle\sum_{\lambda\in\Lambda} (g_\lambda \sigma_+ \otimes B_\lambda + \bar{g}_\lambda \sigma_- \otimes B^+_\lambda)]$$

and H_1 is described in interaction representation using H_{1o} as "free part" and H as "interaction part". This leads to the unitary cocycle $U_t = U_{o,t}$ defined by the equation

$$\frac{d}{dt} U_t = -iU_t H_\Lambda(t) \quad ; \quad U_o = 1$$

$$H_\Lambda(t) = \textstyle\sum_{\lambda\in\Lambda} (g_\lambda \sigma_+ \otimes B_\lambda + \bar{g}_\lambda \sigma_- \otimes B^+_\lambda) = \sigma_+ \otimes B_\Lambda(t) + \sigma_- \otimes B^+_\Lambda(t)$$

where

$$B_\Lambda(t) = \textstyle\sum_{\lambda\in\Lambda} g_\lambda \sigma_+ \otimes B_\lambda \cdot \ell^{-i\omega_\lambda t}$$

The commutator between $B_\Lambda(t)$ and $B^+_\Lambda(s)$ is :

$$[B_\Lambda(t) \, , \, B^+_\Lambda(s)] = \textstyle\sum |g_\lambda|^2 \ell^{-i\omega_\lambda(t-s)} = K_\Lambda(t-s)$$

while all the other commutators vanish.

Introducing on $\mathscr{B}(\mathscr{F}(\mathbb{C}^{|\Lambda|}))$ the quasi-free state $\mathscr{E}$ characterized by

$$\mathscr{E}(B_\lambda B_\mu) = \mathscr{E}(B^+_\lambda B^+_\mu) = 0$$
$$\mathscr{E}(B^+_\lambda B_\mu) = \delta_{\lambda\mu} \theta_\lambda$$

(θ_λ a physical constant), one finds

$$\mathscr{E}(B_\Lambda(t) \cdot B_\Lambda(s)) = \mathscr{E}(B^+_\Lambda(t) B^+_\Lambda(s)) = 0$$
$$\mathscr{E}(B_\Lambda(t) \cdot B^+_\Lambda(s)) = \textstyle\sum |g_\lambda|^2 (1 + \theta_\lambda) \ell^{-i\omega_\lambda(t-s)}$$

The Wigner-Weisskopf approximation is obtained, from the rotating wave approximation, by replacing :

$$K_\Lambda(t-s) = x \cdot \delta(t-s) \quad (x \in C)$$

$$\theta_\lambda = \theta = \frac{\exp(-h\omega_o/KT)}{1 - \exp(-h\,\omega_o/KT)}$$

This means that one substitutes for $B_\Lambda(t)$ and $B^+_\Lambda(t)$ two operators $F(t)$, $F^+(t)$ satisfying :

$$[F(t)\ ,\ F(s)] = [F^+(t)\ ,\ F^+(s)] = 0 \tag{6.4}$$

$$[F(t)\ ,\ F^+(s)] = x\delta(t-s) \tag{6.5}$$

and on the algebra generated by the family $\{F(t),\ F^+(t)\}$ one introduces the quasi-free state characterized by

$$\mathscr{E}(F(t)\cdot F(s)) = \mathscr{E}(F^+(t)\cdot F^+(s)) = 0 \tag{6.6}$$

$$\mathscr{E}(F(t)\cdot F^+(s)) = (1+\theta)\cdot\delta(t-s) \tag{6.7}$$

With these approximations the equation for the unitary cocycle becomes

$$\frac{d}{dt} U_t = -iU_t\cdot H(t) \quad ; \quad U_o = 1 \tag{6.8}$$

$$H(t) = \sigma_+ \otimes F(t) + \sigma_- \otimes F^+(t) \tag{6.9}$$

Equation (6.8) is purely formal because, due to (6.5), (6.7) and (6.9), $H(t)$ is not a well defined operator but an operator valued distribution. In analogy with the classical procedure von Waldenfels [9] introduced to methods for the solution of equation (6.8) :

I.) The "Stratonovich method", corresponding to the "singular coupling limit method" in the physical literature, consisting in three steps :

i) regularize the covariance with the substitution, in (6.5) and (6.7) $\delta(t-s) \longrightarrow K_\varepsilon(t-s)$ for some smooth function $K_\varepsilon(\cdot)$.

ii) solve the corresponding ordinary differential equation, finding a cocycle $U_\varepsilon(0,t)$.

iii) determine the limit of $U_\varepsilon(0,t)$ - and of the associated process (Section (5)) as $\varepsilon \longrightarrow 0$ and $K_\varepsilon(t-s) \longrightarrow \delta(t-s)$.

II.) The "multiplicative Ito integral method", (corresponding to the approximation methods in classical probability) in which - instead of the covariance - you regularize the fields. This can be done in several ways. In [9] one considers for each fixed $T \in \mathbb{R}_+$ a partition $z = \{0=t_o < t_1 < \ldots < t_n = T\}$ of the interval $[0,T]$ and introduces the piecewise constant fields :

$$F_z(t) = \frac{1}{t_{k+1}-t_k}\int_{t_k}^{t_{k+1}} F(\tau)d\tau = F(\chi_{(t_k,t_{k+1}]}) \quad : t_k < t \leq t_{k+1}$$

One then solves the ordinary differential equation :

$$\frac{d}{dt} U_z(t) = -iU_z(t)\cdot H_z(t)$$

and studies the limit of $U_z(t)$ (and of the corresponding process) as $|z| = \max_k (t_{k+1}-t_k) \longrightarrow 0$.

For the Wigner-Weisskopf model the existence of the limiting cocycle (and of the corresponding process) was established by von Waldenfels [9] in both cases (I.) and (II.). A third possibility, is to interpret (6.8) as a quantum stochastic differential equation and use the results of R.Hudson and K.R. Parthasarathy [6] to establish the existence, uniqueness and unitarity of the cocycle U(t). Namely, one considers the Hilbert space

$$\Gamma(L^2(\mathbb{R}^+,dt)) \otimes \Gamma(L^2(\mathbb{R}^+,dt)^-) = \mathcal{H}$$

where $\Gamma(H)$ denotes the (boson) Fock space of H and H^- denotes the conjugate Hilbert space of H. On this Hilbert space one considers the representation of the CCR with creation and annihilation operators given by :

$$F(t) = \sqrt{\gamma}\cosh\Phi\cdot a(\chi_{[o,t]}) \otimes 1 + \sqrt{\gamma}\sinh\Phi\cdot 1 \otimes a^+(\bar{\chi}_{[o,t]})$$

$$F^+(t) = \sqrt{\gamma}\cosh\Phi\cdot a^+(\chi_{[o,t]}) \otimes 1 + \sqrt{\gamma}\sinh\Phi\cdot 1 \otimes a(\bar{\chi}_{[o,t]})$$

where $a(\cdot)$ and $a^+(\cdot)$ are the annihilation and creation operators over $\Gamma(L^2(\mathbb{R}^+))$, and by definition, $\gamma = 2\,\mathrm{Re}\,x$, and :

$$\cosh^2\Phi = \frac{1}{1-\exp(-\omega_o/KT)} \quad ; \quad \sinh^2\Phi = \frac{\exp(-\omega_0/KT)}{1-\exp(-\omega_o/KT)} = \theta$$

With these notations the unitary (markovian) cocycle U_t is defined as the solution of the quantum stochastic differential equation

$$dU_t = U_t \{ (-i\sigma_+ \otimes dF(t) - i\sigma_- \otimes dF^+(t)) - \\ - \gamma/_2 (\cosh^2\Phi\sigma_+\sigma_- \otimes 1 + \sinh^2\Phi\,\sigma_-\sigma_+ \otimes 1)dt \tag{6.10}$$

Denoting $E_{o]}$ the conditional expectation characterized by :

$$E_{o]} : x \otimes (Y \otimes Z) \in \mathcal{B}(\mathbb{C}^2) \otimes \mathcal{B}(\Gamma(L^2(\mathbb{R}_+)) \otimes \Gamma(L^2(\mathbb{R}_+)^-)) \longrightarrow \\ \longrightarrow (x \otimes 1 \otimes 1)\langle\Omega, Y\Omega\rangle\langle\Omega, Z\bar{\Omega}\rangle$$

where Ω (resp. $\bar{\Omega}$) denotes the Fock vacuum in $\Gamma(L^2(\mathbb{R}_+))$ (resp. $\Gamma(L^2(\mathbb{R}_+)^-)$) and applying the theory outlined in Section (4), one obtains a semi-group on $\mathcal{B}(\mathbb{C}^2) = \{2\times2 \text{ matrices}\}$ via the prescription:

$$x \in \mathcal{B}(C^2) \longrightarrow E_{o]}(U_t\cdot(x \otimes 1 \otimes 1)U_t^+) \in \mathcal{B}(C^2) \otimes 1 \otimes 1 \cong \mathcal{B}(C^2)$$

whose generator is :

$$L(x) = -\tfrac{1}{2}\cosh^2\Phi\cdot\gamma\{\sigma_+\sigma_-,x\} + \cosh^2\Phi\cdot\gamma\cdot\sigma_+\cdot x\cdot\sigma_- +$$
$$-\tfrac{1}{2}\sinh^2\Phi\cdot\gamma\{\sigma_-\sigma_+,x\} + \sinh^2\Phi\cdot\gamma\cdot\sigma_-\cdot x\cdot\sigma_+$$

($x \in \mathscr{B}(\mathbb{C}^2)$). Referring the algebra of 2×2 complex matrices; $\mathscr{B}(\mathbb{C}^2)$ to the standard basis, we find for L the matrix :

$$\begin{pmatrix} -\gamma\theta & \gamma(\theta+1) & 0 & 0 \\ \gamma\theta & -\gamma(\theta+1) & 0 & 0 \\ 0 & 0 & -\tfrac{1}{2}(2\theta+1)\gamma & 0 \\ 0 & 0 & 0 & \tfrac{1}{2}(2\theta+1)\gamma \end{pmatrix} \qquad (6.11$$

which is exactly the formula found by von Waldenfels via the "multiplicative Ito method" [9] (in his notations γ = 2 Rex). To obtain the formula found by von Waldenfels via the "Stratonovich method" instead of (6.10) one has to look for the solution of the quantum stochastic differential equation :

$$dU_t = U_t\cdot\{-i\sigma_+\otimes dF(t) - i\sigma_-\otimes dF^+(t) - [\gamma/_2(\cosh^2\Phi\sigma_+\sigma_-\otimes 1 +$$
$$-\sinh^2\Phi\sigma_-\sigma_+\otimes 1)]dt - i\beta/_2(2\theta+1)[\cosh^2\Phi\sigma_+\sigma_-\otimes 1 + \sinh^2\Phi\sigma_-\sigma_+\otimes 1]dt\}$$

where, in von Waldenfels notations: γ = 2Rex, β = 2Imx. The connection between the multiplicative Ito (i.e. singular coupling) method and quantum stochastic differential equations was suggested by Frigerio and Gorini [4] and the explicit form of the semi-group obtained in the Wigner-Weisskopf model in the "multiplicative Ito" case (i.e. corresponding to equation (10)) has been independently obtained by H. Maassen [8].

REFERENCES

1.) L. Accardi. On the quantum Feynman-Kac formula. Rendiconti del Seminario Matematico e Fisico di Milano 48 (1978), 135-180

2.) L. Accardi. A quantum formulation of the Feynman-Kac formula. In: Colloquia Mathematica Societatis Janos Bolyai, 27. Random Fields, Esztergom (Hungary) 1979.

3.) L. Accardi, A. Frigerio, J.T. Lewis. Quantum stochastic processes. Publ. Res. Inst. Math. Sci., Kyoto University 18 (1982) 97-133

4.) A. Frigerio, V. Gorini. On stationary Markov dilations of quantum dynamical semi-groups. In: Quantum Probability and applications to the quantum theory of irreversible processes. Ed. by L. Accardi,

A. Frigerio, V. Gorini. Springer LNM, N° 1055

5.) A. Frigerio. Covariant Markov dilations of quantum dynamical semigroups. Preprint (1984)

6.) R. Hudson, K.R. Parthasarathy. Construction of quantum diffusions. In: Quantum Probability and applications to the quantum theory of irreversible processes. Ed. by L. Accardi, A. Frigerio, V. Gorini. Springer LNM, N° 1055.

7.) B. Kummerer, W. Schroder. On the structure of unitary dilations. Semesterbericht Funktionalanalysis Tubingen, Wintersemester 1983-84, 177-225

8.) H. Maassen. The construction of continuous dilations by solving quantum stochastic differential equations. Preprint (1984)

9.) W. von Waldenfels. Ito solution of the linear quantum stochastic differential equation describing light emission and absorption. In: Quantum Probability and applications to the quantum theory of irreversible processes. Ed. by L. Accardi, A. Frigerio, V. Gorini. Springer LNM, N° 1055

ABSOLUTELY CONTINUOUS INVARIANT MEASURES FOR SOME MAPS OF THE CIRCLE

P. M. Blecher and M. V. Jakobson

1. <u>Statement of results</u>. We consider the two-parameter family of maps on the circle

$$f_{q,\omega} : x \mapsto x+\omega+(q/2\pi) \cdot \sin 2\pi x \ , \quad x \in S^1 = \mathbb{R}/\mathbb{Z}$$

and we find a set $M = \{(q,\omega)\}$ of positive Lebesgue measure such that $(q,\omega) \in M$ implies the stochastic behaviour of $f_{q,\omega}$. We present analytical and numerical results which describe the structure of M as follows.

There exists a sequence of points $A_k = (q_k,\omega_k)$, $k\in\mathbb{N}$ converging to the limit $A_\infty = (q_\infty,\omega_\infty)$, where $q_\infty = 1,169701...$, $\omega_\infty = q_\infty/2\pi$, satisfying

Theorem 1. For any k there exists a set $M_k \subset \mathbb{R}^2$ of positive Lebesgue measure, such that A_k is the density point of M_k , and if $(q,\omega) \in M_k$ then the map $f_{q,\omega}: S^1 \to S^1$ has an absolutely continuous invariant probability measure $\mu_{q,\omega}$. The map $f_{q,\omega}$ cyclically permutes k adjacent intervals $\ell^{(i)}_{q,\omega}$, $i \in [0,k-1]$, $\bigcup_{i=0}^{k-1} \ell^{(i)}_{q,\omega} = S^1$. The support of $\mu_{q,\omega}$ consists of k intervals $s^{(i)}_{q,\omega} \subset \ell^{(i)}_{q,\omega}$ of equal measure. For any i the map $f^k_{q,\omega}$ is an exact endomorphism on the measure space $(s^{(i)}_{q,\omega}, \mu_{q,\omega})$, and its natural extension is a Bernoulli automorphism.

In order to prove Theorem 1 for a given k it suffices to verify some conditions of non-degeneracy, see Sect. 3. For $k=1$ these conditions are verified analytically. For $2\leq k\leq 7$ they were verified with the help of a computer.

If the non-degeneracy conditions hold, then Theorem 1 is proved by using the argument of [1].

The limit map $f_{q_\infty \omega_\infty}$ is characterized by simple geometric features, see Sect. 5.

The full picture is structurally stable in the space of two-parameter families $x \mapsto x + \omega + (q/2\pi)h(x)$, where $h(x)$ has period 1 and is C^3-close to $\sin 2\pi x$ (Theorem 3).

For any family under consideration the rate of convergence of A_k to A_∞ is the same: $|A_k - A_\infty| \sim \frac{\text{const}}{k^2}$.

Our work was motivated by a question of Ya. G. Sinai concerning the appearance of stochasticity in the family $f_{q,\omega}$ for $q>1$.

2. Transition from $q<1$ to $q>1$. Let us consider the following map of the cylinder $r \in (-\infty,\infty)$, $x \in \mathbb{R}/\mathbb{Z}$:

$$P_{\lambda,q,\omega} : \begin{pmatrix} r \\ x \end{pmatrix} \mapsto \begin{pmatrix} 1 + \lambda(r-1) + (q/2\pi)\sin 2\pi x \\ x+\omega+\lambda(r-1) + (q/2\pi)\sin 2\pi x \end{pmatrix}$$

If $\lambda=1$ $P_{1,q,\omega}$ is the so-called standard map, which has been studied as a model for certain dynamical systems, see [2], [3]. If $\lambda<1$, $P_{\lambda,q,\omega}$ is dissipative, and for $\lambda=0$ (the case of infinite contraction) it reduces to the map of the circle:

$$f_{q,\omega} : x \mapsto x + \omega + (q/2\pi)\sin 2\pi x \ , \ x \in \mathbb{R}/\mathbb{Z} = S^1 .$$

If $q \leq 1$ then $f_{q,\omega}$ is a homeomorphism of the circle. Such homeomorphisms were studied intensively in connection with the problem of disappearance of invariant tori in the KAM theory, see [4], [5]. We shall study $f_{q,\omega}$ first of all for $q>1$, but first we recall some facts about $f_{q,\omega}$ with the critical value of $q=1$.

For any ω the homeomorphism $f_{1,\omega}$ has a rotation number $\rho(\omega)$ (for definition see e.g. ref. [5]) depending continuously on ω. If $\rho(\omega) = m/k$ is rational then $f_{1,\omega}$ has a periodic orbit of period k. For all rational numbers $\frac{m}{k}$ there exists an interval $I_{m/k}$ such that for $\omega \in I_{m/k}$, $\rho(\omega) = \frac{m}{k}$.

Considering $f_{1,\omega}$ as a map of the Riemann sphere $z \mapsto z + \omega + (1/2\pi)\sin 2\pi z$, $z \in \mathbf{C}$, depending on the complex parameter ω, one can prove by the methods of [6] (see also [7]) the following proposition, which is not directly used below, but seems to be of some interest.

<u>Proposition</u>. For any $I_{m/k}$ defined above there exists a unique parameter value $\omega_0 \in I_{m/k}$ such that f_{1,ω_0} has a superstable periodic trajectory of period k.

The homeomorphism f_{1,ω_0} permutes cyclically k intervals: $\ell^{(0)} \ni \frac{1}{2}$, $\ell^{(1)} = f_{1,\omega_0}\ell^{(0)}, \ldots, \ell^{(k-1)} = f^{k-1}_{1,\omega_0}\ell^{(0)}$ and the graph of $f^k_{1,\omega_0}|\ell^{(0)}$ looks like figure 1^a.

For $q>1$ the degenerate critical point $c=1/2$ bifurcates into two simple critical points

$$c_1 = (1/2\pi)\arccos(-1/q) < 1/2$$

and

$$c_2 = 1 - (1/2\pi)\arccos(-1/q) > 1/2 .$$

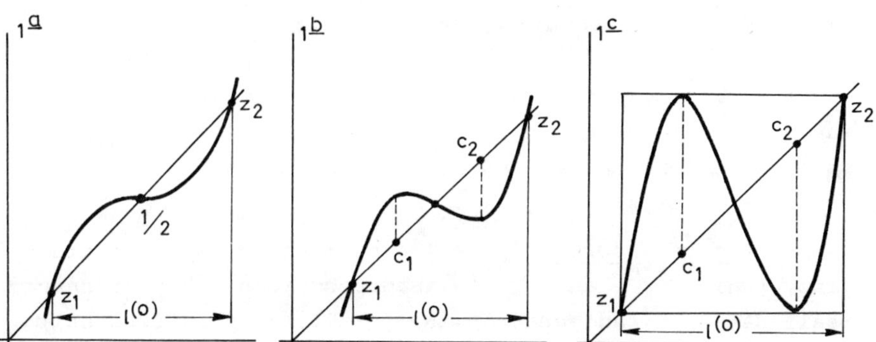

Figure 1 .

For q close to $q_0 = 1$ and ω close to ω_0, $f_{q,\omega}$ still permutes k intervals $\ell^{(i)}_{q,\omega}$, and the graph of $f^k_{q,\omega}|\ell^{(0)}_{q,\omega}$. looks like figure 1^b. A natural idea is to find q and ω characterized by the most chaotic dynamics represented in figure 1^c.

3. <u>Non-degeneracy conditions and proof of Theroem 1'.</u>

Let us denote by $F_{q,\omega}$ the restriction of $f^k_{q,\omega}$ to $\ell^{(0)}_{q,\omega} = [z_1(q,\omega), z_2(q,\omega)]$, by $c_1(q,\omega)$ the point of maximum, and by $c_2(q,\omega)$ the point of minimum of $F_{q,\omega}$.

<u>Non-degeneracy conditions</u>

1) The equation $F_{q,\omega}(c_1(q,\omega)) - z_2(q,\omega) = G(q,\omega) = 0$ defines a smooth curve $\gamma^{(1)}$ on the plane (q,ω), and similarly $F_{q,\omega}(c_2(q,\omega)) - z_1(q,\omega) = H(q,\omega) = 0$ defines a smooth curve $\gamma^{(2)}$.

2) $\gamma^{(1)}$ and $\gamma^{(2)}$ intersect each other at $A_k = (q_k, \omega_k)$, and the Jacobian

$$\det \begin{pmatrix} G_q & G_\omega \\ H_q & H_\omega \end{pmatrix}_{q=q_k,\ \omega=\omega_k} \neq 0 .$$

<u>Theorem 1'</u>. If the non-degeneracy conditions are satisfied then the Lebesgue measure of the set $M_k = \{(q,\omega) \mid F_{q,\omega} : \ell^{(0)}_{q,\omega} \to \ell^{(0)}_{q,\omega}$ has an absolutely continuous invariant probability measure $\mu'_{q,\omega}\}$ is positive, and A_k is the density point of M_k. For $(q,\omega) \in M_k$ $(F_{q,\omega}, \mu'_{q,\omega})$ is an exact endomorphism, and its natural extension is a Bernoulli automorphism.

<u>Proof of Theorem 1'</u>. a) The non-degeneracy conditions imply that $\gamma^{(1)}$ and $\gamma^{(2)}$ intersect each other transversally at A_k, and generate the partition of sufficiently small neighbourhoods of A_k into four quadrants. The inclusions $F_{q,\omega}(c_1(q,\omega)) \subset \ell^{(0)}_{q,\omega}$, $F_{q,\omega}(c_2(q,\omega)) \subset \ell^{(0)}_{q,\omega}$ are are satisfied for (q,ω) belonging to one of these quadrants.

Let us consider a ray $r = \{r(s)\}$ in this quadrant with $r(o)=A_k$. The maps $F_{q,\omega}$ with $(q(s), \omega(s)) = r(s)$

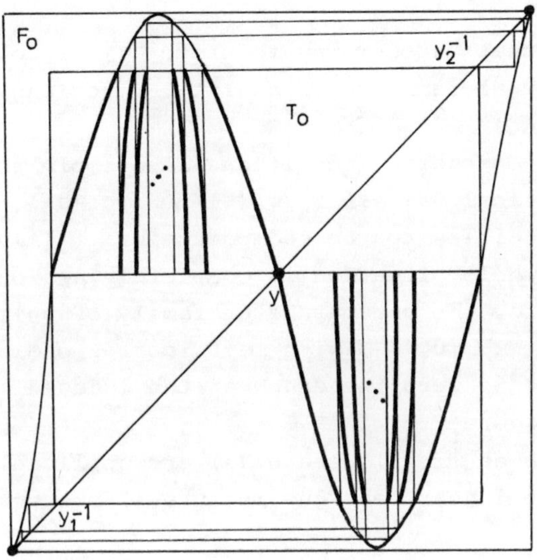

Figure 2.

form a one-parameter family of maps such that the graph of $F_0 = F_{q(0),\omega(0)}$ looks like figure 2. Let $y = F_0(y)$ be the fixed point different from the end-points of $\ell^{(0)}$, and let y_1^{-1}, y_2^{-1} be the preimages of y. We denote by T_0 the first return map induced by F_0 on the interval y_1^{-1}, y_2^{-1}, (see fig. 2.).

b) A straightforward calculation shows that for $q>1$ the mappings $f_{q,\omega}$ have a negative Schwarzian derivative:

$$Sf = f''' ,f' - 3/2(f'')^2 < 0 .$$

This implies (see e.g. [8] about the properties of maps with negative Schwarzian) $SF_0 < 0$. From $SF_0 < 0$ it follows (see [1], [8]) the so-called <u>expanding property</u>:

There exist an $m \in \mathbb{N}$ and $c>1$ such that

$$|T_0^{m\,'}(x)| > c \quad \text{for any} \quad x \in [y_1^{-1}, y_2^{-1}] \tag{1}$$

Notice that another way to prove (1) is to use the theory of normal families of analytic functions due to Montel.

The non-degeneracy conditions also imply that for $s=0$ the critical values $F_s(c_1(s))$, $F_s(c_2(s))$ move with non-zero velocities to the end-points of $\ell^{(0)}(s)$.

c) The whole situation is completely analogous to that which holds for the one-parameter family of unimodal maps $x \longmapsto ax(1-x)$, $x \in [0,1]$, with $s=0$ corresponding to $a=4$. For s close to zero the graph of the induced map T_s looks like Figure 3.

The methods of [1] (see § 13) are applicable. They give that the linear Lebesgue measure of the subset $M(r)$

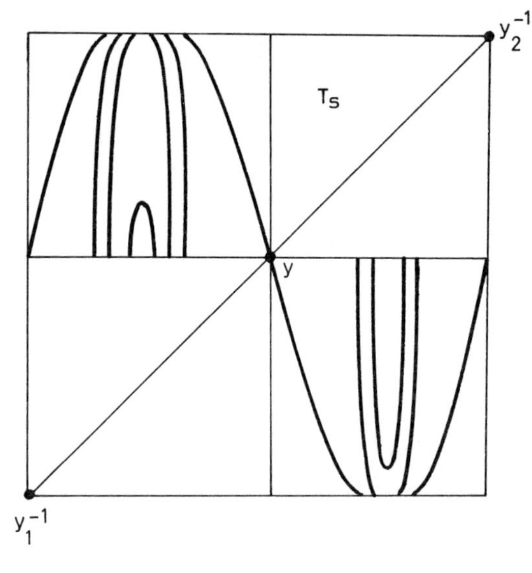

Figure 3.

of the ray r , defined by $M(r) = \{(q(s), \omega(s)) = r(s) : F_{q(s),\omega(s)}$ has an absolutely continuous invariant probability measure $\mu'_{q,\omega}\}$, is positive, and $A_k = r(0)$ is the density point of $M(r)$.

d) The same arguments are valid for any ray in the distinguished quadrant, and thus we obtain, by using the Fubini theorem that $\operatorname{mes} M_k > 0$, and A_k is the density point of M_k in this quadrant.

Ergodic properties of endomorphisms $(F_{q,\omega}, \mu'_{q,\omega})$ and of their natural extensions follow from the results of Ledrappier [9].

4. The choice of the sequence A_k and the proof of Theorem 1

Consider the values q,ω characterized by the following properties (see Figure 4.):

$$f^k(z_1) = z_1 \ ; \ f(z_1) = f(c_2) \ ; \ f(c_1) = f^2(c_2) \ .$$

Then setting $z_2 = f(z_1)$, the interval $\ell^{(0)} = [z_1, z_2]$ triply covers its image $\ell^{(1)} = [z_2, f(z_2)]$ under the action of f . The subsequent iterations of f map $\ell^{(1)}$ homeomorphically on the intervals $\ell^{(2)},\dots,\ell^{(k-1)}$ which are consecutive, adjacent, and the end of $\ell^{(k-1)} = [f^{k-1}(z_1), f^k(z_1)]$ coincides with z_1 (see fig. .4, where $k=3$). Besides the graph of $f^k|[z_1,z_2]$ looks like fig. 1^c. We take these q,ω for q_k, ω_k from Section 1. If the non-degeneracy conditions are satisfied then according to Theorem 1' there exists for $(q,\omega) \in M_k$ an $f^k|\ell^{(0)}$ invariant measure $\mu'_{q,\omega} \ll dx$. To $\mu'_{q,\omega}$ there corresponds a unique f invariant measure $\mu_{q,\omega}$ on the circle. If $q=q_k$, $\omega=\omega_k$ then the support of $\mu'_{q,\omega}$ coincides with $\ell^{(0)}$, and the support of $\mu_{q,\omega}$ coincides with S^1 . If $(q_k,\omega_k) \neq (q,\omega) \in M_k$ then $\ell^{(0)} \supsetneq [F(c_2),F(c_1)] = \operatorname{supp} \mu'_{q,\omega}$, hence $\operatorname{supp} \mu_{q,\omega}$ is the union of k disjoint intervals. The ergodic properties of $\mu_{q,\omega}$ follow from those of $\mu'_{q,\omega}$ which finishes the proof of Theorem 1.

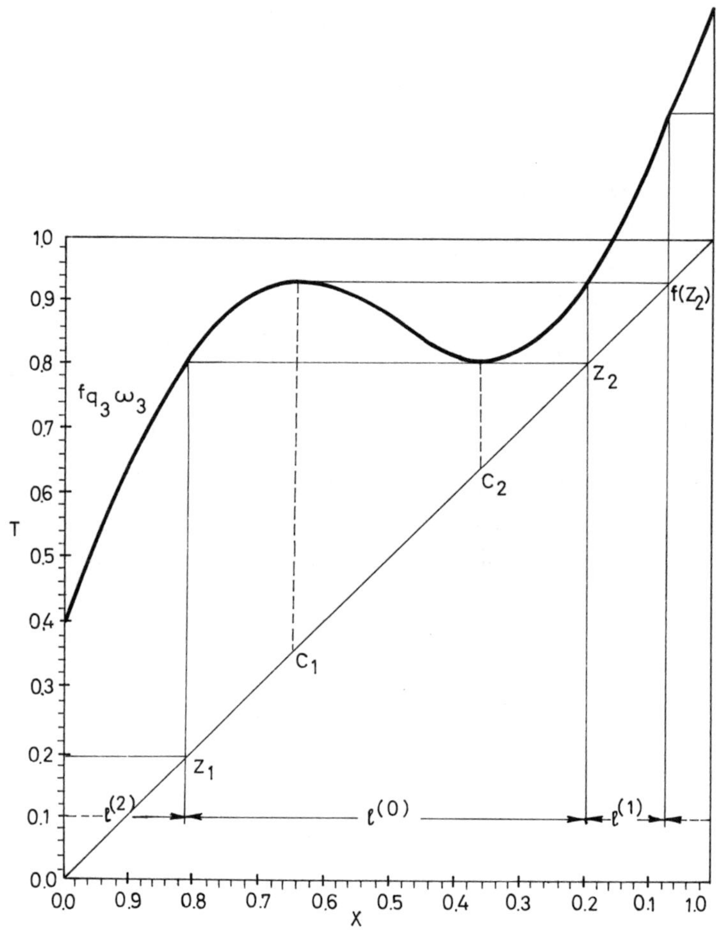
1.0
0.9
0.8
0.7
0.6
0.5
0.4
0.3
0.2
0.1
0.0
T
$f_{q_3\omega_3}$
C_1
C_2
Z_1
Z_2
$f(Z_2)$
$\ell^{(2)}$
$\ell^{(0)}$
$\ell^{(1)}$
0.0 0.9 0.8 0.7 0.6 0.5 0.4 0.3 0.2 0.1 1.0
X

Figure 4.

5. The properties of the map $f_{q_\infty,\omega_\infty}$; the construction of the sets M_k

The values (q_k,ω_k) satisfy the following conditions:

$$f^{2k}(c_i) = f^k(c_i) \ , \ i=1,2; \quad f^2(c_2) = f(c_1) \ .$$

The sequence $A_k = (q_k,\omega_k)$ converges to $A_\infty = (q_\infty,\omega_\infty)$ and the map $f_{q_\infty,\omega_\infty}$ may be characterized by:

(i) $f_{q_\infty,\omega_\infty}$ has a fixed point u such that

$$f_{q_\infty,\omega_\infty}(u) = 1 \ , \ f''_{q_\infty,\omega_\infty}(u) > 0 \ .$$

(ii) $f^2_{q_\infty,\omega_\infty}(c_2) = f_{q_\infty,\omega_\infty}(c_1)$.

Denote by $z_{1\infty}$, $z_{2\infty}$ the endpoints of the interval which is triply mapped by $f_{q_\infty,\omega_\infty}$ on its image $[f_{q_\infty,\omega_\infty}(c_1), f_{q_\infty,\omega_\infty}(c_1)]$ (figure 5.). When $k\to\infty$ then the sequence of intervals $\ell^{(0)}_{q_k,\omega}$ defined above converges to $\ell^{(0)}_{q_\infty,\omega_\infty} = [z_{1\infty}, z_{2\infty}]$.

Theorem 2. When $k\to\infty$ then the sequence of maps $f^k_{q_k,\omega_k}|\ell^{(0)}_{q_k,\omega_k}$ converges in C^1 -topology to the limit map $f^\infty_{q_\infty,\omega_\infty}$ which maps $\ell^{(0)}_{q_\infty,\omega_\infty}$ triply on itself.

Condition (i) defines the line $q = 2\pi\omega$ in the (q,ω)-plane.

Condition (ii) defines the curve Γ which intersects the line $q=2\pi\omega$ at the point (q_∞,ω_∞) . The points A_k lie on the curve Γ (see figure 6.). The curves $\gamma_k^{(1)}$ and $\gamma_k^{(2)}$ defined respectively by $f^{2k}(c_1) = f^k(c_1)$, and $f^{2k}(c_2) = f^k(c_2)$, have A_k as the points of transversal intersection. It is natural to suggest that as $A_k \to A_\infty$ the curves $\gamma_k^{(1)}$ and $\gamma_k^{(2)}$ become parallel to the line $q=2\pi\omega$, and the angles between $\gamma_k^{(1)}$ and $\gamma_k^{(2)}$ tend to zero.

Fig. .6 shows the points $A_k = (q_k,\omega_k)$ and the curves $\gamma_k^{(1)}$, $\gamma_k^{(2)}$ for $3\le k\le 7$. Notice that if $k=1$, then $\omega_1=0$ and q_1 satisfies the equation: $q\cdot\cos\sqrt{q^2-1} = -1$. The curves $\gamma_1^{(1)}$ and $\gamma_1^{(2)}$ are defined by

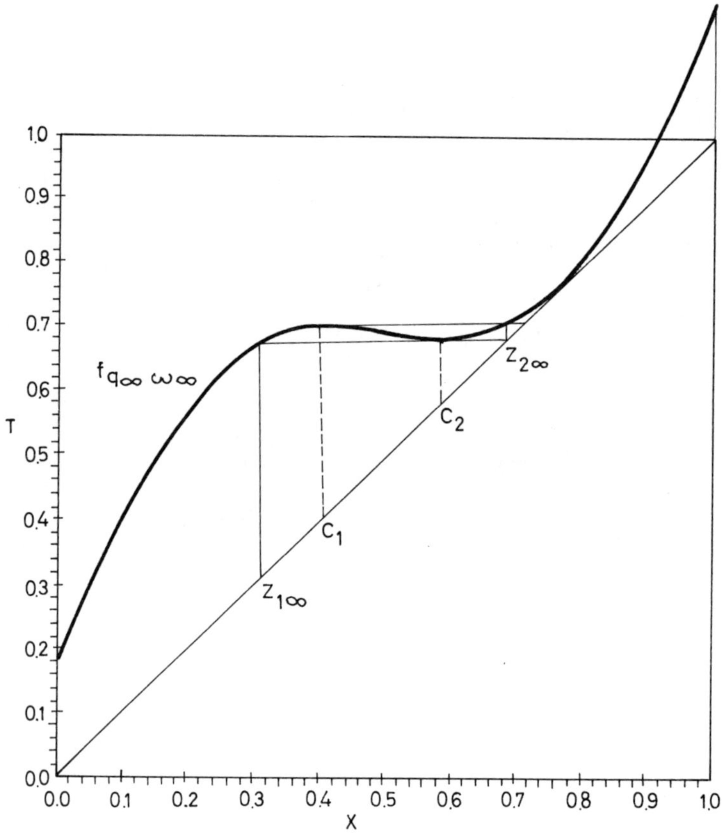

1.0
0.9
0.8
0.7
0.6
0.5
0.4
0.3
0.2
0.1
0.0
T
0.0
0.1
0.2
0.3
0.4
0.5
0.6
0.7
0.8
0.9
1.0
X
$f_{q\infty}\,\omega\infty$
$Z_{1\infty}$
C_1
C_2
$Z_{2\infty}$

Figure 5.

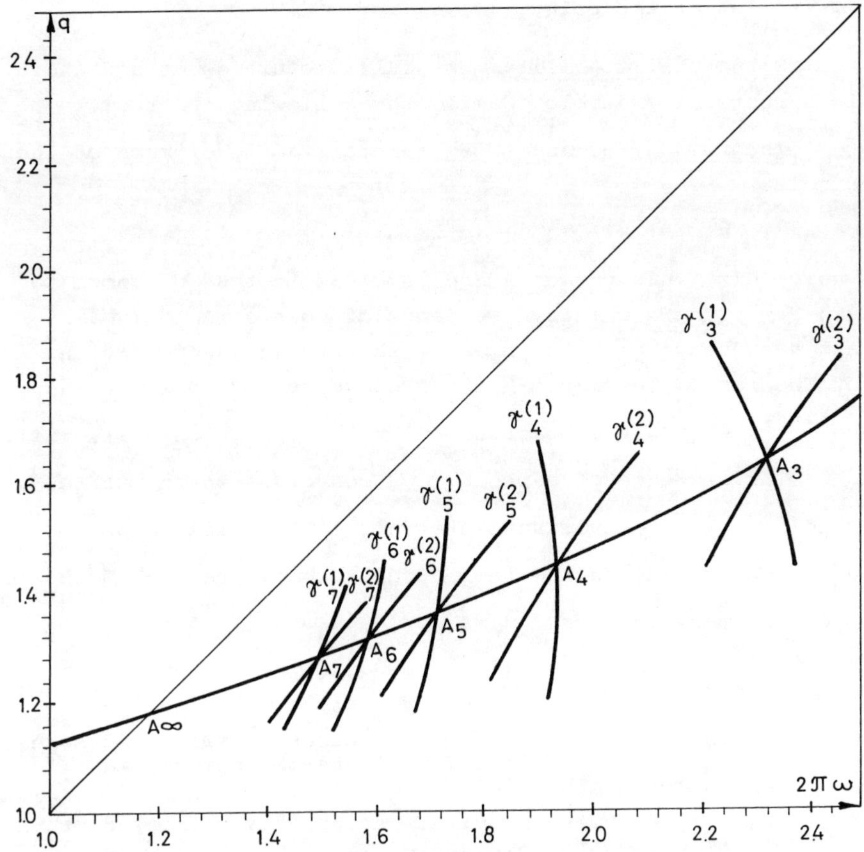

Figure 6.

$$\arc\cos(-1/q) + \sqrt{q^2 - 1} = 2\pi - (\arc\sin \omega/q) - \omega$$

and

$$\arc\cos(-1/q) + \sqrt{q^2 - 1} = 2\pi + (\arc\sin \omega/q) + \omega \ .$$

One can check that $\gamma_1^{(1)}$ and $\gamma_1^{(2)}$ have a transversal intersection at the point $(q_1, 0)$.

6. Structural stability. The full picture described above is structurally stable. Namely the following is true:

Theorem 3. For any two-parameter family of maps of the circle

$$h_{q,\omega} : x \mapsto x + \omega + (q/2\pi) h(x) \ ,$$

where $h(x)$ has period 1 and is close in the C^3 norm to $\sin 2\pi x$, there exists a sequence of points $A_k = (q_k, \omega_k)$ and a sequence of sets $M_k \ni A_k$, satisfying the conditions of Theorem 1. The maps h_{q_k,ω_k} converge to the limit map $h_{q_\infty,\omega_\infty}$, which is characterized by the conditions (i), (ii).

The sequence of maps $h^k_{q_k \omega_k}$ converges on the corresponding $h^k_{q_k \omega_k}$ -invariant intervals to the limit map $h^\infty_{q_\infty \omega_\infty}$ depending on the family under consideration. Moreover, for any such family we have $|A_k - A_\infty| \sim \mathrm{const}/k^2$.

References

[1] Jakobson M. V. Abolutely continuous invariant measures for one-parameter families of one-dimensional maps. Commun. Math. Phys. 81, 39-88 (1981).

[2] Chirikov B. V. A universal instability of many-dimensional oscillator systems. Phys. Rep. 52, 5, 263 (1979).

[3] Mackay R. S. A renormalization approach to invariant circles in area-preserving maps. Physica 7 D, 283-300 (1983).

[4] Feigenbaum M. J., Kadanoff L. P., Shenker S. J. Quasi-periodicity in dissipative systems: a renormalisation analysis. Physica 5 D, 370 (1982).

[5] Ostlund S., Rand D., Sethna J., Siggia E. Universal properties of the transition from quasiperiodicity to chaos in dissipative systems. Physica 8 D, 303-342 (1983).

[6] Douady A., Hubbard J. Iterations des polynomes quadratiques complexes. C. R. Acad. Sc. Paris, 294, 123-126 (1982).

[7] Eremenko A. E., Ljubic M. Yu. Iteration of entire functions. Preprint of Phys.-Techn. Inst. of Low Temp. UkrSSR Acad. of Sci. Kharkov, N 6 (1984).

[8] Guckenheimer J. Sensitive dependence on initial conditions for one-dimensional maps. Commun. Math. Phys. 70, 133-160 (1979).

[9] Ledrappier F. Some properties of absolutely continuous invariant measures on an interval. Ergod. Th. and Dyn. Syst. 1, 77-94 (1981).

SOME PROBLEMS IN VORTEX THEORY

C. Marchioro

1. Introduction

In this lecture we want to study the following Hamiltonian dynamical system, called "vortex system"

$$a_i \dot{x}_i = \frac{\partial H}{\partial y_i} \qquad\qquad (1.1)$$

$$a_i \dot{y}_i = - \frac{\partial H}{\partial x_i}$$

$0 < i \leqslant N \quad ; \quad z_i = (x_i, y_i) \qquad D \subset R^2$

$a_i \in R$ is called vortex intensity

and where H has the form

$$H = \frac{1}{2} \sum_{\substack{i,j \\ i=j}}^{N} a_i\, a_j\, g_D(z_i, z_j) + \frac{1}{2} \sum_{i=1}^{N} a_i^{\,2}\ \gamma_D(z_i, z_i) \qquad (1.2)$$

where g_D is the fundamental solution of the Poisson equation (i.e. $\Delta_z\, g_D(z,z') = -\ \delta(z-z')$, $g_D(z,z')=0$ if z or z' D) and γ_D is its regular part

$$\gamma_D(z,z') = g_D(z,z') + \frac{1}{2\pi} \ln| z-z'| \qquad (1.3)$$

The last term in (1.2) takes into account the interaction between the vortices and the boundary of the domain D (roughly speaking the interaction with the images). Of course if $D=R^2$

$$g_D(z,z') = \frac{1}{2\pi} \ln | z - z'| \qquad (1.4)$$

and the last term in (1.2) vanishes.

We will discuss , shortly, the following problems:

i) the meaning of the dynamical system
ii) its behavior during the time
iii) the statistical mechanics
iv) a mean field limit
v) stochastic vortex system.

2. The meaning of the dinamical system.

The dynamical system (1.1) is strictly related with the fluid mechanics [1] . Let us consider a non viscous incompressible fluid moving in a two dimensional domain D. The Euler equation for the vorticity are

$$\frac{\partial}{\partial t}\omega(z,t) + (u\cdot\nabla)\,\omega(z,t) = 0$$

$$\nabla\cdot u = 0$$

$$\omega(z,t) = \text{curl}\, u(z,t) = \left(\frac{\partial u_2}{\partial x}, -\frac{\partial u_1}{\partial y}\right)$$

$$\omega(z,0) = \omega_o \tag{2.1}$$

$$u\cdot n = 0 \qquad , n = \text{normal of } \partial D$$

$$z \in D \subset R^2$$

$$u = (u_1,u_2) \in R^2 \qquad \text{velocity}$$

The relation between the vorticity $\omega(z)$ and the velocity u (equation $(2.1)_3$) can be inverted and it becomes

$$u(z) = \nabla^{\perp}\int_D g_D(z,z')\,\omega(z')\,dz' \tag{2.2}$$

where

$$\nabla^{\perp} = \left(\frac{\partial}{\partial y} , -\frac{\partial}{\partial x} \right)$$

We suppose now the vorticity concentrated in N points z_i

$$\omega(z) = \sum_{i=1}^{N} a_i \ \delta\,(z-z_i) \tag{2.3}$$

Using (2.2) the velocity field is

$$u(z) = \sum_{i=1}^{N} a_i \ \nabla^{\perp} g_D(z,z_i) \tag{2.4}$$

When $z \to z_i$

$$\nabla \ g_D(z,z_i) \simeq \frac{1}{2\pi} \ \frac{\tau}{|z - z_i|} \tag{2.5}$$

where τ is a unit vector orthogonal to $z-z_i$. That is u becomes singular when z approaches z_i. We evolve ω (z) via the Euler equation (in the weak form), we neglect the singular term and we obtain in a formal way

$$\omega(z,t) = \sum_{i=1}^{N} a_i \ \delta\,(z - z_i(t)) \tag{2.6}$$

where $z_i(t)$ satisfy (1.1). It means that the vorticity remains concentrated and the center moves in the velocity field produced by other vortices but not by itself.

This is a formal proof only. To make it rigorous we consider N disjoint blobs of vorticity

$$\omega_\varepsilon(z,0) = \sum_{i=1}^{N} \omega_\varepsilon^i\,(z,0) \tag{2.7}$$

where

$$\omega_\varepsilon^i\,(z,0) = a_i \ \varepsilon^{-2} \ \chi_{\Lambda_{i,\varepsilon}}$$

Here $\chi_{\Lambda_{i,\varepsilon}^{\varepsilon}}$ is the characteristic function of the set $\Lambda_{i,\varepsilon}$, $\text{meas}\Lambda_{i,\varepsilon} = \varepsilon^2$, diam. $\Lambda_{i,\varepsilon} \simeq \varepsilon$, $\Lambda_{i,\varepsilon} \ni z_i$.

We define center of vorticity of the blob i

$$M^i(t) = \varepsilon^{-2} \int z \; \omega_\varepsilon^i(z,t) \, dz \tag{2.8}$$

where $\omega_\varepsilon^i(z,t)$ is the Euler evolution of $\omega_\varepsilon^i(z,o)$.

We must prove that

$$\lim_{\varepsilon \to o} M^i(t) = z_i(t) \tag{2.9}$$

Equation (2.9) has been proved for every N only for short time [2] and for every time for N=1 [3] and N = 2, sign a_1 = sign a_2 [4]. The general global problem is yet open. As a remarke we note that the difficulty of a rigorous proof is related with the singularity of the velocity field in the limit $\varepsilon \to 0$, limit in which the paths of the particles do not converge. A cancellation of the singular term allows us to obtain (2.9).

3. Behaviour during the time.

We want to study the dynamical system (1,1). It is easily to recognize the existence of some first integral of motion. In fact the energy H itself is a constant of motion. Moreover, if H is translationally invariant, then

$$M = \sum_{i=1}^{N} a_i \, z_i = \text{const} \tag{3.1}$$

If H is rotationally invariant, then

$$I = \sum_{i=1}^{N} a_i \, z_i^2 = \text{const} \tag{3.2}$$

When we deal with two vortices in R^2, the above first integral are sufficient to determine the motion. If $a_1 = a_2$ they rotate with uniform speed around the center of vorticity. If $a_1 = - a_2$ they move on parallel straight lines with uniform velocity.

For more vortices (or complicated domain) the problem is more hard. Not even we are sure of the existence and the uniqueness of the solutions since the logarithmic divergence of the Green function may generate singularities in finite time not prevented by the first integral of motion. It is possible to show that special vorticities and initial conditions lead to a catastrophe in a finite time [5] . These collapse exclude the possibility to give a general theorem of existence and uniqueness of the solution of (1.1). However we can hope that these collapse are exceptional in the sense that the set of the bad initial conditions have null Lebesgue measure in D^N. Equations (1.1) so define a dynamical system, i.e. an evolution of every measurable set. This property has been proved adding some technical hypothesis on the vorticity and/or on the domaid D [6 , 7] .

Let us discuss now the qualitative behavour of the vortex motion. If $D = R^2$ for $N \leq 3$ the system is integrable, while for $N > 4$ there is a numerical and a theoretic evidence that a caotic motion may happen for many initial conditions [8] . Neverthless it is possible to exhibit a positive measure set of initial conditions for which a quasiperiodic motion takes place [9] . We skecth the main idea. We suppose $N = 4$, $D = R^2$, $a_1 \neq -a_2$, $a_3 \neq -a_4$, $a_1 + a_2 \neq -(a_3 + a_4)$. The Hamiltonian H can be written as $H_o + V$ where H_o is the interaction of the two pairs of vortices and the interaction of their centers of vorticity and $V = H - H_o$ that is a small perturbation if the two pairs are far enough. The thesis follows by an application of the KAM theorem [10] since the imperturbed motion given by H_o is quasiperiodic. It is then easy to generalize to arbitrary N by induction.

The previous idea can be extended to a bounded domain D.

In this case of course we cannot put the pairs very far. However [7] we can constract weakly interacting clusters using the singularity of the interaction and scalong the problem

$$z \to \alpha z \qquad ; \quad t \to \alpha^2 t \qquad ; \quad \alpha \text{ large}$$

4. The statistical mechanics.

The statistical mechanics of system (1.1) has been introduced by Onsager with the aim to discuss the turbolence [11]. Formally the problem is very simple: system (1.1) is Hamiltonian, the Liouville theorem holds and we can introduce a "Gibbs measure"

$$\exp(-\beta H) \ / \ \text{normalisation.}$$

Here β plays the role of a factor which can be called "temperature" in analogy with the usual statistical mechanics. Of course a priori β can be positive or negative. If we consider a gas of vortices in a box when β is positive the particles spread down in all the region, while when β is negative two cluster of different vorticity will be formed. In fact vortices interact as a coulomb system, i.e. via a logarithmic interaction and different β correspond in some sense to attraction or repulsion of vortices of the same sign. Of course in the first case only there is a thermodynamical limit as it has been proved rigorously in [12].

For the Euler flow H is not the only first integral of the motion. In fact every integral on D of a funtion of the vorticity is a constant of motion. In particular

$$S = \frac{1}{2} \int_D \omega^2 \, dz$$

called enstrophy plays an important role. We do not enter in this theory and we send the reader to the review paper [13].

5. Mean field limit.

In this section we want to discuss an other relation between system (1.1) and the Euler equations. Suppose that $\omega_o \in L_1 \cap L_\infty$ is an initial profile of vorticity and it is approximated at time zero by

$$\omega_o^N = \frac{1}{N} \sum_{i=1}^{N} \delta_{x_i}$$

where $\{ x_1, \ldots, x_N \}$ are suitable chosen points.
We call $x_i(t)$ the evolution of these points according to the vortex dynamics (1.1) and $\omega(t)$ the evolution of ω_o via the Euler equation.
The question is:

$$\omega_t^N = \frac{1}{N} \sum_{i=1}^{N} \delta_{x_i(t)}$$

is an approximation of $\omega(t)$?
More precisely if

$$\omega_o^N = \frac{1}{N} \sum_{i=1}^{N} \delta_{x_i} \xrightarrow[N \to \infty]{\text{weak}} \omega_o$$

$$\omega_o \in L_1 \cap L_\infty$$

then

$$\frac{1}{N} \sum_{i=1}^{N} \delta_{x_i(t)} \xrightarrow[N \to \infty]{\text{weak}} \omega(t) \qquad ?$$

The answer is positive if we add an essential technical divice: we smooth the kernel and we remove the cutoff simultaneously of the limit $N \to \infty$ [14, 15, 16, 17] .

This result, interesting per se, gives a numerical algorithm for the study of the Euler flow in two dimension. The rapidity of the convergence depends in a deep way on a care choice of the cutoff and on the regularity of the initial condition ω_o.

We note that this mean field theorem is analogous to the relation in kinetic theory between the molecular dynamics and the Vlasov equation, the so called Vlasov limit [18,19,20] .

6. Stochastc vortex model.

We want to approximate by a similar procedure using a vortex model the Navier-Stokes equation for a viscous incompressible two dimensional fluid . The equation are

$$\frac{\partial \omega}{\partial t} + (u \cdot \nabla) \omega = \nu \Delta \omega$$

$$\nabla \cdot u = 0$$

$$\nu > 0$$

We try to take into account the viscosity ν of the problem adding to equations (1.1) a Brownian motion.

In this case the stochastic vortex model is described in a region without boundary (R^2 or T^2) by the stochastic equations

$$d\, x_i(t) = \sum_{\substack{k=1 \\ k=i}}^{N} a_k \nabla^{\perp} g(x_i(t),x_k(t))\, dt + \sigma \, db_i(t) \tag{6.1}$$

where $\{b_i\}_1^N$ are N independent Brownian motions.

First we must prove that equations (6.1) define a dynamical system, that is we must show that the probability of a collapse is negligible. This has been done in T^2 in [6]

Then, we want approximate (as in the previous section) a solution of the Navies-Stokes equations by a measure

$$\tilde{\omega}_t^N(dx) = \frac{1}{N} \sum_{i=1}^{N} \delta \quad (x - x_i^N(t))\, dx$$

where $x_i^N(t)$ are the solutions of (6.1) with $a_i = 1/N$.

We can perform this program [16] smoothing the kernel, removing the cutoff when $N \to \infty$ and choosing $\nu = \sigma^2/2$.

We note that this problem is more complicate that the deterministic case. In fact in this case the vortex model is in some sense (neglecting an infinite term !) a weak solution of the Euler equation and the mean field limit is a continuity theorem with respect to the initial data. Viceversa (6.1) is <u>not</u> a weak solution of the Navier-Stokes equations. When $N \to \infty$ the large number law allow us to obtain the result.

When the boundaries are present the problem is more complicate. In fact the boundaries can produce vorticity and this important effect must be take into account. Chorin [21] introduced a numerical algorithm based on a stochastic vortex model useful to appoximate numerically the Navier-Stokes equations for high Reynolds number. Its rigorous justification is yet missing. For a partial result see [22] .

For a more detailed review of the argument of Section 2,3,5,6 see [7] .

References.

[1] H.Helmholtz, Phil.Mag.,33, p.485 (1867)
G.Kirchhoff, Vorlesungen ueber Math.Phys. Teubener Leipzig (1883)
H.Poincaré, Théories des Turbillons, George Carré (1983)
Lord Kelvin, Mathematical and Physical Papers, Vol IV (No.10,12) Cambridge Univ.Press p.563
C.C.Lin, On the motion of Vortices in Two Dimension, Toronto University Press p.39

[2] C.Marchioro ,M.Pulvirenti, Commun.Math.Phys.,91,563 (1983)

[3] B.Turkington, On the evolution of a concentrated vortex in a ideal fluid (preprint 1984)

[4] C.Marchioro,E.Pagani, Euler evolution of two concentrated vortices moving in a bidimensional bounded region. (in preparation).

[5] H.Aref, Physics of Fluid, 22 ,393 (1979)

[6] D.Dürr,M.Pulvirenti,Commun.Math.Phys. 85, 265 (1982)

[7] C.Marchioro,M.Pulvirenti, Lecture Notes in Phys. 203, Springer-Verlag (1984)

[8] H.Aref, Ann.Rev.Fluid Mech.,15 (1983)

[9] K.M.Khanin,Physica D 4, 261 (1982)

[10] A.N.Kolmogorov,Dokl.Akad.Nauk. SSSR 98,527 (1954)
V.I.Arnold, Usp.Mat.Nauk, 18, 13 (1963)
J.Moser, Math.Ann.,169, 136 (1967)

[11] L.Onsager, Suppl. Nuovo Cimento, 6, "79 (1949)

[12] J.Fröhlich, D.Ruelle, Commun.Math.Phys.,87,1 (1982)

[13] R.H.Kraichnan, D.Montgomery, Rep.Prog.Phys., 43, 547 (1980)

[14] O.Hald, V.Mauceri Del Prete, Math.Comp., 32, 791 (1978)

[15] O.Hald, SIAM J.Numer.Anal., 16, 726 (1979)

[16] C.Marchioro, M.Pulvirenti, Commun.Math.Phys., 84, 483 (1982)

[17] J.T.Beale, A.Majda, Math.Comp., 39, 1 (1982), 39, 29 (1982)

[18] W.Braun, K.Hepp, Commun.Math.Phys., 56, 101 (1977)

[19] R.L.Dobrushin, Sov.J.Funct.Anal. 13, 115 (1979)

[20] H.Neunzert, Lect.Notes in Math. 1048 (1984)

[21] A.J.Chorin, J.Fluid Mech., 57, 785 (1973)

[22] G.Benfatto, M.Pulvirenti, Generation of vorticity near the boundary in two dimensional flows (Comm.Math.Phys 1984)

THE MAXWELL RULE AND PHASE SEPARATION IN THE HIERARCHICAL VECTOR-VALUED ϕ^4 -MODEL

P. M. Blecher

Summary. We investigate Dyson's hierarchical vector-valued ϕ^4 -model at $\beta>\beta_c$ where β_c is the critical inverse temperature. The "Gaussian" case, $\frac{3}{2} > a > 1$, is considered. We prove that the thermodynamical potential is constant in the coexistence region of phases (the Maxwell rule), and find the principal non-constant term of the asymptotics of the small partition function when $V\to\infty$ in this region. We also study the phase separation at $\beta>\beta_c$.

1. Introduction

As it is well-known, in classical ferromagnet spin models, such as the d -dimensional Ising model for $d\geq 2$, and the d -dimensional Heisenberg model for $d\geq 3$, various spatially homogeneous pure phases exist at low temperatures. The phenomenon of phase separation for Ising type models consists in the fact that under fixed density ρ of particles any typical configuration of spins represents a "drop" of particles of a pure phase which is submerged into another phase if ρ is such that $\rho_1<\rho<\rho_2$, where ρ_1 , ρ_2 are densities of particles in the pure phases. The phase separation with finite number of pure phases was investigated in Minlos, Sinai [1], Gallavotti, Miracle-Sole [2], Kuroda [3] and others. For models with continuous symmetry group such as the classical Heisenberg model, where the number of pure phases is infinite, as it was proved in the paper [4] of Fröhlich, Simon, Spencer, there is no description of phase separation at present time as well as no sufficiently constructive description of the pure phases themselves.

In such a situation it is natural to try to study phase separation in simpler models with continuous symmetry group. Such a study will be done in the present paper where we consider phase separation in the hierarchical vector-valued ϕ^4 -model. Moreover we shall establish rigorously the Maxwell rule for this model according to which the thermodynamical potential is constant in the phase coexistence region (see, e.g. [5]), and find the principal term of the asymptotics for the small partition function when $V\to\infty$. We use essentially the results of the paper [6] of Major and the author, where pure phases of the hierarchical vector-valued ϕ^4 -model were constructed, furthermore we use some ideas of the paper [7] of the author, where the Maxwell rule and phase separation in the hierarchical scalar ϕ^4 -model was studied. It is noteworthy that phase separation in the scalar and vector-valued models is similar but not the same.

2. Description of the model and formulation of results

The hierarchical model was introduced by Dyson (see [8]). We use a slightly modified definition of the hierarchical model. We follow the paper [6].

Let $V = \{1,2,\dots\}$, $V_{kn} = \{j\in V \mid (k-1)2^n<j\le k\cdot 2^n\}$; $k,n = 1,2,\dots$. Put $V_{1n}=V_n$. Define for $i,j\in V$

$$n(i,j) = \min\{n \mid \exists k \text{ such that } i,j \in V_{kn}\} .$$

The hierarchical distance is defined as

$$d(i,j) = 2^{n(i,j)-1} \text{ if } i\neq j \ ; \ d(i,i) = 0 .$$

In a vector-valued hierarchical model the spin variables $\sigma(i)$, $i\in V$, take values in the m -dimensional Euclidean space $\mathbb{R}^m$, $m\geq 2$. The energy of a configuration $\sigma = \{\sigma(i),\ i \in V_{kn}\}$ is defined by the Hamiltonian

$$H_{kn}(\sigma) = \sum_{\substack{(i,j):i\neq j \\ i,j \in V_{1n}}} \mathcal{U}(i,j)\ (\sigma(i);\sigma(j)) , \tag{2.1}$$

where $U(i,j) = -d^{-a}(i,j)$ and $(\cdot;\cdot)$ denotes the scalar product in $\mathbb{R}^m$. In particular,

$$H_n(\sigma) = \sum_{\substack{(i,j):i\neq j\\ i,j\in V_{kn}}} U(i,j)\,(\sigma(i);\sigma(j)) .$$

The value a, $2>a>1$, is the parameter of the model.

Given a probability measure ν on $\mathbb{R}^m$ such that $\int_{\mathbb{R}^m} \exp(A(x;x))\,\nu(dx) < \infty$ for any $A>0$, we define the Gibbs distribution in the volume V_{kn} at inverse temperature $\beta = T^{-1}$ by the formula

$$\mu_{kn}(d\sigma) = \mu_{kn}(d\sigma;\beta,\nu) = \frac{\exp[-\beta H_{kn}(\sigma)]}{\Xi_{kn}} \prod_{i\in V_{kn}} \nu(d\sigma(i)) , \tag{2.2}$$

$$\Xi_{kn} = \Xi_{kn}(\beta,\nu) = \int \exp[-\beta H_{kn}(\sigma)] \prod_{i\in V_{kn}} \nu(d\sigma(i)) . \tag{2.3}$$

In this paper we consider the vector-valued ϕ^4-model where the measure ν is defined as

$$\nu(dx) = L^{-1}\exp\left(-\frac{u}{4}(x;x)^2 - \frac{(x;x)}{2}\right)dx , \tag{2.4}$$

$$L = \int_{\mathbb{R}^m} \exp\left(-\frac{u}{4}(x;x)^2 - \frac{(x;x)}{2}\right)dx .$$

Here $u>0$ is a parameter of the model. In the classical spin model $\nu(dx) = L^{-1}\delta(|x|-1)dx$, $L = \int_{\mathbb{R}^m} \delta(|x|-1)dx$.

For the classical hierarchical model Dyson proved in [8] that if β is large enough then there exists some $b>0$ such that $\langle(\sigma(i);\sigma(j))\rangle_n = \int(\sigma(i);\sigma(j))\mu_n(d\sigma;\beta,\nu) > b$ for all $n=1,2,\ldots$ and $i,j\in V_n$. Dyson's estimate means the existence of long-range order. In paper [6] a detailed study of the structure of the pure phases in the hierarchical vector-valued ϕ^4-model was carried out for $1<a<\frac{3}{2}$. Let us recall some results obtained there, which we shall use in the following.

We say that a function $\phi(s)$, $s\in\mathbb{R}^m$, belongs to the space $C^{k+\varepsilon}(\mathbb{R}^m)$, $1>\varepsilon>0$, if $\phi(s)\in C^k(\mathbb{R}^m)$ and for any

$R>0$,

$$\sup_{|s|\le R;\ 0<|t|\le R} \frac{|D^{\alpha}\phi(s+t)-D^{\alpha}\phi(s)|}{|t|^{\varepsilon}} < C_R ,$$

where $\alpha = (\alpha_1,\dots,\alpha_m)$ is any multiindex such that $|\alpha| = \alpha_1+\dots+\alpha_m = k$ and $D^{\alpha} = \partial^{|\alpha|}/\partial s_1^{\alpha_1}\dots\partial s_m^{\alpha_m}$. We say that a function $\phi(s)$, $s\in\mathbb{R}^m$ belongs to the class S , if the following conditions are fulfilled:

1) $\phi(s) = \phi(|s|)$ (isotropy) ;
2) $\phi(s) \in C^{2+\varepsilon}(\mathbb{R}^m)$, $\varepsilon>0$;
3) $\exists M>0$ such that $\phi'(M)=0$ and $\phi''(r) > D > 0$, if $r > \frac{4}{5}M$.

Note that here and later we use the convention to write any function $\phi(s)$ which depends on $|s|$ both as $\phi(s)$ and as $\phi(|s|)$.

Let

$$Z_n(s) = Z_n(s;\beta,u) = \int\delta\left(2^{-n}\sum_{i\in V_n}\sigma(i)-s\right)\exp[-\beta H_n(\sigma)] \times$$

$$\times \prod_{i\in V_n}\{L^{-1}\exp[-\tfrac{u}{4}(\sigma(i);\sigma(i))^2-\tfrac{1}{2}(\sigma(i);\sigma(i))]d\sigma(i)\}$$

be the small partition function of the hierarchical vector-valued ϕ^4 -model. Note that the function $p_n(s) = Z_n(s)/\Xi_n$ is the density of the probability distribution of the random variable $2^{-n}\sum_{i\in V_n}\sigma(i)$ with respect to the Gibbs measure $\mu_n(d\sigma)$.

In [6] a theorem was proved on the asymptotics of the small partition function $Z_n(s)$ when $n\to\infty$. We slightly strengthen now this theorem. Note that it implies a local limit theorem with large deviation asymptotics for the random variable $2^{-n}\sum_{i\in V_n}\sigma(i)$ with respect to $\mu_n(d\sigma)$, when $n\to\infty$. Let $\chi(x) \in C^{\infty}(\mathbb{R}^1)$ be an arbitrary function such that

$$\chi(x) = \begin{cases} 0.1 & \text{when } x \le 0.1 , \\ x & \text{when } x \ge 0.2 , \end{cases}$$

and $0.1 < \chi(x) < 0.2$ when $0.1 < x < 0.2$. Write the function $Z_n(s)$ in the form

$$Z_n(s) = \exp\left[-2^{n(2-a)} \frac{\beta a_0}{2} |s|^2 - 2^n \phi_n(s)\right] , \tag{2.5}$$

where $a_0 = (1-2^{1-a})^{-1}$.

Theorem A. Let $1<a<\frac{3}{2}$ and $\varepsilon>0$ be arbitrary numbers such that $\varepsilon < (a-1)(\frac{3}{2}-a)$. Then $\exists u_0$ such that $\forall u$, $0 < u < u_0$, $\exists \beta_c>0$ such that $\forall\ \beta > \beta_c$ $\exists$ function $\phi(s) = \phi(s;\beta,u) \in S$, $\phi'(M) = 0$, $M \sim (\beta-\beta_c)/u$, $\phi''(M) \sim \beta-\beta_c$ such that the asymptotic formula

$$\phi_n(s) = \phi(s) + \psi_n(s) + \mathcal{O}(2^{n(a-3/2)/2}) \tag{2.6}$$

holds when $n\to\infty$ and $|s| \geq T^{\varepsilon}_{n-1}$. Here T^{ε}_{n-1} is a solution of the equation

$$\phi'_n(T^{\varepsilon}_{n-1}) + 2^{(n-1)(1-a)}\beta a_1(1-\varepsilon)T^{\varepsilon}_{n-1} = 0$$

such that $\varphi''_n(r) > \alpha_0 > 0$ when $r \geq T^{\varepsilon}_{n-1}$ where α_0 does not depend on n and

$$\psi_n(s) = -n \frac{\ln 2}{2} [1+(m-1)(2-a)] + \frac{m}{2}\ln \pi - \frac{1}{2}\ln \phi''(s) -$$

$$- \frac{m-1}{4} \sum_{j=0}^{\infty} 2^{-j} \ln[\chi(\beta a_1 + 2^{(n+j)(a-1)}\phi'(|s|)/|s|] .$$

When $|s| \leq T^{\varepsilon}_{n-1}$, the estimate

$$\exp[-2^n\phi_n(|s|)] \leq \exp[-2^n\phi_n(T^{\varepsilon}_{n-1}) + 2^{(n-1)(2-a)} \frac{\beta a_1}{2}(1-\varepsilon) \times$$

$$\times (|s|^2 - (T^{\varepsilon}_{n-1})^2)]$$

holds.

The notation $f = \mathcal{O}(g)$ used here and in the following means that $f \leq Cg$ when C depends only on a . The notation $f\sim g$ means that $C_1 f \leq g \leq C_2 f$.

Theorem A gives the asymptotics of the small partition function $Z_n(s)$ in the domain $|s| \geq T^{\varepsilon}_{n-1}$ and its estimate in the domain $|s| \leq T^{\varepsilon}_{n-1}$. It slightly strengthens the result of the paper [6] where the asymptotics of $Z_n(s)$ was proved in a somewhat smaller domain $|s| \geq M(1-0{,}01\cdot 2^{n(1-a)})$. We can verify that a local property ensuring the validity of the asymptotic formula (2.6) holds in the domain $|s| \geq T^{\varepsilon}_{n-1}$ and it enables us to extend formula (2.6) to the above domain.

In the present work we obtain the asymptotic behavior of $Z_n(s)$ in the domain $|s|<M$.

Theorem 1. Let $1 < a < 3/2$, $0 < u < u_0$, $\beta>\beta_c$. Then for $|s|<M$

$$\lim_{n\to\infty} \frac{\phi_n(s)-\phi_n(M)}{2^{n(1-a)}} = a_1(M^2-|s|^2),$$

where $a_1 = 1+(1-2^{1-a})^{-1}$.

In particular this theorem implies the phenomenological Maxwell rule for the investigated model. According to this rule the thermodynamical potential $\phi(s) = -\lim_{n\to\infty}|V_n|^{-1}\ln Z_n(s)$ is a constant function in the domain $|s|\leq M$. Indeed,

$$\phi(s) = \lim_{n\to\infty} 2^{-n}\left\{2^{n(2-a)}\frac{\beta a_0}{2}|s|^2+2^n\phi_n(s)\right\} = \lim_{n\to\infty}\phi_n(s) =$$

$$= \lim_{n\to\infty}[\phi_n(M)+2^{n(1-a)}a_1(M^2-|s|^2)] = \lim_{n\to\infty}\phi_n(M) = \phi(M).$$

In Theorem 2 we investigate phase separation in the hierarchical vector-valued ϕ^4-model. Denote

$$s_{kn}(\sigma) = 2^{-n}\sum_{j\in V_{kn}}\sigma(j), \quad s_n(\sigma) = s_{1n}(\sigma),$$

$$\Pr_n\{A \mid s_n(\sigma)=s\} = \int_A \mu_n(d\sigma \mid s_n(\sigma)=s),$$

where

$$\mu_n(d\sigma|s_n(\sigma)=s) = Z_n^{-1}(s)\ \delta(s_n(\sigma)-s)\exp[-\beta H_n(\sigma)] \times$$

$$\times \prod_{j\in V_n}\{L^{-1}\exp[-\tfrac{u}{4}(\sigma(j);\sigma(j))^2 - \tfrac{1}{2}(\sigma(j);\sigma(j)]\}$$

is the small Gibbs ensemble. The value $\Pr_n\{A|s_n(\sigma)=s\}$ is the probability of the event A in the small Gibbs ensemble.

Theorem 2. Let $1 < a < 3/2$, $0<u<u_0$, $\beta>\beta_c$, $s\in\mathbb{R}^m$, $|s|<M$, $\varepsilon>0$, and

$$A_\varepsilon = \{\sigma=\{\sigma(i),\ i\in V_n\} \mid \, ||s_{k,n-1}(\sigma)|-M|<\varepsilon,\ k=1,2\} .$$

Then $\lim_{n\to\infty} \Pr_n\{A_\varepsilon \mid s_n(\sigma)=s\} = 1$.

Roughly speaking Theorem 2 means that when $n\to\infty$, then the small Gibbs ensemble $\mu_n(d\sigma \mid s_n(\sigma)=s)$ is the mixture of pure phases in the subvolumes $V_{1,n-1}$ and $V_{2,n-1}$.

3. Proofs of Theorems 1, 2

For the sake of simplicity we consider the case $m=2$. The extension of our considerations to the case of arbitrary $m\geq 2$ is straightforward. In the proof we use constants C , C_0 , C_1 etc., which will be positive and they depend only on the parameter a of the hierarchical model. They may differ in different estimates. The notation $f = O(g)$ has the usual sense: $f \leq C_g$, the notation $f\sim g$ means that $C_0 f \leq g \leq C_1 g$.

The functions $\phi_n(s)$ satisfy the recurrent equations

$$\phi_{n+1}(s) = -2^{-(n+1)}\ln\{4 \int_{\mathbb{R}^2} \exp[-2^{n(2-a)}\beta a_1(v;v) - 2^n \phi_n(s+v)-2^n \phi_n(s-v)]dv\} \tag{3.1}$$

(see [6]). The main step of our proof will be the demonstration that the integral at the RHS have these equations is essentially concentrated in the neighbourhood of some maximum points which we shall find. Denote

$$\bar{\phi}_n(s) = \begin{cases} \phi_n(s) , \text{ when } |s| \geq T^\varepsilon_{n-1} ; \\ \phi_n(T^\varepsilon_{n-1})+2^{(n-1)(1-a)} \frac{\beta a_1}{2} (1-\varepsilon)((T^\varepsilon_{n-1})^2-|s|^2) , \\ \qquad \text{when } |s| \leq T^\varepsilon_{n-1} , \end{cases}$$

and $\bar{F}_n(v;s) = 2^{n(1-a)}\beta a_1(v;v)+\bar{\phi}_n(s+v)+\bar{\phi}_n(s-v)$. One can see easily that $\bar{\phi}_n(s) \in C^1(\mathbb{R}^2)$ as T^ε_{n-1} is a solution of the equation $\phi_n'(T_{n-1}^{\varepsilon})+2^{(n-1)(1-a)}\beta a_1(1-\varepsilon)T_{n-1}^{\varepsilon} = 0$. Let $s = (r,0)$, $r = |s|$, and

$$\min_{v\in\mathbb{R}^2} \bar{F}_n(v;s) = \bar{F}_n(v^*;s) .$$

Lemma 3.

$$v^* = \begin{cases} 0 , & \text{when } |s| \geq T_n ; \\ (0, \pm\sqrt{T_n^2 - |s|^2}), & \text{when } |s| \leq T_n , \end{cases}$$

where T_n is a solution of the equation

$$\phi_n'(T_n) + 2^{n(1-a)}\beta a_1 T_n = 0 .$$

Proof. Denote

$$\zeta_n(|s|) = \bar{\phi}_n(|s|) + 2^{n(1-a)} \frac{\beta a_1}{2}|s|^2 .$$

Then $\bar{F}_n(v;s) = \zeta_n(|s+v|)+\zeta_n(|s-v|)-2^{n(1-a)}\beta a_1|s|^2$. As the value $2^{n(1-a)}\beta a_1|s|^2$ does not depend on v , we have to find only the minima of the function $\bar{F}_n^0(v;s) = \zeta_n(|s+v|)+$ $+\zeta_n(|s-v|)$. First we show that the T_n is a minimum point of the function $\zeta_n(r)$. Indeed,

$$\zeta_n'(T_n) = \phi_n'(T_n) + 2^{n(1-a)}\beta a_1 T_n = 0 ;$$

$$\zeta_n''(r) = \phi_n''(r)+2^{n(1-a)}\beta a_1 \geq \alpha_0+2^{n(1-a)}\beta a_1>0, \quad \text{if} \quad r\geq T_{n-1}^{\varepsilon} ;$$

$$\zeta_n(r) = \phi_n(T_{n-1}^{\varepsilon})+2^{(n-1)(1-a)} \frac{\beta a_1}{2}(1-\varepsilon)((T_{n-1}^{\varepsilon})^2-r^2)+2^{n(1-a)} \frac{\beta a_1}{2} r^2 =$$

$$=\zeta_n(T_{n-1}^{\varepsilon})+2^{n(1-a)} \frac{\beta a_1}{2} [2^{a-1}(1-\varepsilon)-1]((T_{n-1}^{\varepsilon})^2-r^2) , \quad \text{if} \quad 0 \leq r \leq T_{n-1}^{\varepsilon} .$$

By the above relations we have that

$$\zeta_n(r) \geq \zeta_n(T_n)+(\alpha_0+2^{n(1-a)}\beta a_1) \frac{(r-T_n)^2}{2} , \quad \text{if} \quad r\geq T_{n-1}^{\varepsilon} ; \tag{3.2}$$

$$\zeta_n(r) \geq \zeta_n(T_{n-1}^{\varepsilon})+\alpha_1((T_{n-1}^{\varepsilon})^2-r^2) , \quad \text{if} \quad 0 \leq r \leq T_{n-1}^{\varepsilon} , \tag{3.3}$$

where $\alpha_1 = \frac{\beta a_1}{2}[2^{a-1}(1-\varepsilon)-1] > 0$, so T_n is the minimum point.

Consider the circumferences $C_1 = \{v \mid |s+v|=T_n\}$, $C_2 = \{v \mid |s-v|=T_n\}$. If $|s|\le T_n$, they intersect each other in the points $v^{1,2} = \left(0, \pm\sqrt{T_n^2-|s|^2}\right)$ so

$$\min_v \bar{F}_n^o(v;s) \ge \min_v \zeta_n(|s+v|)+\min_v \zeta_n(|s-v|) \ge$$

$$\ge 2\,\zeta_n(T_n) = \bar{F}_n^o(v^{1,2};s) \text{ , as we claimed.}$$

If $|s| > T_n$ the circumferences C_1, C_2 do not intersect each other. Now we show that the minimum point is $v=0$. As the function $\bar{F}_n^o(v;s)$ is even in v_1 and in v_2 , where $v = (v_1,v_2)$, we may assume that $v_1\ge 0$, $v_2\ge 0$. Let $v^o = (|s|-T_n,0)$. If $|s+v| > |s+v^o|$ then

$$\bar{F}_n^o(v;s) = \zeta_n(|s+v|)+\zeta_n(|s-v|) > \zeta_n(|s+v^o|)+\zeta_n(T_n)=\bar{F}_n^o(v^o;s)$$

so the minimum point does not lie in the domain $|s+v| > |s+v^o|$.

If $|s+v| \le |s+v^o|$, $v = (v_1,v_2)$, $v_1\ge 0$, then $|s|+v_1 \le |s+v| \le |s|+|v^o|$ so $v_1\le |v^o|$ and $\bar{F}_n^o(v;s) \ge \bar{F}_n^o(\bar{v};s)$, where $\bar{v} = (v_1,0)$, as $|s\pm v| \ge |s\pm\bar{v}| \ge T_n$ and so $\zeta_n(|s\pm v|) \ge \zeta_n(|s\pm\bar{v}|)$. Thus the minimum point lies on the segment $[0,v^o]$. On this segment $|s\pm v| \ge T_n > T_{n-1}^{\varepsilon}$ so

$$\frac{\partial^2}{\partial v_1^2}\bar{F}_n^o(v;s) = \zeta_n''(|s+v|)+\zeta_n''(|s-v|) \ge \alpha_0+2^{n(1-a)}\beta a_1 ,$$

hence

$$\bar{F}_n^o(v;s) \ge \bar{F}_n^o(0;s) + (\alpha_0+2^{n(1-a)}\beta a_1)\frac{v^2}{2}$$

i.e. $v=0$ is the minimum point. Lemma 3 is proved.

Let V^* denote the set of the minimum points of the function $\bar{F}_n(v;s)$ in v . By Lemma 3 $V^*=\{0\}$ when $|s| \ge T_n$ and $V^* = \{v^1,v^2\}$ where $v^{1,2} = \left(0,\pm\sqrt{T_n^2-|s|^2}\right)$ when $0<|s|\le T_n$. Denote

$$\rho(v;V^*) = \inf_{v'\in V^*} |v-v'| .$$

Lemma 4. For any $\varepsilon'>0$ there exists a constant $C_{\varepsilon'}>0$ such that for $|s|\geq\varepsilon'$ and $v^*\in V^*$,

$$\bar{F}_n(v;s)-\bar{F}_n(v^*;s) \geq \begin{cases} C_\varepsilon \ , \ \rho^4(v;V^*) \ , & \text{if} \quad \rho(v;V^*) \leq 1 \ ; \\ C_\varepsilon \ , \ \rho^2(v;V^*) \ , & \text{if} \quad \rho(v;V^*) \geq 1 \ . \end{cases}$$

Proof. By the estimates (3.2), (3.3),

$$\zeta_n(r) - \zeta_n(T_n) \geq C(r-T_n)^2 \ . \tag{3.4}$$

If $\varepsilon' \leq |s| \leq T_n$, then simple geometrical considerations show that

$$\max\{\big||s+v|-T_n\big|,\big||s-v|-T_n\big|\} \geq \begin{cases} C(\varepsilon')\rho^2(v;V^*) \ , & \text{if} \quad \rho(v;V^*) \leq 1 \ ; \\ C_o \ \rho(v;V^*) \ , & \text{if} \quad \rho(v;V^*) \geq 1 \ . \end{cases}$$

Therefore

$$\bar{F}_n(v;s)-\bar{F}_n(v^*;s) = \zeta_n(|s+v|)+\zeta_n(|s-v|)-\zeta_n(|s+v^*|)-\zeta_n(|s-v^*|) =$$

$$=\zeta_n(|s+v|)-\zeta_n(T_n)+\zeta_n(|s-v|)-\zeta_n(T_n) \geq \begin{cases} C_1(\varepsilon')\rho^4(v;V^*), & \text{if } \rho(v;V^*) \leq 1 \ ; \\ C_1\rho^2(v;V^*), & \text{if} \quad \rho(v;V^*) \geq 1 \ . \end{cases}$$

The case $|s|\geq T_n$ is considered similarly. Lemma 4 is proved.

Denote $V^s = \{v \mid \big||v|-T_n\big| < 2|s|\}$.

Lemma 5. If $0.1 > \varepsilon' > 0$ then for $|s|<\varepsilon'$ we have

$$\bar{F}_n(v;s) - \bar{F}_n(v^*;s) \geq C \ \rho^2(v;V^s) \ .$$

Proof. The desired inequality is a consequence of (3.4) and the simple geometrical inequality

$$\big||s+v| - T_n\big| \geq \tfrac{1}{2}\rho(v;V^s) \ .$$

Indeed,

$$\bar{F}_n(v;s)-\bar{F}_n(v^*;s) = \zeta_n(|s+v|)-\zeta_n(T_n)+\zeta_n(|s-v|)-\zeta_n(T_n) \geq$$

$$\geq C(|s+v|-T_n)^2 \geq \frac{C}{4}\rho^2(v;V^s) .$$

Lemma 5 is proved.

Proof of Theorem 1. We estimate the function $\phi_n(s)$ from above and below.

Estimate $\phi_n(s)$ from below. We shall prove by induction the inequality

$$\phi_n(s) \geq \bar{\phi}_n(s) = \phi_n(T^{\varepsilon}_{n-1}) + 2^{(n-1)(1-a)}\frac{\beta a_1}{2}(1-\varepsilon)((T^{\varepsilon}_{n-1})^2-|s|^2), \quad (3.5)$$

when $|s| \leq T^{\varepsilon}_{n-1}$. For $n=0$ this inequality is verified easily with the help of the explicit formula $\phi_0(s) = = \frac{u}{4}|s|^4 + \frac{1-\beta a_0}{2}|s|^2$. Indeed, the function $\phi_0(r)-\bar{\phi}_0(r)$ is a polynomial of fourth degree and by construction it has second order zeroes at the points $r = \pm T^{\varepsilon}_{-1}$, and it has no other zeroes, hence $\phi_0(r)-\bar{\phi}_0(r) \geq 0$.

Assume that inequality (3.5) holds for the function $\phi_n(s)$. Prove it for the function $\phi_{n+1}(s)$. By the recurrent formula (3.1),

$$\phi_{n+1}(s) \geq -2^{-(n+1)}\ln\left\{4\int_{\mathbb{R}^2}\exp\left[-2^{n(2-a)}\beta a_1|v|^2 - 2^n\bar{\phi}_n(s+v) - \right.\right.$$
$$\left.\left. - 2^n\bar{\phi}_n(s-v)\right]dv\right\} = -2^{-(n+1)}\ln\left\{4\int_{\mathbb{R}^2}\exp[-2^n\bar{F}_n(v;s)]dv\right\} .$$

By Lemma 4 when $|s|\geq\varepsilon'$ and by Lemma 5 for $|s|\leq\varepsilon'$,

$$\int_{\mathbb{R}^2}\exp[-2^n\bar{F}_n(v;s)]dv \leq C\exp[-2^n\bar{F}_n(v^*;s)] .$$

If $|s|\leq T_n$,

$$\bar{F}_n(v^*;s) = 2\phi_n(T_n) + 2^{n(1-a)}\beta a_1(T_n^2-|s|^2) ,$$

so

$$\phi_{n+1}(s) \geq -2^{-(n+1)}\ln\{4C\exp[-2^{n+1}\phi_n(T_n)-2^{n(2-a)}\beta a_1 \times$$
$$\times(T_n^2-|s|^2)]\} = \phi_n(T_n)+2^{n(1-a)}\frac{\beta a_1}{2}(T_n^2-|s|^2)-2^{-(n+1)}\ln(4C)\geq \quad (3.6)$$
$$\geq \phi_n(T_n)+2^{n(1-a)}\frac{\beta a_1}{2}(1-\varepsilon)(T_n^2-|s|^2)-2^{-(n+1)}\ln(4C) .$$

Let us estimate now the RHS of this inequality. In the proof of Theorem A the following estimates were established:

$$\phi_n(T_n^\varepsilon) - \phi_{n+1}(T_n^\varepsilon) = O(n\ 2^{-n}) \ , \tag{3.7}$$

$$\phi'_n(T_n^\varepsilon) - \phi'_{n+1}(T_n^\varepsilon) = O(T_n^\varepsilon\ u\ 2^{-n}) \ .$$

We use now these estimates. Introduce the function

$$\zeta_{n,\varepsilon}(r) = \phi_n(r) - \phi_n(T_n^\varepsilon) - 2^{n(1-a)}\frac{\beta a_1}{2}(1-\varepsilon)((T_n^\varepsilon)^2 - r^2).$$

It has the following properties:

$$\zeta_{n,\varepsilon}(T_n^\varepsilon) = 0 \ ,$$

$$\zeta'_{n,\varepsilon}(T_n^\varepsilon) = \phi'_n(T_n^\varepsilon) + 2^{n(1-a)}\beta a_1(1-\varepsilon)T_n^\varepsilon =$$

$$= \phi'_n(T_n^\varepsilon) - \phi'_{n+1}(T_n^\varepsilon) = O(T_n^\varepsilon\ u\ 2^{-n}) \ ,$$

$$\zeta''_{n,\varepsilon}(r) = \phi''_n(r) + 2^{n(1-a)}\beta a_1(1-\varepsilon) \ .$$

Hence by the Taylor formula

$$\zeta_{n,\varepsilon}(T_n) = (T_n - T_n^\varepsilon)O(T_n^\varepsilon\ u\ 2^{-n}) + \frac{(T_n - T_n^\varepsilon)^2}{2}(\phi''_n(\theta) + 2^{n(1-a)}\beta a_1(1-\varepsilon)) \ ,$$

where $\theta \in [T_n, T_n^\varepsilon]$, or in a more detailed form

$$\phi_n(T_n) = \phi_n(T_n^\varepsilon) + 2^{n(1-a)}\frac{\beta a_1}{2}(1-\varepsilon)((T_n^\varepsilon)^2 - T_n^2) + (T_n - T_n^\varepsilon) \times$$
$$\times\ O(T_n^\varepsilon\ u\ 2^{-n}) + \frac{(T_n^\varepsilon - T_n)^2}{2}(\phi''_n(\theta) + 2^{n(1-a)}\beta a_1(1-\varepsilon)) \ .$$

Substituting this formula in the RHS of (3.6) we get

$$\phi_{n+1}(|s|) \geq \phi_n(T_n^\varepsilon) + 2^{n(1-a)}\frac{\beta a_1}{2}(1-\varepsilon)((T_n^\varepsilon)^2 - |s|^2) -$$

$$-2^{-(n+1)}\ln(4C) + (T_n - T_n^\varepsilon)O(T_n^\varepsilon\ u\ 2^{-n}) + \frac{(T_n^\varepsilon - T_n)^2}{2}(\phi''_n(\theta) +$$

$$+\ 2^{n(1-a)}\beta a_1(1-\varepsilon)) \ ,$$

or taking into account (3.7),

$$\phi_{n+1}(|s|) \geq \phi_{n+1}(T_n^\varepsilon) + 2^{n(1-a)} \frac{\beta a_1}{2}(1-\varepsilon)((T_n^\varepsilon)^2 - |s|^2) +$$

$$+ \mathcal{O}(n2^{-n}) + \frac{(T_n^\varepsilon - T_n)^2}{2}(\phi_n''(\theta) \div 2^{n(1-a)}\beta a_1(1-\varepsilon)) \quad . \tag{3.8}$$

It remains to estimate $T_n^\varepsilon - T_n$. We have:

$$\phi_n''(\theta_0)(T_n^\varepsilon - T_n) = \phi_n'(T_n^\varepsilon) - \phi_n'(T_n) = \phi_{n+1}'(T_n^\varepsilon) - \phi_n'(T_n) +$$

$$+ \mathcal{O}(T_n^\varepsilon u\, 2^{-n}) = -2^{n(1-a)}\beta a_1(1-\varepsilon)T_n^\varepsilon + 2^{n(1-a)}\beta a_1 T_n +$$

$$+ \mathcal{O}(T_n^\varepsilon u\, 2^{-n}) = 2^{n(1-a)}\beta a_1 \varepsilon T_n^\varepsilon + 2^{n(1-a)}\beta a_1(T_n - T_n^\varepsilon) +$$

$$+ \mathcal{O}(T_n^\varepsilon u\, 2^{-n}) \quad ,$$

where $\theta_0 \in [T_n, T_n^\varepsilon]$. Hence

$$T_n^\varepsilon - T_n = \frac{2^{n(1-a)}\beta a_1 \varepsilon T_n^\varepsilon + \mathcal{O}(T_n^\varepsilon u\, 2^{-n})}{\phi_n''(\theta_0) + 2^{n(1-a)}\beta a_1} \quad . \tag{3.9}$$

Thus

$$\frac{(T_n^\varepsilon - T_n)^2}{2}(\phi_n''(\theta) + 2^{n(1-a)}\beta a_1(1-\varepsilon)) \sim 2^{-n} \times \frac{2^{n(3-2a)}\beta^2 (T_n^\varepsilon)^2 \varepsilon^2}{\phi_n''(\theta_0) + 2^{n(1-a)}\beta a_1} \sim$$

$$\sim 2^{-n} \times \frac{2^{n(3-2a)}\beta^2 M^2 \varepsilon^2}{\phi''(M)} \quad .$$

As $M^2 \sim (\beta - \beta_c)/u$, $\phi''(M) \sim \beta - \beta_c$ we get finally that

$$\frac{(T_n^\varepsilon - T_n)^2}{2}(\phi_n''(\theta) + 2^{n(1-a)}\beta a_1(1-\varepsilon)) \sim 2^{-n} \times \frac{2^{n(3-2a)}\beta^2 \varepsilon^2}{u} \quad .$$

As $2^{n(3-2a)}\beta^2\varepsilon^2/u \gg n$ then by the inequality (3.8) we have now

$$\phi_{n+1}(|s|) \geq \phi_{n+1}(T_n^\varepsilon)+2^{n(1-a)}\frac{\beta a_1}{2}(1-\varepsilon)((T_n^\varepsilon)^2-|s|^2) .$$

We proved the estimate (3.5) for $|s|\leq T_n$. It remains to prove it for $T_n < |s| \leq T_n^\varepsilon$. Here we shall consider two cases: 1) $T_n < |s| \leq (T_n+T_n^\varepsilon)/2$; 2) $(T_n+T_n^\varepsilon)/2 < |s| \leq T_n^\varepsilon$. In the first case we use the equality $F_n(v^*;s) = 2\phi_n(|s|)$ and similarly (3.6) we get the estimate

$$\phi_{n+1}(|s|) \geq \phi_n(|s|)-2^{-(n+1)}\ln(4C) .$$

Further estimations may be proven in the same way as in the case $|s|\leq T_n$ and we omit them.

In the second case, when $(T_n+T_n^\varepsilon)/2 < |s| \leq T_n^\varepsilon$, the function $\phi_{n+1}(|s|)$ satisfies the estimates of Theorem A and according to these estimates

$$\phi''_{n+1}(r) > 0 > \bar{\phi}''_{n+1}(r) , \text{ when } r \in [(T_n+T_n^\varepsilon)/2, T_n^\varepsilon] .$$

Since by definition $\phi_{n+1}(T_n^\varepsilon) = \bar{\phi}_{n+1}(T_n^\varepsilon)$, $\phi'_{n+1}(T_n^\varepsilon) = \bar{\phi}'_{n+1}(T_n^\varepsilon)$ hence the last inequalities imply the estimate $\phi_{n+1}(r) \geq \bar{\phi}_{n+1}(r)$. The proof of estimate (3.5) is completed.

Estimate of $\phi_n(s)$ from above. Let us prove the estimate

$$\phi_{n+1}(|s|) \leq \phi_n(T_n)+2^{n(1-a)}\frac{\beta a_1}{2}(T_n^2-|s|^2)+Cn\ 2^{-n} \qquad (3.10)$$

when $|s|\leq T_n$. For this purpose we shall use the estimate

$$\zeta_n(|s\pm v|)-\zeta_n(T_n) = \zeta_n(|s\pm v|)-\zeta_n(|s\pm v^*|) \leq \frac{A}{2}|v-v^*|^2 ,$$

when $|v-v^*| \leq T_n-T_{n-1}^\varepsilon$, where $A = \sup \zeta''_n(|s|) = \sup(\phi''_n(|s|)+2^{n(1-a)}\beta a_1)$ and sup is taken with respect to $T_{n-1}^\varepsilon \leq |s| \leq 2\ T_n-T_{n-1}^\varepsilon$. Substituting this estimate into the recurrent equation (3.1) we get:

$$\phi_{n+1}(|s|) \leq -2^{-(n+1)}\ln\Big\{4 \int_{|v-v^*|<T_n-T_{n-1}^\varepsilon} \exp[2^{n(2-a)} \times$$

$$\times\beta a_1|s|^2-2^n\zeta_n(|s+v|)-2^n\zeta_n(|s-v|)]dv\Big\} \le -2^{n(1-a)}\frac{\beta a_1}{2}|s|^2 +$$

$$+ \zeta_n(T_n)-2^{-(n+1)}\ln\Big\{4 \int\limits_{|v-v^*|<T_n-T^{\varepsilon}_{n-1}} \exp(-2^nA|v-v^*|^2)dv\Big\} .$$

Note that $(2^nA)^{1/2}(T_n-T^*_{n-1}) \gg 1$ (see (3.9)), so the last relation implies:

$$\phi_{n+1}(|s|) \le \phi_n(T_n)+2^{n(1-a)} \frac{\beta a_1}{2}(T_n^2-|s|^2)-2^{-(n+1)}\ln\left\{4\sqrt{\frac{3}{2^nA}}\right\} \le$$

$$\le \phi_n(T_n)+2^{n(1-a)} \frac{\beta a_1}{2}(T_n^2-|s|^2)+Cn2^{-n} .$$

The estimate (3.10) is proved.

Let us show now that

$$\lim_{n\to\infty} \frac{\phi_n(T_n)-\phi_n(M)}{2^{n(1-a)}} = 0 .$$

As $\phi_n'(M) = O(M\cdot 2^{-n}u)$, $\phi_n'(T_n) = -2^{n(1-a)}\beta a_1T_n$ then $M-T_n \sim 2^{n(1-a)}\beta a_1M / \phi''(M)$. So $\phi_n(T_n)-\phi_n(M) \sim \phi''(M)(T_n-M)^2 \sim$ $\sim 2^{2n(1-a)}\beta^2M^2 / \phi''(M) \sim 2^{2n(1-a)}\beta^2 / u$. Hence

$$\frac{\phi_n(T_n)-\phi_n(M)}{2^{n(1-a)}} \sim 2^{n(1-a)}\beta^2/u \to 0$$

when $n\to\infty$ what was stated. Similarly

$$\lim_{n\to\infty} \frac{\phi_n(T^{\varepsilon}_{n-1})-\phi_n(M)}{2^{n(1-a)}} = 0 .$$

We get now by the estimate (3.5),

$$\varliminf_{n\to\infty} \frac{\phi_n(|s|)-\phi_n(M)}{2^{(n-1)(1-a)}} \ge \frac{\beta a_1}{2}(M^2-|s|^2)(1-\varepsilon) .$$

During the proof we used an estimate of ε from below only once when we assumed that $2^{n(3-2a)}\beta^2\varepsilon^2 / u \gg n$. Hence it is clear that we may consider a sequence $\varepsilon = \varepsilon_n \to 0$ when $n\to\infty$.

Then the last relation gives

$$\underline{\lim}_{n\to\infty} \frac{\phi_n(|s|)-\phi_n(M)}{2^{(n-1)(2-a)}} = \frac{\beta a_1}{2}(M^2-|s|^2) .$$

Similarly with the help of (3.10) we get

$$\overline{\lim}_{n\to\infty} \frac{\phi_n(|s|)-\phi_n(M)}{2^{(n-1)(2-a)}} = \frac{\beta a_1}{2}(M^2-|s|^2)$$

hence

$$\lim_{n\to\infty} \frac{\phi_n(|s|)-\phi_n(M)}{2^{(n-1)(2-a)}} = \frac{\beta a_1}{2}(M^2-|s|^2) .$$

Theorem 1 is proved.

<u>Proof of Theorem 2</u>. We use the formula

$$\Pr_n\{A_\varepsilon \mid s_n(\sigma)=s\} = \frac{\int_{||s\pm v|-M|<\varepsilon} \exp[-2^n F_n(v;s)]dv}{\int_{\mathbb{R}^2} \exp[-2^n F_n(v;s)]dv} \tag{3.11}$$

where $F_n(v;s) = 2^{n(1-a)}\beta a_1|v|^2+\phi_n(s+v)+\phi_n(s-v)$ (see [7]). Using Lemmata 4, 5 we get:

$$\int_{\mathbb{R}^2\setminus\{||s\pm v|-M|<\varepsilon\}} \exp[-2^n F_n(v;s)]dv \le$$

$$\le \int_{\mathbb{R}^2\setminus\{||s\pm v|-M|<\varepsilon\}} \exp[-2^n \bar{F}_n(v;s)]dv \le$$

$$\le C \exp [-2^n \bar{F}_n(v^*;s)-2^n C_0\varepsilon^4] =$$

$$= C \exp [-2^n F_n(v^*;s)-2^n C_0\varepsilon^4] .$$

On the other hand using the estimate (3.10) we get:

$$\int_{\mathbb{R}^2} \exp [-2^n F_n(v;s)]dv \ge C_1 \exp [-2^n F_n(v^*;s)-C_2 n] .$$

Hence

$$\Pr_n\{A_\varepsilon | s_n(\sigma)=s\} \geq 1 - \frac{C\exp[-2^n F_n(v^*;s)-2^n C_0 \varepsilon^4]}{C_1\exp[-2^n F_n(v^*;s)-C_2 n]} \geq$$

$$\geq 1 - C_3 \exp(-2^n C_0 \varepsilon^4 + C_2 n) \to 1$$

when $n\to\infty$. Theorem 2 is proved.

References

[1] Minlos R. A., Sinai Ya. G. Phenomenon of "phase separation" at low temperatures in some lattice models of gase. I. Matem. Sborn. 73, No 3, 375-448 (1967), II. Trans. Moscow Mathem. Soc. 19, 113-178 (1968).

[2] Gallavotti G., Miracle-Sole S. Equilibrium states of the Ising model in the two phase region. Phys. Rev. B5, 2555 (1972).

[3] Kuroda K. The probabilistic treatment of phase separations in lattice models composed of more than two types of particles. Publ. RIMS Kyoto Univ. 18, No 1, 275-305 (1982).

[4] Fröhlich J., Simon B., Spencer T. Infrared bounds, phase transitions and continuous symmetry breaking. Commun. Math. Phys. 50, 79-95 (1976).

[5] Stanley H. E. Introduction to phase transitions and critical phenomena. Clarendon Press, Oxford, 1971.

[6] Blecher P. M., Major P. Renormalization of Dyson's hierarchical vector-valued ϕ^4 -model at low temperatures, preprint Math. Inst. Hungar. Ac. Sci. 1984, Commun. Math. Phys. (in press).

[7] Blecher P. M. Phenomenon of phase separation in ϕ^4 - hierarchical model. Theoret. Math. Phys. (in press).

[8] Dyson F. J. Existence of a phase transition in a one-dimensional Ising ferromagnet. Commun. Math. Phys. 12, No 2, 91 (1969).

CONSTRUCTIVE CRITERION FOR THE UNIQUENESS OF GIBBS FIELD

R. L. Dobrushin and S. B. Shlosman

1. Introduction

In this report we consider ν -dimensional lattice systems with a given translation-invariant potential U . Our main result is the construction of a set of conditions C_V on the potential U , where $V \subset \mathbf{Z}^\nu$ is any finite volume, such that if for some V the condition C_V holds for the interaction U , then the Gibbs state with this interaction is unique. The complexity of the conditions C_V increases, of course, with the volume V . The one-point criterion, $C_{\{t\}}$, was introduced earlier by one of us [1]. It was intensively used later (see [2]-[7]).

Under the conditions C_V , some property of exponential decay of correlations can be proven. This property is Very Strong (see (3.10)), though not the Strongest one (see (3.11)). (In this report only finite range potentials are treaten - otherwise the exponential decay statement has to be replaced by a slower one. The infinite-range potentials will be treated in a separate publication.)

To discuss the effectiveness of our conditions C_V , we restrict ourselves to the models with finite single-spin spaces. In this case the set $\mathbf{A}_0$ of the interactions, for which for some V the condition C_V holds, includes the set of those interactions, for which the unicity is accompanied by the Strongest exponential decay of correlations mentioned above. In a subsequent paper [8] we show that this last property follows from a natural property of analyticity of the partition functions. This and other speculations make plausible the following

Conjecture: For a wide class of potentials the boundary of the set A_0 is the surface of phase transitions.

But from our definitions it follows that $A_0 = \bigcup_{n=1}^{\infty} A_n$, where $A_1 \subset A_2 \subset \ldots$ and the boundary of the set A_n can be specified by a finite number $N(n)$ of arithmetical and logical operations. This is why we call our criterion "effective". Of course, $N(n) \to \infty$ with $n \to \infty$. Nevertheless, our criterion provides one of the first possibilities to locate the phase-transition threshold for a general situation, in principle with any given accuracy (compare, however, with [9]).

This has to be compared with many wellknown numerical methods for exploring the phase-transition region, utilized by mathematical physicists (for example, MC-simulation). The main difference lies in the following. While both the traditional methods and ours are finite-volume ones, the results given by the former are necessarily approximate when attributed to infinite-volume systems; so their implications about phase transitions cannot be justified with mathematical rigour. On the other hand, the criterion described in this paper can be used as a basis for computer-assisted rigorous proof of the absence of phase transition. This is because our result has the form of the following

Theorem: If Something $(=C_V)$ happens in some finite volume $(=V)$, then surely there is no phase transition (in infinite volume)!

We are going to finish computer-assisted proofs of unicity for certain models in the near future.

The content of this report is the following: in Section 2 we state our conditions C_V and prove that C_V implies unicity. The next Section 3 deals with various definitions of correlation decay, and we prove the exponential decay, provided the condition C_V holds. The setting of both of the Sections 2, 3 is abstract - we study random fields with a specification given. We turn ourselves to the Gibbs case in Section 4, in which in addition to the general results of Sections 2, 3, some specific results can be proven - using ground state structures. The final Section 5 deals with the effectiveness of the criterion presented. We show that the set A_0 is effectively enumerable and present an algorithm, which enables one to check uniqueness for any given potential. We finish this section by fixing some notations used throughout this paper:

$\mathbb{Z}^\nu$ - ν-dimensional lattice, $\nu \geq 1$,

S - single-spin space,

$\tilde{B}$ - σ-algebra of subsets of S,

$\sigma, \bar{\sigma}, \tau, \ldots : \mathbb{Z}^\nu \to S$ - maps, or "configurations on $\mathbb{Z}^\nu$",

Ω - the set of all configurations,

$V, W, \Lambda, \ldots \subset \mathbb{Z}^\nu$ - subsets,

$|V|$ - the cardinality of V,

$V^c = \mathbb{Z}^\nu \setminus V$,

$\mathcal{D}_n = \{t \in \mathbb{Z}^\nu : -n \leq t^i \leq n, \ i=1,\ldots,\nu\}$ - the n-cube, centered in the origin,

Ω_V - the set of all configurations on $V: \{\sigma_V : V \to S\}$,

$\sigma_V = \sigma|_V$ - the restriction of $\sigma \in \Omega$ or $\sigma \in \Omega_W$ on $V \subset W$,

$\sigma_{V_1} \cup \sigma_{V_2} \in \Omega_{V_1 \cup V_2}$ is defined by $(\sigma_{V_1} \cup \sigma_{V_2})|_{V_i} = \sigma_{V_i}$, $i=1,2$ for $V_1, V_2 \subset \mathbb{Z}^\nu$, $V_1 \cap V_2 = \emptyset$, $\sigma_{V_i} \in \Omega_{V_i}$, $i=1,2$,

$\mathcal{B}_V$ - σ-algebra of the subsets of Ω_V, generated by the cylinders $\{\sigma_V \in \Omega_V : \sigma_t \in B_t, \ B_t \in \tilde{B}, \ t \in V\}$

$\mathcal{B} = \mathcal{B}_{\mathbb{Z}^\nu}$,

$\partial V = \partial_r V = \{t \in V^c : \operatorname{dist}(t,V) \leq r\}$, where $r \geq 0$.

2. Unicity

Let $(S,\tilde{B})$ be a measurable Polish space (or finite space, if preferred), and $\rho(\cdot,\cdot)$ be a metric on S, such that the function $\rho(\sigma,\tau)$ on $S\times S$ is measurable with respect to the σ-algebra $\tilde{B}\times\tilde{B}$. Consider a pair μ_1, μ_2 of probability measures on S. By a joint distribution of μ_1, μ_2 we call any measure μ on $S\times S$ such that

$$\begin{aligned} \mu(B\times S) &= \mu_1(B) \\ \mu(S\times B) &= \mu_2(B) \end{aligned} \tag{2.1}$$

for all $B \in \mathcal{B}$. The set of all these μ will be denoted by $K(\mu_1,\mu_2)$. For any measure μ on $S \times S$ and any measures μ_1, μ_2 on S let

$$\bar{R}_\rho(\mu) = \int_{S\times S} \rho(\sigma,\tau)\mu(d\sigma,d\tau) \tag{2.2}$$

$$R_\rho(\mu_1',\mu_2) = \inf_{\mu\in K(\mu_1,\mu_2)} \bar{R}_\rho(\mu) .$$

This $R_\rho(\mu_1,\mu_2)$ can be shown to be a metric on the space $M(\mu_0)$ of the measures μ with $R_\rho(\mu,\mu_0) < \infty$ for some μ_0 (see [1]). The metric R_ρ is called Kantorovich metric, or Kantorovich-Rubinstein, or Kantorovich-Rubinstein-Ornstein-Vasserstein and so on (see a recent historical-oriented review [10], which contains about 150 references beginning with Monge's paper published in 1781).

For

$$\rho(\sigma,\tau) = \begin{cases} 1, & \sigma\neq\tau , \\ 0, & \sigma=\tau , \end{cases} \tag{2.3}$$

$R_\rho(\cdot,\cdot)$ coincides with the variation distance $\mathrm{Var}(\cdot,\cdot)$.

For any $V \subset \mathbb{Z}^\nu$, any probability measure P on $(\Omega_V,\mathcal{B}_V)$ and any $W \subset V$ we denote P_W the projection of P on $(\Omega_W, \mathcal{B}_W)$, given by

$$P_W(A) = P(\{\sigma\in\Omega_V : \sigma_W \in A\}) , \ A \in \mathcal{B}_W . \tag{2.4}$$

By a random field we mean any probability measure P on $(\Omega,\mathcal{B})$.

By an r-specification $(r\geq 0)$ we mean a system of functions $Q = \{Q_V : V\subset\mathbb{Z}^\nu, |V|<\infty\}$, $Q_V = Q_V(A|\bar\sigma)$, where $A\in\mathcal{B}_V$, $\bar\sigma\in\Omega$, such that

1) for any A the function $Q_V(A|\cdot)$ is $\mathcal{B}_{\partial_r V}$-measurable,

2) for any $\bar\sigma$ the function $Q_V(\cdot|\bar\sigma)$ is a probability measure on $(\Omega_V,\mathcal{B}_V)$.

Sometime we shall think of $Q_V(A|\cdot)$ as of a function on $\Omega_{\partial_r V}$, or even on $\Omega_{\partial_r V\cap W}$ for some $W \subset \mathbb{Z}^\nu$. This is done for notational convenience and is justified by 1). We denote by $Q_{V,\Lambda}(\cdot|\bar{\sigma})$ the projection of $Q_V(\cdot|\bar{\sigma})$ on a set $\Lambda\subset V$.

A random field P is V-consistent with a specification Q for a set $V\subset \mathbb{Z}^\nu$, $|V|<\infty$, if the conditional distribution of P with respect to the σ-algebra $\mathcal{B}_{V^c}$, $P_V(\cdot|\cdot)$, coincides with $Q_V(\cdot|\cdot)$ almost everywhere.

A specification Q is called self-consistent, if for any finite $V\subset \mathbb{Z}^\nu$ and $\bar{\sigma}\in\Omega$ the random field $Q_V^{\bar{\sigma}}$, defined by the following conditions

1) $(Q_V^{\bar{\sigma}})_V$ coincides with $Q_V(\cdot|\bar{\sigma})$,

2) $(Q_V^{\bar{\sigma}})_{V^c}(\bar{\sigma}) = 1$,

is W-consistent with Q for any $W\subset V$.

A random field P is called consistent with a self-consistent specification Q, (or Q-consistent) if P is V-consistent with Q for all finite $V\subset \mathbb{Z}^\nu$.

The main example of self-consistent specifications are Gibbs specifications (see Sect. 4).

In what follows, we shall consider only translation-invariant specifications. This is done solely in order to simplify notations.

We say the random field P to be of exponential growth, if for some $\tau_0\in S$, g, $G<\infty$

$$\rho_t(P) = \int_\Omega \rho(\sigma_t, \tau_0)P(d\sigma) \le G\exp\{g|t|\},\ t\in \mathbb{Z}^\nu . \tag{2.5}$$

We denote the set of all Q-consistent fields P, satisfying (2.5) for some $\tau_0\in S$ and $G<\infty$ by $\mathcal{P}_Q(g)$.

Theorem 2.1. Suppose that for a given r-specification Q there exists a volume V, such that the following condition holds:

Condition C_V: there exists a function $k_t\ge 0$, $t\in\partial V$ with the following properties:

1) for any $t\in\partial V$ and any $\bar{\sigma}^1, \bar{\sigma}^2\in\Omega$, $\bar{\sigma}^1_s = \bar{\sigma}^2_s$ for $s\ne t$

$$R_{\rho_V}(Q_V(\cdot|\bar{\sigma}^1),\ Q_V(\cdot|\bar{\sigma}^2)) \le k_t \rho(\bar{\sigma}_t^1,\ \bar{\sigma}_t^2)\ , \tag{2.6}$$

where

$$\rho_V(\sigma^1,\sigma^2) = \sum_{t\in V} \rho(\sigma_t^1,\sigma_t^2)\ ,\quad \sigma^1,\sigma^2 \in \Omega_V \tag{2.7}$$

for $|V|<\infty$.

2)

$$|V|^{-1} \sum_{t\in\partial V} k_t = \gamma < 1\ . \tag{2.8}$$

Then there exists a value $g_0 = g_0(V,r,\gamma) > 0$, such that the set $P_Q(g_0)$ contains no more than one element.

Remarks

2.1. In the case $|V|=1$ Theorem 2.1 was proven by one of us in [1]. This particular case is called "mean-field" bound (see [3]), so our theorem is an improvement of the latter bound.

2.2. In the case when S has finite diameter, the theorem implies the unicity in the class of all Q -consistent fields. On the other hand, for S unbounded, the restriction (2.5) is natural (see [11] for Gaussian case), and the theorem fails without it.

2.3. Suppose that

$$d = \text{diam } S = \sup_{\sigma,\tau\in S} \rho(\sigma,\tau) \le 1 \tag{2.9}$$

(as in the case (2.3) of the variation distance).

Then the following more simple condition implies unicity:

For some finite $V\subset\mathbb{Z}^\nu$ and any $\bar{\sigma}^1,\bar{\sigma}^2\in\Omega$ different exactly in one point $t\in\mathbb{Z}^\nu$,

$$R_{\rho_V}(Q_V(\cdot|\bar{\sigma}^1),\ Q_V(\cdot|\bar{\sigma}^2)) \le (1-\varepsilon)\frac{|V|}{|\partial V|} \tag{2.10}$$

with some $\varepsilon>0$.

As for the proof of the above statement, one has only to put $k_t = |V|/|\partial V|$ for $t\in\partial V$ and to apply Theorem 2.1.

The proof of Theorem 2.1 is based on the following lemmas.

Lemma 2.1. Suppose that condition C_V holds for the specification Q. Let W be any given volume, and

$$T(W) = \{t \in \mathbb{Z}^{\nu} : (V \cup \partial V) + t \subset W\} . \tag{2.11}$$

Let P^1, P^2 be two random fields, $(V+t)$ -consistent with Q for all $t \in T(W)$.

Then, for any $\delta > 0$ there exists an element $\mu \in K(P^1_W, P^2_W)$ (which is a measure on $\Omega_W \times \Omega_W$), such that for

$$f_t = \int_{\Omega_W \times \Omega_W} \rho(\sigma^1_t, \sigma^2_t) \mu(d\sigma^1, d\sigma^2) , \quad t \in W \tag{2.12}$$

and for all $s \in T(W)$

$$\sum_{t \in V} f_{t+s} \le \sum_{t \in \partial V} k_t f_{t+s} + \delta . \tag{2.13}$$

Proof of Lemma 2.1. It is enough to take for the measure μ any element of the family $K(P^1_W, P^2_W)$ with the property

$$\sum_{t \in W} f_t \le R_{\rho_W} (P^1_W, P^2_W) + \delta/2 \tag{2.14}$$

(the definitions (2.2) and (2.7) implies the existence of such a μ). To see (2.13) holds for this measure, let us fix some $s_0 \in T(W)$ and check (2.13) for it. To simplify the notations, suppose $s_0 = 0 \in \mathbb{Z}^{\nu}$. From (2.6) it follows that for any $\bar{\sigma}^1, \bar{\sigma}^2 \in \Omega$ there exists a measure $\hat{\mu}(\cdot, \cdot | \bar{\sigma}^1, \bar{\sigma}^2) \in K(Q_V(\cdot | \bar{\sigma}^1), Q_V(\cdot | \bar{\sigma}^2))$ on $\Omega_V \times \Omega_V$ such that

$$\int_{\Omega_V \times \Omega_V} \rho_V(\sigma^1, \sigma^2) \hat{\mu}(d\sigma^1, d\sigma^2 \mid \bar{\sigma}^1, \bar{\sigma}^2) \le \sum_{t \in \partial V} k_t \rho(\bar{\sigma}^1_t, \bar{\sigma}^2_t) + \delta/2 . \tag{2.15}$$

Now we are going to apply to the measure μ on $\Omega_V \times \Omega_V$ the following "surgery" construction (see [1], [12]). For any $W_1 \subset W$ denote by $\bar{B}_{W_1}$ the σ -subalgebra of $B_W \times B_W$ generated by the pairs $B' \times B'' \subset \Omega_W \times \Omega_W$, where $B', B'' \in B_{W_1}$. Now the V -surgery, applied to μ , results in a new measure $\tilde{\mu}$ on $\Omega_W \times \Omega_W$, which is defined by the following two properties:

1) $\tilde{\mu}$ equals μ on $\bar{\mathcal{B}}_{W \setminus V}$;

2) its conditional distribution with respect to $\bar{\mathcal{B}}_{W \setminus V}$ is equal μ -everywhere to $\hat{\mu}(\cdot,\cdot|\cdot,\cdot)$.

From the Q -consistency of P^1, P^2 it follows that $\tilde{\mu} \in K(P^1_W, P^2_W)$, hence

$$\sum_{t \in W} \tilde{f}_t \geq R_{\rho_W}(P^1_W, P^2_W) , \tag{2.16}$$

where $\tilde{f}_t$ is defined by (2.12) with $\tilde{\mu}$ instead of μ . By definition,

$$f_t = \tilde{f}_t \quad , \quad t \in W \setminus V . \tag{2.17}$$

From (2.15) it follows that

$$\sum_{t \in V} \tilde{f}_t = \int_{\Omega_W \times \Omega_W} \tilde{\mu}(d\bar{\sigma}^1, d\bar{\sigma}^2) \int_{\Omega_V \times \Omega_V} \rho_V(\sigma^1,\sigma^2)\hat{\mu}(d\sigma^1, d\sigma^2 | \bar{\sigma}^1, \bar{\sigma}^2) \leq$$

$$\leq \sum_{t \in \partial V} k_t \tilde{f}_t + \delta/2 . \tag{2.18}$$

From (2.14), (2.16), (2.17) it follows that

$$\left| \sum_{t \in V} f_t - \sum_{t \in V} \tilde{f}_t \right| = \left| \sum_{t \in W} f_t - \sum_{t \in W} \tilde{f}_t \right| \leq \delta/2 , \tag{2.19}$$

hence from (2.18)

$$\sum_{t \in V} f_t \leq \sum_{t \in \partial V} k_t \tilde{f}_t + \delta \tag{2.20}$$

which together with (2.17) implies (2.13).

Lemma 2.2. Let condition 2) of Theorem 2.1 hold (see (2.8)). Let a volume $W \subset \mathbb{Z}^\nu$, $|W| < \infty$, and a positive function f_t , $t \in W$ be given, such that for all $s \in T(W)$ the bounds (2.13) hold with some $\delta > 0$. Let $\Lambda \subset W$ be given and for some $\bar{g} > 0$ let us define the function

$$C(t) = C_{\Lambda,\bar{g}}(t) = \exp\{-\bar{g} \operatorname{dist}(t,\Lambda)\} , \quad t \in \mathbb{Z}^\nu . \tag{2.21}$$

Then for some value $\bar{g} = \bar{g}(\gamma,r,V)$, $\bar{C} = \bar{C}(\gamma,r,V)$

$$\sum_{t\in W} f(t)c(t) \le \bar{C} \sum_{t\in\bar{\partial}_V W} f_t c(t) + \delta|W| \tag{2.22}$$

where

$$\bar{\partial}_V W = \{t\in W : \operatorname{dist}(t,W^c) \le \operatorname{diam}(V\cup\partial V)\} . \tag{2.23}$$

Proof of Lemma 2.2: After multiplying each of the bounds (2.13) by the corresponding value $c(s)$ and summing them over all $s \in T(W)$ one arrives to the following estimate:

$$\sum_{t\in W} f_t\left(\sum_{s\in\mathbb{Z}^\nu:t-s\in V} c(s) - \sum_{s\in\mathbb{Z}^\nu:t-s\in\partial V} k_{t-s}c(s)\right) \le \tag{2.24}$$

$$\le \sum_{t\in W} f_t\left(\sum_{s\in(T(W))^c:t-s\in V} c(s) - \sum_{s\in(T(W))^c:t-s\in\partial V} k_{t-s}\, c(s)\right) + \delta|W| .$$

Let

$$m_1 = \max_{t,s:t-s\in V\cup\partial V} \frac{c(s)}{c(t)} , \quad m_2 = \min_{t,s:t-s\in V\cup\partial V} \frac{c(s)}{c(t)} . \tag{2.25}$$

Then the left-hand factor in (2.24) can be estimated using (2.8) as follows:

$$\sum_{s\in\mathbb{Z}^\nu:t-s\in V} c(s) - \sum_{s\in\mathbb{Z}^\nu:t-s\in\partial V} k_{t-s}\, c(s) \ge c(t)|V|(m_2-\gamma m_1) \tag{2.26}$$

which in turn is larger than $c(t)|V|\kappa$ provided $0<\kappa<1-\gamma$ and $\bar{g}=\bar{g}(\kappa)$ is small enough. The analogous factor in the right-hand side of (2.24) is non-zero only for $t\in\bar{\partial}_V W$. It can be estimated from above by $2|V|$, provided $\bar{g}$ is so small that $m_1 \le 2$. From this and (2.26) follows (2.22).

Proof of Theorem 2.1: Let $P^1,P^2 \in \mathcal{P}_Q(g)$. To prove the theorem it is enough to show that for any finite $\Lambda\subset\mathbb{Z}^\nu$ $P^1_\Lambda = P^2_\Lambda$. Let $\Lambda\subset W$ for some finite $W\subset\mathbb{Z}^\nu$. From Lemmas 2.1, 2.2 it follows the existence of a measure $\mu\in K(P^1_W,P^2_W)$ such that

$$\sum_{t\in W} f_t c(t) \le \bar{C} \sum_{t\in\bar{\partial}_V W} f_t c(t) + \delta |W| \ . \tag{2.27}$$

Using definitions (2.21), (2.12), (2.7) one has

$$\sum_{t\in W} f_t c(t) \ge \sum_{t\in\Lambda} f_t \ge R_{\rho_\Lambda}(P^1_\Lambda, P^2_\Lambda) \ . \tag{2.28}$$

On the other hand from (2.5) and (2.12) it follows that

$$\begin{aligned} f_t &\le \int_{\Omega_W\times\Omega_W} (\rho(\sigma^1_t,\tau)+\rho(\sigma^2_t,\tau))\mu(d\sigma^1,d\sigma^2) = \\ &= \rho_t(P^1)+\rho_t(P^2) \le 2\ G\exp\{g|t|\} \ . \end{aligned} \tag{2.29}$$

Now take W to be the n-cube $\mathcal{D}_n$ (defined in Sect. 1). It is clear that for $g\le g_0<\bar{g}$ and for a suitably chosen $\delta=\delta(n)$ the right-hand side of (2.27) goes to zero as $n\to\infty$. From (2.27) and (2.28) it follows then that $R_{\rho_\Lambda}(P^1_\Lambda,P^2_\Lambda)=0$. But R_{ρ_Λ} is a metric, hence $P^1_\Lambda = P^2_\Lambda$.

3. The decay of correlations

In this section we shall show that for any Q-consistent random field P the correlations decay exponentially, provided condition C_V of Theorem 2.1 holds. In what follows we suppose for simplicity that condition (2.9) on the diameter of S holds.

Theorem 3.1. Let Q be a self-consistent r-specification and the conditions of Theorem 2.1 hold. Let P be the (unique) random field consistent with Q. Then there exist constants $G_1<\infty$, $g_1>0$ depending only on γ, r, V such that for any $\Lambda\subset W\subset \mathbb{Z}^\nu$ and any $\bar{\sigma}\in\Omega$

$$R_{\rho_\Lambda}(Q_{W,\Lambda}(\cdot|\sigma),P_\Lambda) \le G_1 \sum_{t\in\bar{\partial}_V W} \exp\{-g_1 \mathrm{dist}(t,\Lambda)\} \ . \tag{3.1}$$

An analogous bound holds for the specification Q itself: for $\bar{\sigma}^1,\bar{\sigma}^2\in\Omega$

$$R_{\rho_\Lambda}(Q_{W,\Lambda}(\cdot|\bar{\sigma}^1),Q_{W,\Lambda}(\cdot|\bar{\sigma}^2)) \le G_1 \sum_{t\in\bar{\partial}_V W} \exp\{-g_1 \mathrm{dist}(t,\Lambda)\} \ . \tag{3.2}$$

Proof: The proof is very similar to the proof of Theorem 2.1. To prove (3.2) one has to consider the measures $P^i = Q_W^{\sigma^i}$, i=1,2 (see Sect. 2). These measures are (V+t) - consistent with Q for $t \in T(W)$, hence Lemmas 2.1, 2.2 are valid for them. The condition $P^i \in P(g)$ follows from (2.9). The inequality (3.1) is proved in the same way.

Now we state the decay property in a more familiar form.

Theorem 3.2. Under the hypothesis of Theorem 3.1 let $\Lambda_i \subset \mathbb{Z}^\nu$ be finite and ψ_i be $\mathcal{B}_{\Lambda_i}$ -measurable for i=1,2. Suppose that ψ_1 is bounded while ψ_2 is Lipschitz: for $C_1, C_2 < \infty$

$$|\psi_1(\sigma)| \le C_1 \ , \ \sigma \in \Omega \tag{3.3}$$

$$|\psi_2(\sigma^1) - \psi_2(\sigma^2)| \le C_2 \rho_{\Lambda_2}(\sigma^1_{\Lambda_2}, \sigma^2_{\Lambda_2}) \ .$$

Then

$$\begin{aligned} &\left|\int_\Omega \psi_1(\sigma)\psi_2(\sigma)P(d\sigma) - \int_\Omega \psi_1(\sigma)P(d\sigma)\int_\Omega \psi_2(\sigma)P(d\sigma)\right| \le \\ &\le G_1 C_1 C_2 \sum_{t \in \bar{\partial}_V(\Lambda_1)^c} \exp\{-g_1 \operatorname{dist}(t,\Lambda_2)\} \ , \end{aligned} \tag{3.4}$$

where G_1, g_1 - are the same as in Theorem 3.1.

<u>Proof</u>: We begin by the following remark due already to Kantorovich: for any function $\psi(\sigma)$, $\sigma \in S$, which is Lipschitz with a constant $C>0$ and for any measures μ_1, μ_2

$$\left|\int_S \psi(\sigma)\mu_1(d\sigma) - \int_S \psi(\sigma)\mu_2(d\sigma)\right| \le C\mathcal{R}_\rho(\mu_1,\mu_2) \tag{3.5}$$

Indeed if $\mu \in K(\mu_1,\mu_2)$ then

$$\left|\int_S \psi(\sigma)\mu_1(d\sigma) - \int_S \psi(\sigma)\mu_2(d\sigma)\right| \le$$

$$\le \int_{S\times S} |\psi(\sigma)-\psi(\tau)|\mu(d\sigma,d\tau) \le C\int_{S\times S} \rho(\sigma,\tau)\mu(d\sigma,d\tau)$$

so (3.5) follows. Now without loss of generality one may assume that

$$\int_\Omega \psi_2(\sigma) P(d\sigma) = 0 \ . \tag{3.6}$$

Let $P^{\Lambda_1}(\cdot|\cdot)$ be the conditional distribution of P with respect to the σ-algebra $\mathcal{B}(\Lambda_1)$. Then

$$\rho_{\Lambda_2}\left[(P^{\Lambda_1}(\cdot|\bar{\sigma}))_{\Lambda_2}, P_{\Lambda_2}\right] \le G_1 \sum_{t\in\bar{\partial}_V(\Lambda_1)^c} \exp\{-g_1 \text{ dist } (t,\Lambda_2)\} \tag{3.7}$$

which follows from (3.1) if one takes $\Lambda=\Lambda_2$, $W=\mathcal{D}_n\setminus\Lambda_1$ and then puts $n\to\infty$. From (3.5), (3.6), (3.7) it follows that

$$\left|\int_\Omega \psi_2(\sigma) P^{\Lambda_1}(d\sigma|\bar{\sigma})\right| \le G_1 C_2 \sum_{t\in\bar{\partial}_V(\Lambda_1)^c} \exp\{-g_1 \text{ dist } (t,\Lambda_2)\} \tag{3.8}$$

But

$$\int_\Omega \psi_1(\sigma)\psi_2(\sigma)P(d\sigma) = \int_\Omega P(d\bar{\sigma})\psi_1(\bar{\sigma})\int_\Omega \psi_2(\sigma)P^{\Lambda_1}(d\sigma|\bar{\sigma}) \ .$$

Hence (3.8) implies (3.4).

Remarks.

3.1. In the case when ρ is given by (2.3) any bounded function ψ on Ω_Λ is Lipschitz with respect to ρ_Λ with Lipschitz constants $C = \sup_{\sigma\in\Omega_\Lambda} \psi(\sigma)$. So in this case the bound (3.4) is applicable to all bounded functions.

3.2. The bound (3.1) is more powerful than (3.4) (because the δ-function is not Lipschitz!).

3.3. One would like to improve the bound (3.4) substituting $\exp\{-g \text{ dist } (\Lambda_1,\Lambda_2)\}$ instead of the right-hand sum. However this "improved" bound does not hold in general. Really note that in the general case the measure $P_{\Lambda_1\cup\Lambda_2}$ is singular with respect to the product measure $P_{\Lambda_1}\times P_{\Lambda_2}$ if for example $\Lambda_i = \{t=(t^1,t^2)\in\mathbb{Z}^2 : t^1=a_i\}$, $i=1,2$ for any $a_1\ne a_2$. It is clear because in the general case the expectations of random variables of the type

$$\sum_{k=-n}^{n} \phi\left(x_{(a_1,k)}, x_{(a_2,k)}\right)$$

are different for both measures and the law of large numbers proves that these variables are almost constant for large n.

3.4. From (3.1) the unicity of a Q-consistent field follows.

3.5. Consider now the case of ρ given by (2.3). Then for the corresponding metric ρ_Λ ("Hamming distance") the following estimate is evident

$$\rho(\sigma^1_\Lambda,\sigma^2_\Lambda) \le \rho_\Lambda(\sigma^1_\Lambda,\sigma^2_\Lambda) \le |\Lambda|\rho(\sigma^1_\Lambda,\sigma^2_\Lambda)$$

where ρ is the metric on Ω_Λ, satisfying (2.3). Hence for any measures μ_1,μ_2 on Ω_Λ

$$\mathrm{Var}(\mu_1,\mu_2) \le R\rho_\Lambda(\mu_1,\mu_2) \le |\Lambda|\mathrm{Var}(\mu_1,\mu_2) \quad . \tag{3.9}$$

In our case the estimate (3.2) implies the "Very Strong Decay Property": for any $\bar\sigma^1,\bar\sigma^2 \in \Omega$

$$\mathrm{Var}(Q_{W,\Lambda}(\cdot|\bar\sigma^1),Q_{W,\Lambda}(\cdot|\bar\sigma^2)) \le G_1 \sum_{t\in\bar\partial_V W} \exp\{-g_1 \mathrm{dist}(t,\Lambda)\} \tag{3.10}$$

It is important to compare (3.10) with the following "The Strongest Decay Property": for some $G_2,g_2>0$ and any $\bar\sigma^1,\bar\sigma^2\in\Omega$ such that $\bar\sigma^1_s = \bar\sigma^2_s$ unless $s\neq t_0$, $t_0\in\partial W$

$$\mathrm{Var}(Q_{W,\Lambda}(\cdot|\bar\sigma^1),Q_{W,\Lambda}(\cdot|\bar\sigma^2)) \le G_2 \exp\{-g_2\, \mathrm{dist}(t_0,\Lambda)\} \tag{3.11}$$

The main difference between (3.10) and (3.11) is the following. In the case of (3.11) the conditional distributions in W corresponding to boundary conditions $\bar\sigma^1,\bar\sigma^2$ different only in a point t_0 differ essentially only in a vicinity of t_0 while in the case of (3.10) these distributions may differ essentially along the whole boundary of W. In [8] we show that (3.11) is equivalent to a natural analyticity condition on the partition function.

Theorem 3.3. Condition (3.11) implies condition C_V of Theorem 2.1 for some V.

Proof: Let $V = \mathcal{D}_n$ and $\Lambda_d = \{t\in\mathcal{D}_n : |t-t_0|\ge d\}$, $\tilde\Lambda_d = \mathcal{D}_n\setminus\Lambda_d$. Let $\tilde\mu\in K(Q_{\mathcal{D}_n,\Lambda_d}(\cdot|\bar\sigma^1),Q_{\mathcal{D}_n,\Lambda_d}(\cdot|\bar\sigma^2))$ be such that

$$\bar{R}\rho_{\Lambda_d}(\tilde{\mu}) \leq 2\, G_2 \exp\{-g_2 d\} \tag{3.12}$$

Consider the measure μ on $\Omega_{\mathcal{D}_n} \times \Omega_{\mathcal{D}_n}$ given by the following two properties

1) its projection on $\Lambda_d \times \Lambda_d$ is equal to $\tilde{\mu}$

2) its conditional distribution on $\tilde{\Lambda}_d \times \tilde{\Lambda}_d$ with respect to the σ-algebra $\bar{B}_{\lambda_d}$ (see the proof of Lemma 2.1) is for any $\tau^1, \tau^2 \in \Omega_{\Lambda_d}$ given by the product

$$Q_{\tilde{\Lambda}_d}\left(\cdot \mid \tau^1 \cup \bar{\sigma}^1_{(\tilde{\Lambda}_d)^c}\right) \times Q_{\tilde{\Lambda}_d}\left(\cdot \mid \tau^2 \cup \bar{\sigma}^2_{(\tilde{\Lambda}_d)^c}\right) .$$

Clearly $\mu \in K(Q_{\mathcal{D}_n}(\cdot \mid \bar{\sigma}^1), Q_{\mathcal{D}_n}(\cdot \mid \bar{\sigma}^2))$ and

$$R_{\rho_{\mathcal{D}_n}}(\mu) \leq |\tilde{\Lambda}_d| + 2G_2 \exp\{-g_2 d\} |\Lambda_d| .$$

It is clear that for some choice of $d=d(n)$ the right-hand side is $o(n)$ and so (2.10) is true for d large enough.

We shall discuss below (see Remark 4.2 and [13] for more details) some examples where (3.10) holds while (3.11) does not. In some of these cases condition C_V holds true for some V and in others they are violated for all V. Note that condition C_V for $|V|=1$ does imply (3.11) (see [1]).

4. Gibbs fields with finite single-spin space

In this and the following sections the space S is assumed to be finite endowed with the metric (2.3).

Let $\mathbb{A}_r$ be the set of all translation-invariant intersections U with radius r, $U = \{U_A(\sigma) = U_A(\sigma_A), A \subset \mathbb{Z}^\nu, |A| < \infty\}$, $U_A(\sigma) \equiv 0$ if $\operatorname{diam} A > r$. For $U \in \mathbb{A}_r$ let

$$\|U\| = \max_{\sigma \in \Omega, A \subset \mathbb{Z}^\nu} |U_A(\sigma)| . \tag{4.1}$$

To each interaction $U \in \mathbb{A}_r$ certain r-specification $Q^U = \{Q^U_V, V \subset \mathbb{Z}^\nu\}$ corresponds, which is given by

$$Q_V^U(\{\sigma\}|\bar{\sigma}) = \exp\{-H_V(\sigma|\bar{\sigma})\}|Z(V|\bar{\sigma}) \quad , \quad \sigma\in\Omega_V \tag{4.2}$$

where

$$Z(V|\bar{\sigma}) = \sum_{\sigma\in\Omega_V} \exp\{-H_V(\sigma|\bar{\sigma})\} \tag{4.3}$$

is called the partition function and

$$H_V(\sigma|\bar{\sigma}) = \sum_{A:A\cap V\neq\emptyset} \mathcal{U}_A(\sigma_{V\cap A} \cup \bar{\sigma}_{V^c\cap A}) . \tag{4.4}$$

Any Q^U -consistent field is called a Gibbs field with the inter U (or a U -Gibbs field). The results of Sections 2, 3 are certainly valid for them. By $A_V \subset A_r$ we denote the set of interactions for which condition C_V of Theorem 2.1 holds. Let $E_r = \bigcup_{V\subset\mathbb{Z}^\nu} A_V$. It is easy to see that E_r is an open subset of A_r.

Let $U\in A_r$ and $U_\beta = \beta U\in A_r$. The factor β is called the inverse temperature. Let $V\subset\mathbb{Z}^\nu$ $|V|<\infty$ and $G_V(\bar{\sigma})\subset$ $\subset\Omega_V$ be the set of "conditional ground state configurations", i.e. of those configurations $\sigma\in\Omega_V$ for which

$$H_V(\sigma|\bar{\sigma}) = \min_{\sigma'\in\Omega_V} H_V(\sigma'|\bar{\sigma}) . \tag{4.5}$$

The ground specification $\bar{Q}^U$ is defined by the formula:

$$\bar{Q}_V^U(\sigma|\bar{\sigma}) = \begin{cases} |G_V(\bar{\sigma})|^{-1} & \text{if } \sigma\in G_V(\bar{\sigma}) \\ 0 & \text{otherwise} \end{cases} \tag{4.6}$$

This is a self-consistent r -specification and

$$\lim_{\beta\to\infty} Q_V^{U_\beta}(\sigma|\bar{\sigma}) = \bar{Q}_V^U(\sigma|\bar{\sigma})$$

for all V , $\sigma\in\Omega_V$ (see [14] for more details).

Let $V\subset\mathbb{Z}^\nu$ be a finite volume and $t_0\in\partial V$. Let $M_{t_0}^U(V)\subset V$ be the set of all points $t\in V$ such that if $\sigma^1\in G_V(\bar{\sigma}^1)$, $\sigma^2\in G_V(\bar{\sigma}^2)$ and $\bar{\sigma}_s^1 = \bar{\sigma}_s^2$ unless $s=t_0$ then necessarily $\sigma_t^1 = \sigma_t^2$. In other words the conditional ground configuration does not depend on the condition at t_0 outside $M^U_{t_0}(V)$.

Theorem 4.1. Suppose that for some finite $V \subset \mathbb{Z}^\nu$

$$\sum_{t_0 \in \partial V} |V \setminus M^U_{t_0}(V)| < |V| . \tag{4.7}$$

Then condition C_V holds for the specification $Q^{U\beta}$, provided β is large enough. In particular the (βU)-Gibbs field is unique and the decay condition (3.10) holds for it.

Proof. Let $\mu_\beta \in K(Q^{U\beta}_V(\cdot|\bar\sigma^1), Q^{U\beta}_V(\cdot|\bar\sigma^2))$ where $\bar\sigma^1_s = \bar\sigma^2_s$ for $s = t_0$. Denoting $M^U_{t_0}(V)$ by M one has

$$\begin{aligned}\int_{\Omega_V \times \Omega_V} \rho_V(\sigma^1_V, \sigma^2_V)\mu_\beta(d\sigma^1_V, d\sigma^2_V) \le |V\setminus M| + \\ + \int_{\Omega_M \times \Omega_M} \rho_M(\sigma^1_M, \sigma^2_M)(\mu_\beta)_M(d\sigma^1_M, d\sigma^2_M)\end{aligned} \tag{4.8}$$

where $(\mu_\beta)_M$ is the projection of μ_β on $\Omega_M \times \Omega_M$. But $(\mu_\beta)_M$ is concentrated on the pairs of identical configurations as $\beta \to \infty$. Hence the right-hand side integral in (4.8) goes to zero as $\beta \to \infty$ and (4.7) implies condition C_V.

Remarks

4.1. Consider the following condition: for some $d > 0$, any finite $V \subset \mathbb{Z}^\nu$ and any $\bar\sigma^1, \bar\sigma^2 \in \Omega$, $\bar\sigma^1_s = \bar\sigma^2_s$ if $s \ne t$ and $\sigma^i \in G_V(\bar\sigma^i)$, $i=1,2$

$$\sigma^1_s = \sigma^2_s \quad \text{if} \quad \text{dist}(s,t) \ge d \tag{4.9}$$

Clearly (4.9) implies (4.7). Under condition (4.9) the unicity is proven in [12], though the condition itself was not clearly formulated there. This condition implies also (see again [12]) the decay property (3.11).

4.2. In [12] it was introduced the condition: for some $d > 0$, all finite $V \subset \mathbb{Z}^\nu$ and $\bar\sigma^1, \bar\sigma^2 \in \Omega$, $\sigma^1 \in G_V(\bar\sigma^1)$, $\sigma^2 \in G_V(\bar\sigma^2)$

$$\sigma^1_s = \sigma^2_s \quad \text{if} \quad \text{dist}(s, V^c) \ge d . \tag{4.10}$$

This condition is also implied by (4.9). The difference between (4.9) and (4.10) is the same as between (3.11) and (3.10). The change of the boundary condition in one point t_0 results in a change of the conditional ground configurations only in a vicinity of the point t_0 in the former case while it may lead to a change along all the boundary ∂V in the latter case.

The first example of an interaction satisfying (4.10), but not (4.9) was constructed by Navratil (Prague), in the context of criticizing the paper [12], where these conditions were confused. The corresponding class of models is studied in details in [13], where they are called "Czech models". In spite of the fact that (3.11) does not hold for them for large β (which erroneously was stated in [12]) the unicity statement for them holds true which is shown in [13] together with the weaker decay property (3.10). Condition (4.10) follows from the statement of unicity of the periodic ground configuration with Peierls (or GPS) condition (see [12]).

These Czech models provide examples mentioned in Remark 3.5. They have many other unusual features. For example, the uniqueness in the whole space $\mathbb{Z}^\nu$ for them can be accompanied by the non-uniqueness in the half-space $\mathbb{Z}_+^\nu = \{t \in \mathbb{Z}^\nu : t^1 \geq 0\}$. Another example of unusual behaviour is that the zeroes of the partition functions $Z(V|\bar{\sigma})$ tend to the real axis as $V \to \mathbb{Z}^\nu$ violating thus the natural analyticity assumptions (see [8]). Although, the limit free energies of the Czech models are complex analytic functions! As a result one is driven to the conclusion that the phase transitions of a special type take place in Czech models. In this sense they do not provide a counterexample to the hypothesis stated in the Introduction of this report.

5. Effectiveness

In this section we discuss the effectiveness property of the conditions of Theorem 2.1. We begin with some elementary facts of algorithm theory. It is generally agreed that the effectively calculable functions are the recursive ones (we are speaking about functions from $(\mathbb{Z}_+)^k$ to $\mathbb{Z}_+$ where $\mathbb{Z}_+ = \{1,2,\ldots\}$ and $k \in \mathbb{Z}_+$). A set $R \subset \mathbb{Z}_+$ is called enumerable, if it is the image of some recursive function. This means the existence of an algorithm which allows one to construct certain finite subsets of the set R , R_n in the n 'th step, in such a way that $R_n \nearrow R$ when $n \to \infty$. At the same time the complement $\mathbb{Z}_+ \setminus R$ is not necessarily enumerable, so, in general, there is no algorithm which allows one to check that $k \notin R$.

Now let us consider the set $A_r' \subset A_r$ of rational potentials: $U \in A_r'$ iff all the values $U_A(\sigma_A)$ are rational. The set A_r' is countable so it is possible to identify it with $\mathbb{Z}_+$. Let $E_r' = E_r \cap A_r'$.

Theorem 5.1. The set E_r' is enumerable.

We give here only a sketch of the proof. Note that the exponent can be approximated by its Taylor series which has rational coefficients. Hence, it is easy to see that one can define the conditional probability distributions $Q_V^{U,n}(\cdot|\cdot)$, $n \in \mathbb{Z}_+$ such that

$$|Q_V^{U,n}(B|\bar{\sigma}) - Q_V^U(B|\bar{\sigma})| \le (s(U,n))^{-1}, \ n \in \mathbb{Z}_+, \ U \in A_r' \tag{5.1}$$

where

1) for $V \subset \mathbb{Z}^\nu$, $B \in \mathcal{B}_V$, $\sigma \in \Omega_\partial$ fixed the function $Q_V^{U,n}(\beta|\bar{\sigma})$ takes on only rational values, and is recursive (the set of rationals was identified with $\mathbb{Z}_+$),

2) the function $s(U,n) \to \infty$ in a monotone way as $n \to \infty$ $s(U,n) \in \mathbb{Z}_+$ and is a recursive function of $(U,n) \in A_r' \times \mathbb{Z}_+$. (Warning! The functions $Q_V^{U,n}(\cdot|\cdot)$ need not to be consistent!)

Consider now the family of pairs $(V, \mu_{\bar{\sigma}^1,\bar{\sigma}^2})$ where $V \subset \mathbb{Z}^\nu$ is finite, $\bar{\sigma}^1, \bar{\sigma}^2 \in \Omega_{\partial V}$ is a pair of boundary conditions which differ only at a point $t \in \partial V$ and, finally, $\mu_{\bar{\sigma}^1,\bar{\sigma}^2}$ is a probability measure on $\Omega_V \times \Omega_V$ with rational values depending on $\bar{\sigma}^1, \bar{\sigma}^2$. Now let us define the function $\phi = \phi(U,n,(V,\mu_{\bar{\sigma}^1,\bar{\sigma}^2}))$ to be equal to U if

1) for all $\bar{\sigma}^1$, $\bar{\sigma}^2$ the measure $\mu_{\bar{\sigma}^1,\bar{\sigma}^2} \in K(Q_V^{U,n}(\cdot|\bar{\sigma}^1), Q_V^{U,n}(\cdot|\bar{\sigma}^2))$ and

2) $$\sum_{t\in\partial V} \max_{\bar{\sigma}^1,\bar{\sigma}^2:\bar{\sigma}^1_s=\bar{\sigma}^2_s, s\neq t} \bar{R}_{\rho_V}(\mu_{\bar{\sigma}^1,\bar{\sigma}^2}) \leq \frac{|V|}{|\partial V|}-\frac{|V|}{s(U,n)}; \quad (5.2)$$

otherwise ϕ is equal to some fixed $U_0' \in E_r'$. The function ϕ is defined on a countable set and after a natural identifying of its domain with $\mathbb{Z}_+$ it is recursive. It is easy to see also that its image is exactly E_r'.

The same idea can be stated otherwise. For each pair (U,n), $U\in A_r'$, $n \in \mathbb{Z}_+$ consider a ball $O(U,n)$ in A_r of of radius n^{-1} centered at U. The set of all these pairs can be identified with $\mathbb{Z}_+$. We call the subset $\tilde{A}\subset A_r$ effectively enumerable if for some enumerable set W of pairs (U,n)

$$\bigcup_{(U,n)\in W} O(U,n) = \tilde{A}. \quad (5.3)$$

Theorem 5.1. The set E_r (of Sect. 4) is effectively enumerable.

The proof is almost the same as above.

Of course, the algorithm described in Theorem 5.1 is not suitable for application: it was constructed exclusively to simplify the proof of the theorem. In real situations the key part of the problem is to look for a measure μ minimizing - or almost minimizing - the integral (2.2) for $\rho=\rho_V$. This is an usual linear programming problem. But usual algorithms are of no help here because the size of the problem increases exponentially fast with $|V|$, and also they do not use specific features of our problem. These are the following: it is intuitively clear that the main contribution to the optimal value of (2.2) comes from those points $s\in V$ which are in a vicinity of the point $t\in\partial V$ where the boundary conditions $\bar{\sigma}^1$, $\bar{\sigma}^2$ differ. Hence to construct an almost optimal measure one has to begin somehow with the construction of a μ which is good in this vicinity.

We finish our report by presenting such an algorithm

which results in the joint distribution sought, i.e. a measure μ on $\Omega_V \times \Omega_V$ which is close enough to optimal. We believe algorithms of such type to be of importance in attempts to prove the unicity by using computers.

The algorithm. Let a volume $V \subset \mathbf{Z}^\nu$ as well as a pair $\bar{\sigma}^1$, $\bar{\sigma}^2$ of boundary conditions be given with $\bar{\sigma}_s^1 = \bar{\sigma}_s^2$ for $s \neq t \in \partial V$. We construct a measure $\mu \in K(Q_V^U(\cdot|\bar{\sigma}^1), Q_V^U(\cdot|\bar{\sigma}^2))$.

We shall use the following notations:

$$F_j = \{s \in V : r(j-1) < \text{dist}(s,t) \leq rj\}$$
$$H_j = \bigcup_{k \leq j} F_k \tag{5.4}$$

The input data for the i-th step of our algorithm are the following:

1) a measure P_{i-1} on $\Omega_V \times \Omega_V$,

2) measures q_{i-1}^1, q_{i-1}^2 on $\Omega_{H_{j-1}}$,

with the following property: for any ξ supp $\in q_{i-1}^1$, $\zeta \in$ supp q_{i-1}^2 and any $j=1,\ldots,i-1$

$$\xi|_{F_j} \neq \zeta|_{F_j} . \tag{5.5}$$

(Warning: In general these measures are not probabilistic!)

The parameter i runs from 1 to $[\text{diam}\, V/r]+1$. As for the initial conditions

$$P_0 \equiv 0$$
$$q_0^1 = q_0^2 = 1$$

(the set $H_0 = \emptyset$ and $\Omega_\emptyset$ consists of one element).

We describe now the i-th step.

Let us consider the set of all pairs (ξ,ζ), $\xi \in$ $\in$ supp q_{i-1}^1, and let us number these pairs by an index $k=1,\ldots,k(i)$ such that if $k_1<k_2$ then $\rho_{H_{i-1}}(\xi^{k_1},\zeta^{k_1}) \leq \rho_{H_{i-1}}(\xi^{k_2},\zeta^{k_2})$. The i-th step is again an inductive procedure on k. The input data for the k-th substep are the following:

1) a measure $P_{i-1,k-1}$ on $\Omega_V \times \Omega_V$,

2) measures $q^1_{i-1,k-1}$, $q^2_{i-1,k-1}$ on Ω_{H_i} .

For $k=1$ the measure $P_{i-1,0} = P_{i-1}$ and

$$q^{\ell}_{i-1,0}(\xi \cup \tau) = q^{\ell}_{i-1}(\xi) Q^U_{V \setminus H_{i-1}, F_i}(\tau \mid \bar{\sigma}^{\ell} \cup \xi) \tag{5.6}$$

$\ell = 1,2$, $\xi \in \Omega_{H_{i-1}}$, $\tau \in \Omega_{F_i}$.

For $k = k(i)$ the measure $P_{i-1,k(i)} = P_i$

$$q^{\ell}_{i-1,k(i)} = q^{\ell}_i , \quad \ell = 1,2 . \tag{5.7}$$

To describe the transformation of the measure let us introduce for $\xi, \eta \in \Omega_{H_{i-1}}$, $\sigma \in \Omega_{V \setminus H_{i-1}}$ the quantity

$$\begin{aligned} &M_{i-1,k-1}(\xi,\zeta,\sigma) = \\ &= \min\left(q^1_{i-1,k-1}(\xi) Q^U_{V \setminus H_i}(\sigma \mid \bar{\sigma}^1_{V^c} \cup \xi), q^2_{i-1,k-1}(\zeta) Q^U_{V \setminus H_i}(\sigma \mid \bar{\sigma}^2_{V^c} \cup \zeta)\right) \end{aligned} \tag{5.8}$$

Now we define for $\sigma^1, \sigma^2 \in \Omega_V$, $\xi, \zeta \in \Omega_{H_{i-1}}$

$$P_{i-1}(\sigma^1,\sigma^2) = \begin{cases} P_{i-1,k-1}(\sigma^1,\sigma^2) + M_{i-1,k-1}(\xi_k,\zeta_k,\sigma) & \text{if} \\ \quad \sigma^1_{H_{i-1}} = \xi_k, \sigma^2_{H_{i-1}} = \zeta_k, \sigma^1_{V \setminus H_{i-1}} = \sigma^2_{V \setminus H_{i-1}} = \sigma \\ P_{i-1,k-1}(\sigma^1,\sigma^2) & \text{otherwise} \end{cases} \tag{5.9}$$

$$q^1_{i-1,k}(\xi) = \begin{cases} q^1_{i-1,k-1}(\xi_k) - \sum\limits_{\tau \in \Omega_{V \setminus H_{i-1}}} M_{i-1,k-1}(\xi_k,\zeta_k,\tau) \\ \qquad\qquad\qquad \text{if } \xi = \xi_k \\ q^1_{i-1,k-1}(\xi) \quad \text{otherwise} \end{cases} \tag{5.10}$$

$$q_{i-1,k}^2(\zeta) = \begin{cases} q_{i-1,k-1}^2(\zeta_k) - \sum_{\tau \in \Omega_{V \setminus H_{i-1}}} M_{i-1,k-1}(\xi_k,\zeta_k,\tau) & \text{if } \zeta = \zeta_k \\ q_{i-1,k-1}^2(\zeta) & \text{otherwise} \end{cases} \tag{5.11}$$

From the definition (5.9)-(5.11) it follows that the inductive condition (5.5) is reproduced.

The algorithm presented allows one to construct a sequence $\mu_i \in K(Q_V^U(\cdot|\bar{\sigma}^1), Q_V^U(\cdot|\bar{\sigma}^2))$ namely

$$\mu_i(\sigma^1,\sigma^2) = P_i(\sigma^1,\sigma^2) + q_{i,0}^1(\sigma^1) q_{i,0}^2(\sigma^2)$$

(see (5.6)). The quantity $\bar{R}_{\rho_V}(\mu_i)$ is estimated as follows:

$$\bar{R}_{\rho_V}(\mu_i) \le \rho_i = \bar{R}_{\rho_V}(P_i) + |V| \left(1 - \sum_{\sigma^1,\sigma^2 \in \Omega_V} P(\sigma^1,\sigma^2) \right) .$$

Hence if for some i $\rho_i < k_t$ where k_t, $t \in \partial V$ satisfy (2.8) one may terminate at the step i. Note, however, that we do not claim the convergence of μ_i to an optimal μ - in general it does not hold!

Finally the following a priori bound holds:

$$\rho_i \le \sum_{j \le i} (\mathrm{Var}_{F_j}) |H_j| + (\mathrm{Var}_{F_{i+1}}) |V| \tag{5.12}$$

where for $\Lambda \subset V$

$$\mathrm{Var}_\Lambda = \mathrm{Var}(Q_{V,\Lambda}^U(\cdot|\bar{\sigma}^1), Q_{V,\Lambda}^U(\cdot|\bar{\sigma}^2)) . \tag{5.13}$$

Indeed if $P_k(\sigma^1,\sigma^2) \ne 0$ then by (5.9) $\rho_V(\sigma^1,\sigma^2) \le |H_k|$. On the other hand for $\ell = 1,2$

$$\sum_{\sigma^1,\sigma^2 \in \Omega_V} P_i(\sigma^1,\sigma^2) = 1 - \sum_{\sigma \in \Omega H_i} q_i^\ell(\sigma) = 1 - \mathrm{Var}_{V \setminus H_i} .$$

At last

$$\mathrm{Var}_{V \setminus H_i} = \mathrm{Var}_{F_{i+1}}$$

because the r -Markov property of the measures $Q_V^U(\cdot|\bar{\sigma}^\ell)$

implies that the conditional distributions, induced by these measures, for the restriction $\sigma_{V\setminus H_{i+1}}$ of σ_V to $V\setminus H_{i+1}$ under the condition that σ_{H_i} is fixed, coincide for $\ell=1$ and $\ell=2$.

If condition (3.11) is true then $\rho_i < |V|/|\partial V|$ for large enough i and V and so the algorithm proves the property C_V .

References

[1] Dobrushin R. L. Prescribing a system of random variables by the help of conditional distributions. Theory Prob. and its Appl. 15, N 3, 469-497 (1970).

[2] Gross L. Decay of correlations in classical lattice models at high temperature. Comm. Math. Phys. 68, N 1, 9-27 (1979).

[3] Simon B. A remark on Dobrushin's uniqueness theorem. Comm. Math. Phys. 68, N 2, 183-185 (1979).

[4] Levin S. L. Application of Dobrushin's uniqueness theorem to N-vector model. Comm. Math. Phys. 78, N 1, 65-74 (1980).

[5] Gross L. Absence of second-order phase transitions in the Dobrushin uniqueness region. J. Stat. Phys. 25, N 1, 57-72 (1981).

[6] Künsch H. Decay of correlations under Dobrushin's uniqueness condition and its applications. Comm. Math. Phys. 84, N 2, 207-222 (1982).

[7] Klein P. Dobrushin's uniqueness theorem and the decay of correlations in continuum statistical mechanics. Comm. Math. Phys. 86, N 2, 227-246 (1982).

[8] Shlosman S. B., Dobrushin R. L. Complete analytical Gibbsian fields. This volume.

[9] Lieb, E. H. A refinement of Simon's correlation inequality. Comm. Math. Phys. 77, N 1, 127-136 (1980).

[10] Rachev S. T. Monge-Kantorovich problem of a mass shift and its application in the stochastic. Theory Prob. and its Appl. (in print).

[11] Dobrushin R. L. Gaussian random fields - Gibbsian point of view. In "Multicomponent random systems", Marcell Dekker, N. Y. - Basel, 119-151 (1980).

[12] Dobrushin R. L., Pecherski E. A. Uniqueness conditions for finitely dependent random fields. In "Random Fields", vol. 1, North-Holland, Amsterdam-Oxford-N.Y., 223-262 (1981).

[13] Dobrushin R. L., Shlosman S. B. The problem of translational invariance in statistical physics. Soviet Scientific Reviews, Ser. C. Vol. 5 (1984).

[14] Shlosman S. B. Meditation on Ĉźech models: uniqueness, half-space non-uniqueness and analyticity properties (in preparation).

COMPLETELY ANALYTICAL GIBBS FIELDS

R. L. Dobrushin and S. B. Shlosman

1. Introduction

In this report we describe the class of interactions $\mathbb{A}$ for which the corresponding Gibbs field is unique and possess every possible virtue one can imagine. We mean by this the well-known regularity properties of Gibbs fields in the high-temperature region. This class of interactions is defined axiomatically by ten (!) very natural properties. But the reader should not be confused by their great amount because all of them turn out to be equivalent! This fact alone shows that the class considered is natural. We call the potentials of the class completely analitical. The boundary of $\mathbb{A}$ corresponds to phase transition surface. Furthermore, the region $\mathbb{A}$ can be defined constructively by a certain algorithm. Here we use the word "constructive" in the same sense, as we used it in report [1], which is conceptually close to this one. For the sake of simplicity, we restrict ourselves to random fields on $\mathbb{Z}^{\nu}$, with finite single-spin space S and finite-range, translation-invariant interactions with finite values. The ideas of the present report and those of [1] can be applied in more complex cases; in particular, we study perturbations of Gaussian fields in [2]. Throughout this report we shall use the notations of [1], §1.

The Ten Conditions, which define our class of interactions $\mathbb{A}$ and which are equivalent, can be divided into three types. Conditions of the first type are those of analyticity. Namely, in one of them we ask for the following estimate to hold:

for $U \in A$, the partition function in any finite volume V, with any boundary condition $\bar{\sigma}$ satisfies the following bound:

$$|V|^{-1} |\ln Z(U+\tilde{U}|\bar{\sigma})| \leq C(U) \quad (1.1)$$

uniformly in V, $\bar{\sigma}$ and $\tilde{U}$, where the latter is any complex (not necessarily translation-invariant) interaction small enough.

Condition (1.1) is inspired by the well-known Lee-Yang method to locate the points of phase transition, which are interpreted as the limit points in the complex plane of the zeros of the partition functions, when the underlying volume V goes to infinity. From (1.1) the analyticity of the free energy in U follows. But condition (1.1) is even stronger than the usual requirement of analyticity of the free energy, because it excludes the cases when the latter is accompanied by accumulation of zeroes of the partition functions near the real axis (see an example in [3]).

Conditions of the second type are stated in terms of the variation distance, $\mathrm{Var}(\cdot,\cdot)$, between the projections $Q_{V,\Lambda}(\cdot|\bar{\sigma})$ of the conditional Gibbs distributions $Q_V(\cdot|\bar{\sigma})$ in the volume V into the subvolume $\Lambda \subset V$ for different boundary conditions $\bar{\sigma}$; one of them is the following:

for any finite $V \subset \mathbb{Z}^\nu$, $\bar{\sigma}^1, \bar{\sigma}^2 \in \Omega_{V^c}$, which differ only in one point $t \in V^c$,

$$\mathrm{Var}(Q_{V,\Lambda}(\cdot|\bar{\sigma}^1),\ Q_{V,\Lambda}(\cdot|\bar{\sigma}^2)) \leq K \exp\{-\alpha \mathrm{dist}(t,\Lambda)\}, \quad (1.2)$$

where $K = K(U) < \infty$, $\alpha = \alpha(U) > 0$.

Finally, conditions of the third type are stated in terms of decay properties of semi-invariants which are similar to the properties studied in the papers [4]-[8].

For completely analytical potentials the central limit theorem holds with various improvements, as well as many other properties of "very good" fields.

Two basic ideas are used in the proof of the equivalence of the Ten Conditions. To show that (1.1) follows from (1.2), we use induction on $|V|$. The same idea was used earlier by one of us ([9, 10]) in the one-dimensional case, but the general case can be treated as well. Since (1.2) for U implies (1.1) for small perturbations of U , we obtain that (1.1) is valid in high-temperature region (taking for U the non-interacting potential). The same is valid for the low-temperature region when the ground-state is unique, as well as for the small activity region at any temperature.

To derive (1.2) from (1.1) we use a modified version of the conform transformation method, which was developed earlier in [4]-[8]. The same modification is used for estimation of the semiinvariants and other aims.

These two methods enable us to obtain all natural properties of the Gibbs fields under consideration without invoking the cumbersome machinery of correlation equations or cluster expansion (cf. [11], [12]). The same techniques can be applied in more general cases (see [2]), where the usual approach requires cumbrous calculations.

2. The main result

Let $\mathcal{A} = \{A_i \subset \mathbb{Z}^\nu,\ i=1,\dots,k\}$ be a finite collection of finite subsets of $\mathbb{Z}^\nu$ and $\Delta = \Delta(\mathcal{A}) = \{A \subset \mathbb{Z} : A = A_i + t$ for some $i = i(A) = 1,\dots,k$, $t=t(A) \in \mathbb{Z}^\nu\}$. We denote by $\mathfrak{A}_\Delta$, $\mathfrak{A}^{\mathbb{C}}_\Delta$ the real and the complex Banach spaces of translation-invariant interactions with support in Δ . An interaction $U \in \mathfrak{A}_\Delta$ $(\mathfrak{A}^{\mathbb{C}}_\Delta)$ is thus a family $U = \{U_A(\sigma) \equiv U_A(\sigma_A),\ A \subset \mathbb{Z}^\nu,\ |A| < \infty,\ \sigma \in \Omega\}$, such that

$$U_A(\sigma) = U_{A+t}(\sigma_{+t}) \ , \text{ where } \ (\sigma_{+t})_s = \sigma_{s+t} \ , \tag{2.1}$$

$$U_A \equiv 0 \quad \text{unless} \quad A \in \Delta(\mathcal{A}) \ . \tag{2.2}$$

The norm of U is given by

$$\|U\| = \sup_{A,\sigma} |U_A(\sigma)| \ . \tag{2.3}$$

The radius of interaction U is the number

$$r = r(U) = \max_{A:U_A \neq 0} \text{diam}(A) .$$

If $\Delta = \{A \subset \mathbb{Z}^\nu : \text{diam } A \leq r\}$, then the corresponding spaces $\mathbb{A}_\Delta$, $\mathbb{A}^{\mathbb{C}}_\Delta$ will be also denoted by $\mathbb{A}_r$, $\mathbb{A}^{\mathbb{C}}_r$. Throughout the most part of this section the families A and Δ will be fixed and so the corresponding index will be omitted, and we shall speak about the spaces $\mathbb{A}$, $\mathbb{A}^{\mathbb{C}}$, with $r = r(\mathbb{A})$ being the maximum radius of interactions $U \in \mathbb{A}$.

For any set $\mathfrak{A} \subset \mathbb{A}$ we define its <u>main component</u> $M(\mathfrak{A})$ to be the maximal open connected subset of $\mathfrak{A}$, which contains the zero interaction $U^0 = \{U^0_A \equiv 0\}$.

We shall denote by $\mathfrak{A}_\alpha$ the set of interactions satisfying condition α - which is one of the Ten Conditions to be formulated below. (α runs through Ia, Ib, IIa, IIb, IIc, IIIa, IIIb, IIIc, IIId, IIIe.)

We present now the main results of our report

Theorem 1.1. The main components $M(\mathfrak{A}_\alpha)$ coincide. (α = Ia to IIIe.)

This common main component is called the set of completely analytical potentials.

Theorem 2.2. The classes of interactions A, B, C, D, E to be defined below, consist of completely analytical interactions.

We now start to formulate our Ten Conditions and five classes of interactions.

Let $\tilde{\mathbb{A}}^{\mathbb{C}}$ be the set of all complex interactions which satisfy (2.2), but not necessarily (2.1), with the norm (2.3). Of course, $\mathbb{A}^{\mathbb{C}} \subset \tilde{\mathbb{A}}^{\mathbb{C}}$. For any U $\tilde{\mathbb{A}}^{\mathbb{C}}$, $V \subset \mathbb{Z}^\nu$ finite and boundary condition $\bar{\sigma} \in \Omega$ let

$$Z_V(U|\bar{\sigma}) = \sum_{\sigma_V \in \Omega_V} \exp\{-H_V(\sigma_V|\bar{\sigma})\} , \tag{2.4}$$

where

$$H_V(\sigma_V|\bar{\sigma}) = \sum_{A:A \cap V \neq \emptyset} U_A(\sigma_V|\bar{\sigma}_{V^c}) . \tag{2.5}$$

Condition Ia. $U \in \mathfrak{A}_{Ia}$ if there exist $C < \infty$ and $\varepsilon > 0$, such that for any V, $\bar{\sigma}$ the function $Z_V(\tilde{U}|\bar{\sigma})$ is holomorphic and nonvanishing in the region

$$\mathcal{O}_\varepsilon(U) = \{\tilde{U} \in \tilde{A}^{\mathbb{C}} : \|\tilde{U}-U\| < \varepsilon\} , \tag{2.6}$$

and moreover,

$$|\ln Z_V(\tilde{U}|\bar{\sigma})| < C|V| \quad \text{for} \quad \tilde{U} \in \mathcal{O}_\varepsilon(U) . \tag{2.7}$$

Remark 1. The function $Z_V(\tilde{U}|\bar{\sigma})$ depends only on those $\tilde{U}_A(\sigma_A)$, for which $A \cap V \neq \emptyset$. So by being holomorphic we mean the usual property of functions of several complex variables.

Remark 2. $\ln Z_V(\tilde{U}|\bar{\sigma})$ is the uniquely defined holomorphic function, which coincides with the usual (real) logarithm for $\tilde{U}$ real. Its analytic continuation to $\mathcal{O}_\varepsilon(U)$ is possible and unique because the latter set is contractible and $Z|_{\mathcal{O}_\varepsilon(U)} \neq 0$.

Condition Ib. $U \in \mathcal{U}_{Ib}$ iff there exist two constants $C<\infty$, $\varepsilon>0$, such that for any V, $\bar{\sigma}$ the function $Z_V(\tilde{U}|\bar{\sigma})$ is holomorphic and nonvanishing in the region $\mathcal{O}_\varepsilon(U)$ and moreover

$$|\ln[Z_V(\tilde{U}|\bar{\sigma})/Z_V(U|\bar{\sigma})]| \leq c|V \cap \operatorname{supp}(\tilde{U}-U)| , \tag{2.8}$$

where for any $\Phi \in \tilde{A}^{\mathbb{C}}$

$$\operatorname{supp} \Phi = \bigcup_{A:\ \Phi_A \neq 0} A . \tag{2.9}$$

We recall now the definition of semiinvariants (see, e.g. [11], [12]). Let $\xi_1,\dots,\xi_m \in S$ be random variables with the joint probability distribution $q = q(x_1,\dots,x_m)$, $x_i \in S$. The semiinvariant of order $(k_1,\dots,k_m)$, where $k_i>0$ is the number

$$\langle \xi_1^{,k_1}, \xi_2^{,k_2},\dots,\xi_m^{,k_m}\rangle_q = \frac{\partial^{k_1+\dots+k_m}}{\partial^{k_1} z_1 \dots \partial^{k_m} z_m} \ln \phi(z_1,\dots,z_m)\Big|_{z_1=z_2=\dots=z_m=0} \tag{2.10}$$

where

$$\phi(z_1,\dots,z_m) = \sum_{x_1,\dots,x_m \in S} \exp\{z_1x_1+\dots+z_mx_m\} q(x_1,\dots,x_m) \tag{2.11}$$

is the generating function, and $z_i \in \mathbf{C}$.

Now let

$$Q_V^U(\sigma_V|\bar{\sigma}) = \exp\{-H_V(\sigma_V|\bar{\sigma})\}/Z_V(U|\bar{\sigma}) \tag{2.12}$$

be the conditional Gibbs distribution in V , and $\psi_1(\sigma_{A_1}),\ldots,\psi_m(\sigma_{A_m})$ be real functions, where $A_i \subset V$ are (not necessarily distinct) subsets, then

$$\begin{aligned}\langle \psi_1^{,k_1},\ldots,\psi_m^{,k_m}\rangle_{Q_V(\cdot\,\bar{\sigma})} &\equiv \langle \psi_1^{,k_1},\ldots,\psi_m^{,k_m}|U,V,\bar{\sigma}\rangle = \\ &= \frac{\partial^{k_1+\ldots+k_m}}{\partial^{k_1}z_1\ldots\partial^{k_m}z_m}\left(\ln Z_V(\tilde{U}(z_1,\ldots,z_m)|\bar{\sigma})\right)\Big|_{z_1=z_2=\ldots=z_m=0} ,\end{aligned} \tag{2.13}$$

where

$$(\tilde{U}(z_1,\ldots,z_m))_A = U_A + \sum_{i:A_i=A} z_i\psi_i . \tag{2.14}$$

Condition IIa. $U \in \mathfrak{A}_{IIa}$ iff for some constants $C>0$, $\varepsilon>0$ and for all V, $\bar{\sigma}$, $\psi_1,\ldots,\psi_m$, $k_1,\ldots,k_m$ with $|\psi_i|\le 1$ and $A_i \in \Delta$, the function $\langle \psi_1^{,k_1},\ldots,\psi_m^{,k_m}|\tilde{U},V,\bar{\sigma}\rangle$, defined by (2.13) for real $\tilde{U}$, can be extended to a holomorphic function on $\mathcal{O}_\varepsilon(U)$, with the following bound to hold:

$$|\langle \psi_1^{,k_1},\psi_2^{,k_2},\ldots,\psi_m^{,k_m}|\tilde{U},V,\bar{\sigma}\rangle| \le (k_1!\ldots k_m!)C^{k_1+\ldots+k_m} \tag{2.15}$$

$\tilde{U} \in \mathcal{O}_\varepsilon(U)$.

Condition IIb. $U \in \mathfrak{A}_{IIb}$ iff there exists a constant $C>0$, such that for all V, $\bar{\sigma}$, $\psi_1,\ldots,\psi_m$, $k_1,\ldots,k_m$ with $|\psi_i|\le 1$ and $A_i \in \Delta$

$$\begin{aligned}&|\langle\psi_1^{,k_1},\ldots,\psi_m^{,k_m}|U,V,\bar{\sigma}\rangle| \le \\ &\le (k_1!\ldots k_m!)C^{k_1+\ldots+k_m}\sum_{\Gamma\in G(A_1,\ldots,A_m)}\prod_{\gamma\in\varepsilon(\Gamma)}\phi(|\gamma|) ,\end{aligned} \tag{2.16}$$

where $G(A_1,\ldots,A_m)$ is the set of all trees Γ with m vertices identified with the sets $A_1,\ldots,A_m$, $E(\Gamma)$ is the set of all bounds $\gamma=(A_{i_\gamma},A_{j_\gamma})$ of Γ, $|\gamma|=\mathrm{dist}(A_{i_\gamma},A_{j_\gamma})$ and finally $\phi(d)>0$ is a decreasing function on $\mathbf{Z}_+$ with

$$\sum_{t\in\mathbf{Z}^\nu}\phi(|t|)|t|^{\nu-1}<\infty \tag{2.17}$$

Condition IIc. $U\in\mathfrak{A}_{IIc}$ iff for some constants $C>0$, $\alpha>0$ and for all V, $\bar\sigma$, $\psi_1,\ldots,\psi_m$, $k_1,\ldots,k_m$ with $|\psi_i|\le 1$ and $A_i\in\Delta$

$$|\langle\psi_1^{,k_1},\ldots,\psi_m^{,k_m}|U,V,\bar\sigma\rangle|\le(k_1!\ldots k_m!)C^{k_1+\ldots+k_m}\exp\{-\alpha d(A_1,\ldots,A_m)\}, \tag{2.18}$$

where $d(A_1,\ldots,A_m)=\min\limits_{B:\,B\cup(A_1\cup\ldots\cup A_m)\text{ is connected}}|B|$,

and connectedness is meant in the sense of the graph $\mathbf{Z}^\nu$ with edges joining nearest neighbours.

For $\Lambda\subset V$ we define

$$Q^U_{V,\Lambda}(B|\bar\sigma)=\sum_{\sigma_V\in\Omega_V:\sigma_\Lambda\in B}Q^U_V(\sigma_V|\bar\sigma)\quad\text{with}\quad B\subset\Omega_\Lambda\,. \tag{2.19}$$

Condition IIIa. $U\in\mathfrak{A}_{IIIa}$ iff for some constants $\delta<1$ and $\rho>0$ and for all finite $V\subset\mathbf{Z}^\nu$, $t\in\partial V$, $\bar\sigma^1,\bar\sigma^2\in\Omega$ with $\bar\sigma^1_s=\bar\sigma^2_s$ for $s\neq t$

$$\mathrm{Var}(Q^U_{V,B(t,\rho,V)}(\cdot|\bar\sigma^1),Q^U_{V,B(t,\rho,V)}(\cdot|\bar\sigma^2))<\delta|B(t,\rho,V)|^{-1}, \tag{2.20}$$

where

$$B(t,\rho,V)=\{s\in V:\rho<|s-t|\le p+r\},\quad r=r(U)\,. \tag{2.21}$$

Condition IIIb. $U\in\mathfrak{A}_{IIIb}$ iff for some decreasing function $\phi(d)$ with

$$\lim_{d\to\infty}\phi(d)d^{\nu-1}=0\,, \tag{2.22}$$

for the same V, t, $\bar\sigma^1$, $\bar\sigma^2$ as in IIIa, and for any $\Lambda\subset V$

$$\mathrm{Var}(Q^U_{V,\Lambda}(\cdot|\bar\sigma^1),Q^U_{V,\Lambda}(\cdot|\bar\sigma^2))\le\sum_{s\in\Lambda}\phi(|s-t|)\,. \tag{2.23}$$

Condition IIIc. $U \in \mathcal{A}_{IIIc}$ iff for some constants $K<\infty$ and $\gamma>0$ and for the same $V, \Lambda, t, \bar{\sigma}^1, \bar{\sigma}^2$ as in IIIb

$$\mathrm{Var}(Q^U_{V,\Lambda}(\cdot|\bar{\sigma}^1), Q^U_{V,\Lambda}(\cdot|\bar{\sigma}^2)) \le K \exp\{-\gamma \operatorname{dist}(t,\Lambda)\} \quad (2.24)$$

Condition IIId. $U \in \mathcal{A}_{IIId}$ iff for some $K<\infty$, $\gamma>0$, for the same $V, \Lambda, t, \bar{\sigma}^1, \bar{\sigma}^2$ as in IIIb and for all $\sigma_\Lambda \in \Omega_\Lambda$

$$\left| \frac{Q^U_{V,\Lambda}(\{\sigma_\Lambda\}|\bar{\sigma}^1)}{Q^U_{V,\Lambda}(\{\sigma_\Lambda\}|\bar{\sigma}^2)} - 1 \right| \le K \exp\{-\gamma \operatorname{dist}(t,\Lambda)\} . \quad (2.25)$$

Condition IIIe (CONSTRUCTIVE condition). $U \in \mathcal{A}_{IIIe}$ iff U satisfies the same conditions as in IIIc but only for volumes $V \subset \mathcal{D}_n$ (see [1] for the definition of $\mathcal{D}_n$), where

$$n = n(K, \gamma, \nu, r) \quad (2.26)$$

is some constant which can be EXPLICITELY estimated.

We define now the five classes of interactions, which were mentioned in Theorem 2.2.

Class A (High-temperature class). $U \in A$ iff for some $U_0 \in \mathbb{A}_0$, $\|U-U_0\| < \varepsilon(\Delta)$, for $U \in \mathbb{A} = \mathbb{A}_\Delta$ and $\varepsilon(\Delta)$ small enough.

Class B (Large magnetic field - or chemical potential - class). $U \in B$ iff U is of the form Φ_μ, where for some $\Phi \in \mathbb{A}$ and $s_0 \in S$

$$(\Phi_\mu)_A(\sigma_A) = \begin{cases} \Phi_{\{t\}}(s_0) - \mu & \text{if } A = \{t\},\ \sigma_t = s_0 \\ \Phi_A(\sigma_A) & \text{otherwise,} \end{cases} \quad (2.27)$$

provided $\mu \ge \bar{\mu}(\|\Phi\|, r(\mathbb{A}), \nu)$, where the latter is a function large enough.

To define class C we recall the $\tilde{B}(d)$-property, introduced in [13]. For $U \in \mathbb{A}_r$, $V \subset \mathbb{Z}^\nu$, $\bar{\sigma} \in \Omega$ denote by $G^U_V(\bar{\sigma})$ the set of all $\sigma \in \Omega_V$ such that

$$H_V(\sigma|\bar{\sigma}) = \min_{\tilde{\sigma} \in \Omega_V} H_V(\tilde{\sigma}|\bar{\sigma}) \quad (2.28)$$

The potential U is called $\tilde{B}(d)$ -potential, with $d>0$, if for any $V \subset \mathbf{Z}^{\nu}$, $t \in \partial V$, $\bar{\sigma}^1, \bar{\sigma}^2 \in \Omega$ with $\bar{\sigma}^1_s = \bar{\sigma}^2_s$ for $s \neq t$ and any $\sigma^i \in G^U_V(\bar{\sigma}^i)$, $i=1,2$

$$\sigma^1_u = \sigma^2_u \tag{2.29}$$

for all $u \in V$ such $\operatorname{dist}(u,t) \geq d$. Let

$$U_\beta = \beta U , \quad \beta > 0 . \tag{2.30}$$

Class C (Low-temperature $\tilde{B}(d)$ -case of a unique ground state). Suppose that a potential $\hat{U} \in \mathbb{A}$ as well as all the potentials U_μ defined by (2.27) have $\tilde{B}(d)$ -property for some $d>0$, $s_0 \in S$ and all μ, $0<\mu<\bar{\mu}(\|\hat{U}\|, r(\mathbb{A}), \nu)$.

$U \in C$ iff $U = \bar{U}_\beta$, defined by (2.30), provided $\beta \geq \bar{\beta}(\|U\|, r(\mathbb{A}), \nu)$ where the latter is a function large enough.

Class $\mathcal{D}$ (Ferromagnet in large external field). $U \in \mathcal{D}$ iff the space $S = \{-1,+1\}$, and

$$U_A(\sigma_A) = \begin{cases} -U_{s-t}\, \sigma_s \sigma_t & \text{for } A = \{s,t\} \\ -h\, \sigma_t & \text{for } A = \{t\} \\ 0 & \text{for other } A \end{cases} \tag{2.31}$$

where $U_{s-t} \geq 0$ and

$$|h| > \sum_{s \in \mathbf{Z}^{\nu} \setminus 0} U_s . \tag{2.32}$$

Class E (One-dimensional and almost one-dimensional interactions). $U \in E$ if $\|U - \Phi\| \leq \varepsilon((\|\Phi\|), r, \nu)$, for some longitudinal potential $\Phi \in \mathbb{A}$, and for some ε small enough. The interaction Φ is longitudinal if $\Phi_A \neq 0$ implies that for any two points $(t_1,\ldots,t_\nu),(s_1,\ldots,s_\nu) \in A$, $s_i = t_i$ for $i = 2,\ldots,\nu$ (see [13]).

Comments

2.1. We believe that all the families $\mathcal{O}_\alpha$ are connected, thus implying $M(\mathcal{O}_\alpha) = \mathcal{O}_\alpha$. However we have no proof of this hypothesis. We need the condition of being in the main component only when proving the implications IIa ⇒ IIc,

Ia ⇒ IIId.

2.2. If $\Delta \subset \Delta'$ then the inclusion $A_\Delta \subset A_{\Delta'}$ induces the inclusion of the corresponding sets of completely analytical potentials: $M(\mathfrak{A}_\alpha(A_\Delta)) \subset M(\mathfrak{A}_\alpha(A_{\Delta'}))$. This follows from the fact that the definition of the class $\mathfrak{A}_{IIIb}$ does not use the value $r(A)$. Our statement is, however, not so innocuous: it means in particular that if the interaction $U \in A_\Delta$ is completely analytic in the sense of Ia, i.e. all partition functions $Z_V(U+\tilde{U}|\bar{\sigma})$ are nonzero and (2.7) holds for the complex perturbations $\tilde{U} \in \tilde{A}_\Delta^{\mathbf{C}}$, then the same is true for perturbations from $\tilde{A}_\Delta^{\mathbf{C}}$, for all V and $\bar{\sigma}$ with any Δ' ! This is by no means à priori evident, and in this place the fact that the interaction U lies in the same component with the zero interaction is again important. Similarly conditions IIa, IIb, IIc can be extended to semiinvariants of functions $\psi_i(\sigma_{A_i})$ with arbitrary finite A_i but with constants C, ε depending on A_i .

2.3. Recall that by Gibbs fields with interaction U in a volume $V \subset \mathbf{Z}^\nu$ which is not necessarily finite), with boundary condition $\bar{\sigma} \in \Omega$ we mean any probability measure P on Ω_V , such that its conditional distribution in any finite $\Lambda \subset V$ subject to the condition $\sigma_{\Lambda^c \cap V} \in \Omega_{\Lambda^c \cap V}$ equals a.e. the Gibbs specification $Q_\Lambda(\cdot|\sigma_{\Lambda^c \cap V} \cup \bar{\sigma}_{V^c})$. Applying now condition IIIc with $V \cap \mathcal{D}_n$ instead of V , and taking the limit $n \to \infty$, we find immediately that the Gibbs state for completely analytical potentials is unique in any volume V .

2.4. The novel features of the analyticity condition Ia are two-fold: we consider all complex perturbations of the interaction U - and not only the translation-invariant ones, and we consider all boundary conditions $\bar{\sigma}$. These features are essential. Of course, using Van Hove's theorem and the compactness criterion for the analytic functions, one can prove that condition Ia implies the existence of the limit (in the sense of Van Hove)

$$-f(\tilde{U}) = \lim_{V \to \infty} |V|^{-1} \ln Z_V(\tilde{U}|\bar{\sigma}) \tag{2.33}$$

for all $\tilde{U} \in \mathcal{O}'_\varepsilon(U) = \mathcal{O}_\varepsilon(U) \cap \mathbb{A}^{\mathbf{C}}$, which is holomorphic on $\mathcal{O}'_\varepsilon(U)$. The main point here is that the converse implication does not hold - see [1], §3, and [3] for details. The potential constructed in [3] is not completely analytic, and the unicity of the Gibbs field in $\mathbb{Z}^\nu$ is accompanied with a non-unicity in the half-space $\mathbb{Z}^\nu_+ = \{t \in \mathbb{Z}^\nu : t^{(1)} \geq 0\}$ for a special boundary condition. So it is natural to think that the corresponding potential exhibits some sort of phase transition.

2.5. Condition Ib improves essentially condition Ia. For example, it implies the following bound. Let $\tilde{\mathbb{A}}^{\mathbf{C}}(U) = \{\tilde{U} \in \tilde{\mathbb{A}}^{\mathbf{C}} : |\mathrm{supp}(U-\tilde{U})| < \infty\}$. For any $V \subset \mathbb{Z}^\nu$, $\bar{\sigma} \in \Omega$, $U \in \mathbb{A}$ and $\tilde{U} \in \tilde{\mathbb{A}}^{\mathbf{C}}(U)$ let

$$\hat{Z}_V(\tilde{U}|\bar{\sigma}) = \int_{\Omega_V} \exp\left\{- \sum_{A: A \cap V \neq \emptyset} (\tilde{U}-U)_A(\sigma_V \cup \bar{\sigma}_{V^c})\right\} P^V_{\bar{\sigma}}(d\sigma_V) , \tag{2.34}$$

where $P^V_{\bar{\sigma}}$ is a Gibbs field in V with interaction U and boundary condition $\bar{\sigma}$. Then $\ln \hat{Z}_V(U|\bar{\sigma})$ is holomorphic in $\mathcal{O}_\varepsilon(U) \cap \tilde{\mathbb{A}}^{\mathbf{C}}(U)$ and

$$|\ln \hat{Z}_V(U|\bar{\sigma})| \leq C_1 |V \cap \mathrm{supp}(U-\tilde{U})| , \tag{2.35}$$

where C_1 does not depend on U, V and $\bar{\sigma}$. Indeed, for finite V (2.35) is the same as (2.8). In general case one has to take $V \cap \mathcal{D}_n$ instead of V in (2.8) and then tend with n to ∞. The bound (2.35) can be used to derive the properties of completely analytical interactions in infinite volumes.

2.6. The bounds on semiinvariants of the type IIa-IIc have occured earlier for specific classes of potentials in many papers (see [6]-[10]). For example, a statement of [8] is close to our IIa ⇔ IIc.

2.7. In the same way as (2.33) for the free energy, follows from IIa the existence of the limits

$$\langle \psi_1'^{k_1}, \psi_2'^{k_2}, \ldots, \psi_m'^{k_m} | \tilde{U} \rangle = \lim_{V \to \infty} \langle \psi_1'^{k_1}, \ldots, \psi_m'^{k_m} | \tilde{U}, V, \bar{\sigma} \rangle \tag{2.36}$$

for $\tilde{U} \in \mathcal{O}_\varepsilon(U)$. They are semiinvariants of the Gibbs field in $\mathbb{Z}^\nu$ for real $\tilde{U}$, and are holomorphic on $\mathcal{O}_\varepsilon(U)$. From our equivalence statement follows that the bound (2.18) holds for the limiting semiinvariants (2.36). We show below (Prop. 4.6) that the convergence in (2.36) is exponential. But the example already discussed in Comment 2.4 shows that the bounds on infinite-volume semiinvariants do not imply the complete analyticity.

2.8. Equivalences IIb ⇔ IIc and IIIb ⇔ IIIc illustrate the general principle that in the finite range case between the exponential- and power-law no intermediate correlation decay can occur (see [13]-[15]). Equivalence IIIb ⇔ IIIc with $\nu^2+\varepsilon$ instead of ν in (2.22), was obtained in [13]. Nevertheless, our result does not cover that of [13], because we are restricted to the main component. Condition IIIc was formulated already in [13], where it was derived from the unicity condition of [13] (see also [1], §3).

2.9. The main peculiarity of condition IIIe is its constructiveness. Using it, one can prove the complete analyticity of a given potential after a finite amount of calculations. Putting it more rigorously, the set of completely analytic interactions is effectively enumerable (see the discussion in [1], §4). However, the proof of equivalence of IIIe and all the other conditions is very involved, and we do not present it in this report. Even our estimate on the number $n(k,\gamma,\nu,r)$ is so awkward at present that we do not demonstrate it here. We hope to improve it in a forthcoming publication.

Let us compare condition IIIe with the unicity criterion of [1]. We showed in [1] that IIIb implies this unicity criterion. On the other hand, the example of [3] shows that the latter can hold together with a non-uniqueness in a half-space. Hence it is valid not only for completely analytical potentials - though these extra interactions are in a sense "exotic". It seems that now only criterion of [1] can be recommended for computer-assisted proofs.

2.10. Consider the random variable

$$S_\Lambda = \sum_{B:B+s\subset\Lambda} \phi(\{\sigma_{t+s},\ t\in B\}) , \tag{2.37}$$

where ϕ is some function on Ω_B , $\Lambda\subset V$, and the distribution of σ_V is the conditional Gibbs distribution $P_{\bar{\sigma}}^V$ in $V \subset \mathbb{Z}^\nu$ with boundary condition $\bar{\sigma}$. Then the characteristic function of S_Λ is

$$\langle e^{zS_\Lambda} \rangle = \hat{Z}_V(U + z\Phi|\bar{\sigma}) \tag{2.38}$$

where

$$\Phi_A(\sigma_A) = \begin{cases} \phi(\sigma_{A-s}) \ , \text{ if } \ A = B+s \subset \Lambda \\ 0 \ \text{otherwise.} \end{cases}$$

From (2.35) follows that its logarithm is holomorphic and bounded for $|z| \leq \varepsilon_0$, where ε_0 is independent of Λ. Hence the general result of Statulevičus [16] together with positive lower bounds on the variance of S_Λ in the limit $|\Lambda| \to \infty$ (see [17]) imply the central limit theorem for S_Λ with various improvements. The local limit theorem in our case follows from the global one (see [18]).

2.11. The interactions of class $\mathcal{A}$ correspond to high temperature, and those of class $\mathcal{B}$ to low density. These classes are highly explored by now, and the properties discussed above can be proven for them using correlation equations and cluster expansion. We think, however, that our approach is more straightforward and easier.

2.12. Class $\mathcal{C}$ corresponds to the case of unique ground state. Condition $\tilde{B}(d)$ is an additional essential restriction. Without it one can construct examples of not completely analytic interactions with unique ground state (see [1] and [3]). We require condition $\tilde{B}(d)$ not only for U itself, but also for U_μ. This is done only to ensure the inclusion of the low-temperature interaction βU in the main component. Conditions of the type III follow for βU even if only U itself is $\tilde{B}(d)$, but we do not know whether βU is contained always in the main component.

2.13. For the case of the ferromagnetic class $\mathcal{D}$, it is well known that the unicity holds for $h \neq 0$. But the known techniques for locating the zeroes of the partition function (beginning with Lee-Yang method), enables one to prove condition Ia only under the additional restriction (2.32).

It is worth mentioning that under this restriction the set $G_V^U(\bar{\sigma})$ (see (2.28)) always consisits of exactly one configuration: $\sigma_V \equiv \text{sign } h$.

The scheme of the proof of the theorems. We shall prove the following implications between the Ten Conditions:

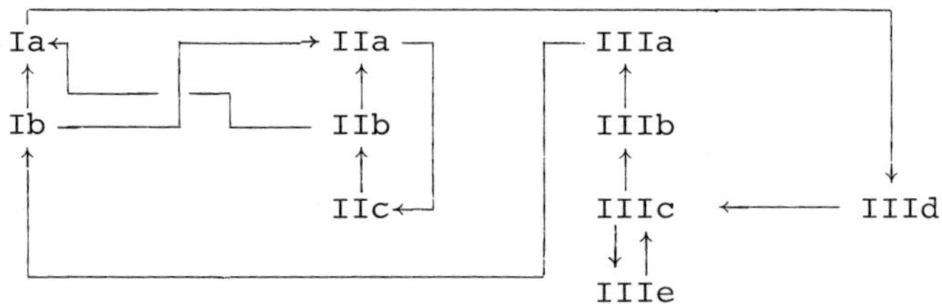

(2.39)

The arrows $X \to Y$ mean that condition Y is satisfied for the main component of the set of interactions with property X . The theorem follows from the fact that in the oriented graph (2.39) there is a circuit passing at least once through each vertex. The implications $\text{Ib} \to \text{Ia}$, $\text{Ib} \to \text{IIa}$, $\text{IIc} \to \text{IIb} \to \text{IIa}$, $\text{IIId} \to \underset{\substack{\downarrow \\ \text{IIIe}}}{\text{IIIc}} \to \text{IIIb} \to \text{IIIa}$ are almost evident. Therefore we make only minor remarks concerning them. To prove $\text{Ib} \to \text{Ia}$ it is useful to note that for U real $|\ln Z_V(U|\bar{\sigma})| \leq C|V|$ with C independent of V, $\bar{\sigma}$. Proving $\text{Ib} \to \text{IIa}$ one notes first that $|\text{supp}(\tilde{U}(z_1, \ldots, z_m) - U)| = |\bigcup_{i=1}^{m} A_i| \leq C_1 m$, where $C_1 = C_1(r, \nu)$, and then applies the Cauchy formula for the derivatives of holomorphic functions. To prove $\text{IIId} \to \text{IIIc}$ one needs to note that for any two probability distributions $Q_1(x)$, $Q_2(x)$ on a finite set X

$$\text{Var}(Q_1, Q_2) = \frac{1}{2} \sum_{x \in X} |Q_1(x) - Q_2(x)| = \frac{1}{2} \sum_{x \in X} Q(x) \left| 1 - \frac{Q_1(x)}{Q_2(x)} \right| \leq$$

$$\leq \frac{1}{2} \max_{x \in X} \left| 1 - \frac{Q_1(x)}{Q_2(x)} \right| .$$

The implication $\text{IIIa} \to \text{Ib}$ is the content of Proposition 3.1, which follows from Theorem 3.1. The proof of $\text{IIIe} \to \text{IIIc}$ follows from Theorem 3.2, the proof of which is, however, not easy, and will be published separately. Propositions 4.5 and 4.4 prove the implications IIa $\rightarrow$ IIc and Ia $\rightarrow$ IIId; they follow from Theorem 4.2. In Section 4 we also explain the implication IIb $\rightarrow$ Ia (Prop. 4.6).

One can prove Theorem 2.2 in many ways, because it is enough to check any one of the Ten Conditions.

The easiest of the conditions to check is condition IIIe. Indeed, one can easily show that it is true for the classes $\mathcal{A}$, $\mathcal{B}$, $\mathcal{C}$ (see the end of §3). We present also other ways to prove the theorem, especially because we do not present in our publication the proof of IIIe $\Rightarrow$ IIIc.

Note that for any interaction $U_0 \in \mathbf{A}_0$ condition IIIa is evidently fulfilled, because the corresponding field is a field of independent random variables. But IIIa $\rightarrow$ Ia, and Ia is valid in an open subset of $\mathbf{A}$, thus the complete analyticity for the class $\mathcal{A}$ is proven.

The specification for the classes $\mathcal{A}$, $\mathcal{B}$ are close to those for independent random variables (for class $\mathcal{B}$ those are the variables, taking the constant value s_0). Thus again condition IIIa holds for $\mathcal{A}$, $\mathcal{B}$. It follows for class $\mathcal{C}$ from the main result of [13]. For $\mathcal{E}$ it was also checked in [13].

On the other hand, one can check Ia straightforwardly. For class $\mathcal{D}$ it can be done using the techniques of the Asano-Ruelle-Slawny method (see [19], Prop. 4.2, 4.1). For class $\mathcal{E}$ it follows from known formulas for partition functions through transfer-matrices ([20]).

To see that all the five classes of ours are in the main component, one has to note that each of them is connected and $U^o \in \mathcal{A} \cap \mathcal{E}$, $\mathcal{B} \cap \mathcal{A} \neq \emptyset$, $\mathcal{C} \cap \mathcal{B} \neq \emptyset$, $\mathcal{D} \cap \mathcal{B} \neq \emptyset$.

3. Estimates on partition functions

In this section we present a new method for estimating partition functions. We shall consider an arbitrary self-consistent r-specification $Q = \{Q_V(\cdot|\bar{\sigma}), V \subset \mathbf{Z}^\nu, |V| < \infty, \bar{\sigma} \in \Omega\}$. For any function ϕ on Ω let

$$\langle\phi\rangle_{V,\bar{\sigma}} = \sum_{\sigma_V \in \Omega_V} \phi(\sigma_V \cup \bar{\sigma}_{V^c}) Q_V(\sigma_V|\bar{\sigma}) \quad . \tag{3.1}$$

For $\Lambda \subset V$ we denote by $Q_{V,\Lambda}(\cdot|\bar{\sigma})$ the projection of $Q_V(\cdot|\bar{\sigma})$ onto Λ ;

$$Q_{V,\Lambda}(\sigma_\Lambda|\bar{\sigma}) = Q_V(\{\sigma_V: (\sigma_V)_\Lambda = \sigma_\Lambda\}|\bar{\sigma}), \quad \sigma_\Lambda \in \Omega_\Lambda . \tag{3.2}$$

Finally, for any $\tilde{U} \in \mathbb{A}_r^{\mathbf{C}}$, finite $V \subset \mathbb{Z}^\nu$ and $\bar{\sigma} \in \Omega$ let (see (2.4))

$$Z_V(Q,\tilde{U}|\bar{\sigma}) = \langle \exp\{-\tilde{H}_V(\cdot|\bar{\sigma})\}\rangle_{V,\bar{\sigma}} . \tag{3.3}$$

Theorem 3.1. Let $\rho > 0$, $\delta < 1$ and $\varepsilon > 0$ be specified. Suppose that

1) For any finite $V \subset \mathbb{Z}^\nu$, $t \in V^c$ and $\bar{\sigma}^1, \bar{\sigma}^2 \in \Omega$ with $\bar{\sigma}_s^1 = \bar{\sigma}_s^2$ for $s \neq t$

$$\begin{aligned} \mathrm{Var}(Q_{V,B(t,\rho,V)}(\cdot|\bar{\sigma}^1), Q_{V,B(t,\rho,V)}(\cdot|\bar{\sigma}^2)) &\leq \\ &\leq \delta |B(t,\rho,V)|^{-1} , \end{aligned} \tag{3.4}$$

where $B(t,\rho,V)$ is defined by (2.18);

2) $\tilde{U} \in \mathcal{O}_\varepsilon$, where

$$\mathcal{O}_\varepsilon = \left\{ \tilde{U} \in \mathbb{A}_r^{\mathbf{C}} : \sup_{V \subset \mathbb{Z}^\nu,\, \sigma_V \in \Omega_V} |\tilde{U}_V(\sigma_V)| < \varepsilon \right\} . \tag{3.5}$$

There exists a function $\varepsilon_0(\delta)$, depending only on ν, r , such that conditions 1), 2) with $\varepsilon \leq \varepsilon_0(\delta)$ imply that the functions $\ln Z_V(Q,\tilde{U}|\bar{\sigma})$ are holomorphic in the variables $\{\tilde{U}_A(\sigma_A);\ \sigma_A \in \Omega_A,\ A \cap V \neq \emptyset\}$. Moreover, there exists a function $C(\varepsilon,\delta,\rho)$, depending only on ν, r , $C(\varepsilon,\delta,\rho) \to 0$ as $\varepsilon \to 0$, such that (cf. (2.9))

$$|\ln Z_V(Q,\tilde{U}|\bar{\sigma})| \leq C(\varepsilon,\delta,\rho)\,|V \cap \mathrm{supp}\,\tilde{U}| . \tag{3.6}$$

The proof is based on the following lemma.

Lemma 3.1. Let conditions 1), 2) of Theorem 3.1 hold, $V \subset \mathbb{Z}^\nu$ be finite, $t \in \partial V$ and

$$\tilde{V} = \{s \in V: |s-t| > \rho + r\} . \tag{3.7}$$

Then for all $\bar{\sigma} \in \Omega$ and $\varepsilon \leq \varepsilon_0(\delta,\rho)$, with some $\varepsilon_0(\delta,\rho)$, the partition functions $Z_V(Q,\tilde{U}|\bar{\sigma})$ and $Z_{\tilde{V}}(Q,\tilde{U}|\bar{\sigma})$ do not vanish. Furthermore:

1) for any $\bar{\sigma}, \tilde{\sigma} \in \Omega$, such that $\bar{\sigma}_s = \tilde{\sigma}_s$ for $s \in \partial V \cap \partial V$ (here and in the sequel $Z_\emptyset = 1$ by definition)

$$Z_V(Q,\tilde{U}|\bar{\sigma})/Z_{\tilde{V}}(Q,\tilde{U}|\tilde{\sigma}) = 1+\theta_1 C_1 \exp\{-\kappa \operatorname{dist}(t, V \cap \operatorname{supp} \tilde{U})\} , \quad (3.8)$$

2) for any $\bar{\sigma}^1, \bar{\sigma}^2 \in \Omega$, such that $\bar{\sigma}^1_s = \bar{\sigma}^2_s$ for $s \neq t$

$$Z_V(Q,\tilde{U}|\bar{\sigma}^1)/Z_V(Q,\tilde{U}|\bar{\sigma}^2) = 1+\theta_2 C_2 \exp\{-\kappa \operatorname{dist}(t, V \cap \operatorname{supp} \tilde{U})\} . \quad (3.9)$$

Here $\kappa = \kappa(\delta)$, $C_j = C_j(\varepsilon) \to 0$ with $\varepsilon \to 0$ are functions, depending only on ν, r, ρ, and $\theta_1 = \theta_1(\sigma, \tilde{\sigma})$, $\theta_2 = \theta_2(\bar{\sigma}^1, \bar{\sigma}^2)$ are such that $|\theta_j| \le 1$, $j=1,2$.

The proof of the lemma goes by an induction on $|V|$.

Supposing (3.8), (3.9) to be valid for all $W \subset \mathbb{Z}^\nu$ with $|W| \le |V|-1$, and some κ , C_j , which will be specified later, we first prove (3.8) for the volume V . Next, we prove (3.9) for V , assuming both (3.8), (3.9) are valid for smaller volumes, while (3.8) is valid for V itself. The initial step of induction is valid automatically, because $Z_\emptyset = 1$. The variables Q, $\tilde{U}$, t, ρ will be fixed throughout the proof and omitted in all notations. We also denote by

$$d = \operatorname{dist}(t, V \cap \operatorname{supp} \tilde{U}) , \quad \tilde{d} = \operatorname{dist}(V \cap \partial\tilde{V}, V \cap \operatorname{supp} \tilde{U}) . \quad (3.10)$$

We begin with (3.8). Let

$$\tilde{Z} = Z_{\tilde{V}}(\tilde{\sigma}) . \quad (3.11)$$

By induction, $\tilde{Z} \neq 0$. By the definition (3.1)

$$Z_V(\bar{\sigma}) = \langle \tilde{Z}(x_{\partial\tilde{V}\cap V})[1+\phi(x_{V\setminus\tilde{V}})] \rangle_{V,\bar{\sigma}} \quad (3.12)$$

where

$$\tilde{Z}(x_{\partial\tilde{V}\cap V}) = Z_{\tilde{V}}\left(x_{\partial\tilde{V}\cap V} \cup \bar{\sigma}_{(\partial\tilde{V}\cap V)^c} \right) \quad (3.13)$$

and

$$\phi(x_{V\setminus\tilde{V}}) = \exp\left\{ - \sum_{\substack{A: A\cap V \neq 0 \\ A \cap \tilde{V} = 0}} \tilde{U}_A\left(x_V \cup \bar{\sigma}_{V^c} \right) \right\} - 1 . \quad (3.14)$$

But the boundary conditions $\tilde{\sigma}$ and $x_{\partial\tilde{V}\setminus V} \cup \bar{\sigma}_{(\partial\tilde{V}\cap V)^c}$ differ only inside $\partial\tilde{V}\cap V\subseteq B$, hence, applying $|\partial\tilde{V}\cap V|$ times (3.9) to the volume $\tilde{V}$ which is valid by the induction hypothesis, we have

$$\tilde{Z}(x_{\partial\tilde{V}\cap V})/\tilde{Z} = (1+\tilde{\theta}C_2 e^{-\kappa\tilde{d}})^{|B|} , \tag{3.15}$$

where $\tilde{\theta} = \tilde{\theta}(x_{\partial\tilde{V}\cap V})$, $|\tilde{\theta}|\leq 1$. Now we have to consider two cases:

$$\text{I)}\quad d > \rho+r \tag{3.16}$$

$$\text{II)}\quad d \leq \rho+r \tag{3.17}$$

In the first case $\phi(x_{V\setminus\tilde{V}}) \equiv 0$, hence from (3.12), (3.15) follows that

$$Z_V(\bar{\sigma})/\tilde{Z} = (1+\hat{\theta}_1 C_2 \exp(-\kappa\tilde{d}))^{|B|} . \tag{3.18}$$

(Here and in the sequel quantities $\hat{\theta}_j$ satisfy $|\hat{\theta}_j|\leq 1$.) But $d \leq \tilde{d}+(\rho+r)$ and $|B| \leq |B(t,\rho,\mathbb{Z}^\nu)|$, hence for some $C_1 = C_1(C_2,\kappa,\rho,r,\nu)$ which goes to zero together with C_2 while κ, ρ, r, ν remain fixed, from (3.18) follows that

$$Z_V(\bar{\sigma})/\tilde{Z} = 1+\hat{\theta}_2 C_1 \exp\{-\kappa d\} . \tag{3.19}$$

As for the second case, from (3.5) and (3.14) follows that

$$|\phi(x_{V\setminus\tilde{V}})| \leq K\varepsilon \tag{3.20}$$

where $K = K(\rho,r,\nu)$. Hence from (3.12), (3.15) we deduce that

$$Z_V(\bar{\sigma})/\tilde{Z} = 1+\theta_3\bar{\bar{C}}_1 \tag{3.21}$$

where $\bar{\bar{C}}_1 = \bar{\bar{C}}_1(C_2,\varepsilon,\kappa,\rho,r,\nu)$ and goes to 0 if $C_2\to 0$, $\varepsilon\to 0$, while κ, ρ, r, ν are fixed. Defining

$$C_1 = \max(\bar{C}_1,\bar{\bar{C}}_1 e^{\kappa(\rho+r)}) , \tag{3.22}$$

we arrive to (3.8). The condition that $C_1\to 0$ together with ε is satisfied, provided the same holds for C_2 .

Now, let us prove (3.9). Evidently, one has

$$Z_V(\bar\sigma^1)/Z_V(\bar\sigma^2) = 1 + \frac{(Z_V(\bar\sigma^1)-Z_V(\bar\sigma^2)\tilde Z^{-1}}{Z_V(\bar\sigma^2)\tilde Z^{-1}} . \tag{3.23}$$

Denote by ϕ_j , $j=1,2$ the functions defined by (3.14) with $\bar\sigma^j$ instead of $\bar\sigma$. Using now (3.12), one can rewrite the numerator of (3.23):

$$\begin{aligned}(Z_V(\bar\sigma^1)-Z_V(\bar\sigma^2))\tilde Z^{-1} &= \\ &= \langle \tilde Z(x_{\partial\tilde V\cap V})\tilde Z^{-1}-1\rangle_{V,\bar\sigma^1} - \langle \tilde Z(x_{\partial\tilde V\cap V})\tilde Z^{-1}-1\rangle_{V,\bar\sigma^2} + \\ &\quad + \langle \tilde Z(x_{\partial\tilde V\cap V})\phi_1(x_{V\setminus\tilde V})\rangle_{V,\bar\sigma^1}\tilde Z^{-1} - \langle \tilde Z(x_{\partial\tilde V\cap V})\phi_2(x_{V\setminus\tilde V})\rangle_{V,\bar\sigma^2} Z^{-1} .\end{aligned} \tag{3.24}$$

Using the inclusion $\partial\tilde V\cap V\subset B$ and the estimate (3.15) (which follows from the induction estimate (3.9)), we have

$$\begin{aligned}&|\langle \tilde Z(x_{\partial\tilde V\cup V})\tilde Z^{-1}-1\rangle_{V,\bar\sigma^1} - \langle\tilde Z(x_{\partial\tilde V\cup V})\tilde Z^{-1}-1\rangle_{V,\bar\sigma^2}| \le \\ &\le \left[(1+C_2e^{-\kappa\tilde d})^{|B|}-1\right]\mathrm{Var}(Q_{V,B}(\cdot|\bar\sigma^1),Q_{V,B}(\cdot|\bar\sigma^2)) .\end{aligned} \tag{3.25}$$

Now let us take functions $\kappa(\delta)$, $\varepsilon_0(\delta)$ to be so small that for $C_2(\varepsilon)=\sqrt{\varepsilon}$ the following estimate would be true:

$$(1+C_2\, e^{-\kappa\tilde d})^{|B|}-1 \le |B|C_2\, e^{-\kappa d}\, \delta^{-1/2} ,$$

provided $d\le\tilde d+(\rho+r)$ and $\varepsilon\le\varepsilon_0(\delta)$. Applying then condition (3.4), one has

$$|\langle\tilde Z(x_{\partial\tilde V\cup V})\tilde Z^{-1}-1\rangle_{V,\bar\sigma^1} - \langle \tilde Z(x_{\partial\tilde V\cup V})\tilde Z^{-1}-1\rangle_{V,\bar\sigma^2}| \le C_2\, e^{-\kappa d}\,\delta^{1/2} . \tag{3.26}$$

Again, one has to consider cases I, II. In case I, $\phi_1\equiv\phi_2\equiv 0$, so the only nontrivial term in (3.24) is already estimated in (3.26). Using the already proven estimate (3.8) and taking $\varepsilon_0(\delta)$ to be small enough, we can estimate the denominator of (3.23) by

$$|Z_V(\bar\sigma^2)Z^{-1}| \ge \delta^{1/4} \tag{3.27}$$

hence (3.26) implies for case I that

$$\left|\frac{(Z_V(\bar\sigma^1)-Z_V(\bar\sigma^2))\tilde Z^{-1}}{Z_V(\bar\sigma^2)\tilde Z^{-1}}\right| \le C_2\, e^{-\kappa d} . \tag{3.28}$$

Thus in case I the estimate (3.9) is inductively reproduced. In case II, from (3.15) and (3.20) follows that for $j=1,2$

$$|\tilde Z^{-1}\langle \tilde Z(x_{\partial\tilde V\cap V})\phi_j(x_{V\setminus\tilde V})\rangle_{V,\bar\sigma^j}| \le K\varepsilon(1+C_2)^{|B|} . \tag{3.29}$$

From (3.24), (3.26), (3.29) and (3.27) follows that

$$\left|\frac{(Z_V(\bar\sigma^1)-Z_V(\bar\sigma^2))\tilde Z^{-1}}{Z_V(\bar\sigma^2)\tilde Z^{-1}}\right| \le C_2 e^{-\kappa d}\delta^{1/4}+2K\varepsilon(1+C_2)^{|B|}\delta^{-1/4} . \tag{3.30}$$

As δ is fixed, $d \le \rho+r$, and $\kappa(\delta)$ can be taken smaller than 1, it follows that for $\varepsilon_0(\delta)$ small enough the right-hand side of (3.30) for $\varepsilon < \varepsilon_0(\delta)$ is less than $C_2\, e^{-\kappa d}$ (because of our choice of C_2 to be equal to $\sqrt{\varepsilon}$). Thus condition (3.9) reproduces itself also for case II. The proof of the lemma is thus finished.

The proof of the theorem. Let $V_0=V$, $V_1,V_2,\dots,V_k=\emptyset$ be a sequence of volumes, such that $V_{i+1}=\tilde V_i$ for some $t=t_i \in \partial V_i$. Consider the boundary conditions $\bar\sigma^i \in \Omega$, which satisfy the relation

$$\bar\sigma^i_t = \begin{cases} \bar\sigma_t & \text{for } t\in\partial V_i\cap\partial V \\ \tau & \text{for } t\in\partial V_i\cap V , \end{cases}$$

where $\tau\in S$ is some fixed point. Then

$$\ln Z_V(Q,\tilde U|\bar\sigma) = \sum_{i=0}^{k-1} \ln \frac{Z_{V_i}(Q,\tilde U|\bar\sigma_i)}{Z_{V_{i+1}}(Q,\tilde U|\bar\sigma_{i+1})} . \tag{3.31}$$

From (3.8) follows that

$$\begin{aligned} |\ln Z_V(Q,\tilde U|\bar\sigma)| &\le 2C_1 \sum_{i=0}^{k-1} \exp\{-\kappa\, \mathrm{dist}\,(t_i, V_i\cap \mathrm{supp}\,\tilde U)\} \le \\ &\le 2C_1 \sum_{t\in V} \exp\{-\kappa\, \mathrm{dist}(t, V\cap\mathrm{supp}\,\tilde U)\} \le \\ &\le 2C_1\hat C|V\cap \mathrm{supp}\,\tilde U| , \end{aligned} \tag{3.32}$$

provided ε , and in turn C_1 , are small enough. Here $\hat{C} = \hat{C}(\kappa,\nu) < \infty$. From (3.32) follows (3.6).

Proposition 3.1. $U \in \mathfrak{A}_{IIIa}$ implies $U \in \mathfrak{A}_{Ib}$.

Proof. Note that for the Gibbs specification Q^U , given by (2.12), one has

$$Z_V(Q^U, \tilde{U}-U|\bar{\sigma}) = Z_V(\tilde{U}|\bar{\sigma})/Z_V(U|\bar{\sigma}) \ , \tag{3.33}$$

where $Z_V(\cdot|\cdot)$ is defined by (2.4). Now, condition I of the theorem follows from (2.20). Condition II is valid for the potential $\tilde{U}-U$, provided $\tilde{U} \in \mathfrak{A}_\varepsilon(U)$ (cf. (2.6) with ε small enough). Hence (2.8) follows from (3.6).

Definition 3.1. Let $M \subset \mathbb{Z}^\nu$ and $\phi(\rho) > 0$ and goes to zero as $\rho \to \infty$. A specification Q is called ϕ -mixing on M , if for any finite $V \subset M$, $t \in \partial V$, $\bar{\sigma}^1, \bar{\sigma}^2 \in \Omega$ such that $\bar{\sigma}^1_s = \bar{\sigma}^2_s$ for $s \neq t$, and any ρ ,

$$\mathrm{Var}(Q_{V,B(t,\rho,V)}(\cdot|\bar{\sigma}^1),\ Q_{V,B(t,\rho,V)}(\cdot|\bar{\sigma}^2)) \leq \phi(\rho) \ . \tag{3.34}$$

If $M = \mathbb{Z}^\nu$ the specification Q is called ϕ -mixing.

Note 3.1. If Q is ϕ -mixing and

$$\lim_{\rho\to\infty} \phi(\rho)\rho^{\nu-1} = 0 \ , \tag{3.35}$$

then condition I of Theorem 3.1 is valid with some ρ .

Theorem 3.2. For

$$\phi(\rho) = C \exp\{-\alpha\rho\} \ , \tag{3.36}$$

where $C>0$, $\alpha>0$, there exists a number $m(C,\alpha,r,\nu)$, such that if the r -specification Q is ϕ -mixing on any cube

$$\mathcal{D}(s,m) = \{t \in \mathbb{Z}^\nu : |t^i - s^i| \leq m,\ i=1,\dots,\nu\} \ , \ s \in \mathbb{Z}^\nu \tag{3.37}$$

for some $m \geq m(C,\alpha,r,\nu)$, then Q is ϕ -mixing.

A proof of Theorem 3.2 together with an estimate on $m(C,\alpha,r,\nu)$ will be published later. Note, however, that an important ingredient of the proof is the algorithm of construction of a certain joint distribution on $\Omega_V \times \Omega_V$, presented in §4 of [1].

Proposition 3.2. Let the translation-invariant specification Q be $C\exp\{-\alpha\rho\}$ -mixing, and the family Q^γ of translation-invarian specifications converge to Q in the sense that for all $V \subset \mathbb{Z}^{\nu}$, $\bar{\sigma} \in \Omega$, $B \subset \Omega_V$ $\lim_{\gamma \to 0} Q_V^{\gamma}(B|\bar{\sigma}) = Q_V(B|\bar{\sigma})$. Then the specification Q^{γ} satisfies condition I of Theorem 3.1, provided γ is small enough.

Proof. For γ small enough the specification Q^{γ} is $2C\exp\{-\alpha d\}$ -mixing on any $\mathcal{D}(s,m)$, with $m = m(2C,\alpha,r,\nu)$. Hence our statement follows from Theorem 3.2 and Note 3.1.

From the last proposition condition IIIa follows for the classes $\mathcal{A}$, $\mathcal{B}$, $\mathcal{C}$. Indeed, the corresponding specifications are close to $C\exp\{-\alpha d\}$ -mixing ones. For classes $\mathcal{A}$, $\mathcal{B}$ these are the specifications of independent random variables, while for class $\mathcal{C}$ it is the ground specification

$$\bar{Q}_V^U(\sigma|\bar{\sigma}) = \begin{cases} |G_V^U(\bar{\sigma})|^{-1} & \text{for } \sigma \in G_V^U(\bar{\sigma}) \\ 0 & \text{otherwise.} \end{cases}$$

In this case the mixing condition follows from $\tilde{B}(d)$.

4. Estimates on the functionals of partition functions

The variant of the conform mapping method we use in this section is based on the following simple property of analytic functions:

Proposition 4.1. Let $G \subset \mathbf{C}^1$ be a bounded, connected and simply-connected open region and $z_0 \in G$. For any point $z_1 \in G$, there exist positive constants $C = C(z_1,z_0,G)$, $\alpha = \alpha(z_1,z_0,G)$, such that for any function f, holomorphic in G, conditions

$$1) \quad |f(z)| \le M, \quad z \in G \tag{4.1}$$

and

$$2) \quad f(z_0) = f^1(z_0) = \dots = f^{(n-1)}(z_0) = 0 \tag{4.2}$$

with some $n>0$

imply that

$$|f(z_1)| \le M\,C\exp\{-\alpha n\}. \tag{4.3}$$

Proof. We begin with the case when $G = \bar{G}_1$ is the unit ball centered in $z_0=0$. Then

$$f(z) = z^n \phi(z) \tag{4.4}$$

where ϕ is again analytic in $\bar{G}_1$. From the maximum principle it follows that

$$|\phi(z)| \le M \tag{4.5}$$

(because $|z|_{\partial\bar{G}_1}| \equiv 1$). Hence from (4.4) follows that

$$|f(z)| \le M|z|^n , \tag{4.6}$$

which implies (4.3). For the general case, one uses first the well-known Riemann theorem to construct a conform map $\psi: \bar{G}_1 \to G$ with $\psi(0) = z_0$ and then applies the statement already proven to $f(\psi(z))$.

We shall use consequently the following theorem:

Theorem 4.1. Let $W \subset A_r$ be open and connected. Let also $U^0 \in W$ where U^0 is the zero interaction. Then there exist functions $\alpha(U) = \alpha(U,\varepsilon) > 0$, $C(U) = C(U,\varepsilon) < \infty$, $U \in W$, such that for any analytic function $g(\tilde{U})$, $\tilde{U} \in \mathcal{O}_\varepsilon(W) \subset \tilde{A}^{\mathbf{C}}$, conditions

$$1) \quad |g(\tilde{U})| \le M , \quad \tilde{U} \in \mathcal{O}_\varepsilon(W) \tag{4.7}$$

and

2) for $n>0$ all the derivatives

$$\left.\frac{\partial^m g(\tilde{U})}{\partial^{\ell_1}\tilde{U}_{B_1} \cdots \partial^{\ell_q}\tilde{U}_{B_q}}\right|_{\tilde{U}=U^0} \equiv 0 , \qquad \begin{array}{l} \text{if } \ell_1+\ldots+\ell_q \equiv m \le n , \\ B_1,\ldots,B_q \subset \mathbf{Z}^\nu \end{array} \tag{4.8}$$

imply the bound

$$|g(U)| \le M\,C(U) \exp\{-n\alpha(U)\} , \quad U \in W \tag{4.9}$$

to hold.

Remark. By the partial derivative (4.8) we mean the family

$$\frac{\partial^m g(\tilde{U})}{\partial^{\ell_1}\tilde{U}_{B_1} \cdots \partial^{\ell_q}\tilde{U}_{B_q}} =$$

$$= \left\{ \frac{\partial_g^m(\tilde{U})}{\partial^{\ell_{1,1}} \tilde{U}_{B_1}(\sigma^{1,1}) \partial^{\ell_{1,2}} \tilde{U}_{B_1}(\sigma^{1,2}) \dots \partial^{\ell_{q,u_q}} \tilde{U}_{B_q}(\sigma^{q,u_q})} , \right. \tag{4.10}$$

$$\left. \ell_{i,j} > 0,\ \ell_{i,1}+\dots+\ell_{i,u_i} = \ell_i,\ \sigma^{i,j} \in \Omega_{B_i},\ j=1,\dots,u_i,\ u_i>0,\ i=1,\dots,q \right\} .$$

Proof. As W is connected, for any $U \in W$ there exists a path $\Gamma(s)$, $0 \le s \le 1$, $\Gamma(0) = U^0$, $\tilde{\Gamma}(1) = U$. Using the Weierstrass theorem, we can assume without loss of generality that all the functions $U_A^s(\sigma_A)$ are polynomials in s , where $\Gamma(s) = U^s \in \mathbb{A}_r$. Let

$$f(s) = g(\Gamma(s)) . \tag{4.11}$$

This function can be continued analytically into the region

$$\{z \in \mathbb{C}: -\delta < \mathrm{Re}\, z < 1+\delta, |\mathrm{Im}\, z| < \delta\} , \tag{4.12}$$

where

$$\delta = \delta(U,\varepsilon) \tag{4.13}$$

is small enough. Now, the function f satisfies the conditions of Proposition 4.1 with $z_0 = 0$, $z_1 = 1$. Thus (4.9) follows from (4.3).

We shall apply Theorem 4.1 to functions of $\tilde{U}$, which depend on partition functions and semiinvariants. In order to simplify the notations, we identify the logarithm of the partition function with the semiinvariant:

$$\ln Z_V(\tilde{U}|\bar{\sigma}) = \langle \psi_1^{,k_1}, \dots, \psi_m^{,k_m} | \tilde{U}, V, \bar{\sigma} \rangle , \tag{4.14}$$

where $m=0$ and the families $\{A_i\}$, $\{\psi_i\}$ are empty.

To check condition (4.8) we shall use the following simple fact.

Definition 4.1. A family $\{C_i \subset \mathbb{Z}^\nu : |C_i| < \infty,\ i=1,\dots,s\}$ is called connected, if for any proper subset $I \subset \{1,\dots,s\}$

$$\left(\bigcup_{i \in I} C_i \right) \cap \left(\bigcup_{i \notin I} C_i \right) \ne \emptyset . \tag{4.15}$$

Proposition 4.2.

I.

$$\left.\frac{\partial^{\ell_1+\dots+\ell_q}\langle \psi_1^{,k_1},\dots,\psi_m^{,k_m}|\tilde{U},V,\bar{\sigma}\rangle}{\partial^{\ell_1}\tilde{U}_{B_1}\dots\partial^{\ell_q}\tilde{U}_{B_q}}\right|_{\tilde{U}=U^0} = 0 \tag{4.16}$$

unless the family $\{A_1,\dots,A_m, B_1,\dots,B_q\}$ is connected.

II. Let $V_j \subset \mathbf{Z}$ and $\bar{\sigma}^j \in \Omega_{\partial V^j}$, $j=1,2$ be such that for $\mathcal{D} = \left(\bigcup_{i=1}^{m} A_i\right) \cup \left(\bigcup_{i=1}^{q} B_i\right)$, $\mathcal{D} \cap V^1 = \mathcal{D} \cap V^2$, $\mathcal{D} \cap \partial V^1 = \mathcal{D} \cap \partial V^2$, and for $s \in \mathcal{D} \cap \partial V_1$ $\bar{\sigma}^1_s = \bar{\sigma}^2_s$. Then for any choice of the other parameters,

$$\left.\frac{\partial^{\ell_1+\dots+\ell_q}\langle\psi_1^{,k_1},\dots,\psi_m^{,k_m}|\tilde{U},V_1,\bar{\sigma}^1\rangle}{\partial^{\ell_1}\tilde{U}_{B_1}\dots\partial^{\ell_q}\tilde{U}_{B_q}}\right|_{\tilde{U}=U^0} =$$

$$= \left.\frac{\partial^{\ell_1+\dots+\ell_q}\langle\psi_1^{,k_1},\dots,\psi_m^{,k_m}|\tilde{U},V_2,\bar{\sigma}^2\rangle}{\partial^{\ell_1}\tilde{U}_{B_1}\dots\partial^{\ell_q}\tilde{U}_{B_q}}\right|_{\tilde{U}=U^0}. \tag{4.17}$$

Proof. The Gibbs field in V with interaction U^0 and any boundary condition $\bar{\sigma}$ is nothing else than a family $\{\xi_t, t\in V\}$ of independent random variables, uniformly distributed in S. The components of the derivative (4.10) are the semiinvariants of the form

$$\langle\psi_1^{,k_1},\dots,\psi_m^{,k_m},\chi_{1,1}^{,\ell_{1,1}},\chi_{1,2}^{,\ell_{1,2}},\dots,\chi_{s,u_s}^{,\ell_{s,u_s}}|U^0,V,\bar{\sigma}\rangle, \tag{4.18}$$

where

$$\chi_{i,j}(\sigma_V) = \begin{cases} 1 & \text{for } \sigma_{V\cap B_i} = \sigma^{i,j}_{V\cap B_i} \\ 0 & \text{otherwise,} \end{cases} \tag{4.19}$$

provided

$$\sigma^{i,j}_{\partial V\cap B_i} = \bar{\sigma}_{\partial V\cap B_i} \quad \text{for all} \quad j=1,\ldots,\ell_i\ , i=1,\ldots,s\ ; \tag{4.20}$$

otherwise the corresponding component is zero, because the partition function $Z_V(\tilde{U}|\bar{\sigma})$ does not depend on $\tilde{U}_{B_i}(\sigma^{i,j})$.

Now statement II follows immediately from (4.18)-(4.20). To prove I, note that if (4.15) is violated with $(C_1,\ldots,C_s)= =(A_1,\ldots,A_m,\ B_1,\ldots,B_q)$, the family $(\psi_1,\ldots,\psi_m,\ \chi_{1,1},\ldots \ldots,\chi_{s,u_s})$ can be divided into two groups, depending on independent sets of variables. Hence the corresponding semi-invariant vanishes. (As in this case the generating function (2.11) is a product of functions depending on disjoint subsets of variables z_i .)

Now we shall apply Theorem 4.1.

Proposition 4.3. $U\in M(\mathfrak{A}_{Ia})$ implies that $U\in\mathfrak{A}_{IIId}$.

Proof. Consider the function (see (2.25)):

$$g(U) = \ln \frac{Q^U_{V,\Lambda}(\{\sigma_\Lambda\}|\bar{\sigma}^1)}{Q^U_{V,\Lambda}(\{\sigma_\Lambda\}|\bar{\sigma}^2)} . \tag{4.21}$$

By definitions (2.12), (2.19)

$$\begin{aligned} g(U) &= \ln Z_V(U|\bar{\sigma}^2) - \ln Z_V(U|\bar{\sigma}^1) + \\ &+ \ln Z_{V\setminus\Lambda}(U|\sigma_\Lambda \cup \bar{\sigma}^1_{\Lambda^c}) - \ln Z_{V\setminus\Lambda}(U|\sigma_\Lambda \cup \bar{\sigma}^2_{\Lambda^c}) . \end{aligned} \tag{4.22}$$

From condition Ia follows that the function $g(U)$ can be continued analytically, and then Theorem 4.1 can be applied. Namely, we define $W=W_{C,\varepsilon}$ to be the main component of the potentials, which satisfy (2.7) with constants $C,\ 2\varepsilon$. Surely, the sets $W_{C,\varepsilon}$ exhaust all $\mathfrak{A}_{Ia}$. From (2.7) follows that

$$|g(\tilde{U})| \le 4C|V| \ ,\ \tilde{U}\in\mathcal{O}_\varepsilon(W) . \tag{4.23}$$

Now, the derivative of the type (4.8) can be nonvanishing only if the family $\{B_1,\ldots,B_q\}$ is connected, with $t\in B_i$ for some i and with $\Lambda\cap B_j\neq\emptyset$ for some j . Indeed, if connectedness does not hold, then the derivative vanishes according to Proposition 4.2, I. If, on the other hand, the family is connected, but $t \notin B_j$ for all j, or all the intersections $B_j \cap \Lambda$ are empty, then, according to statement II of the same proposition, we have the complete cancellation of equal terms with different signs. So, if the derivative is nonzero, then $q \geq q_0 = [r^{-1} \operatorname{dist}(t,\Lambda)]$. Hence condition II of Theorem 4.1 holds with $n = q_0 - 1$. From (4.9), (4.23) follows that

$$|g(U)| \leq c(U)|V| \exp\{-\gamma(U) \operatorname{dist}(t,\Lambda)\}. \qquad (4.24)$$

This is not, however, what we need, because of the extra factor $|V|$. But we can eliminate it by using r-Markov property of Gibbs fields. Note that from (4.21), (4.24) follows that for any $\bar{\sigma}^1, \bar{\sigma}^2 \in \Omega$

$$\left| \ln \frac{Q^U_{V,\Lambda}(\{\sigma_\Lambda\}|\bar{\sigma}^1)}{Q^U_{V,\Lambda}(\{\sigma_\Lambda\}|\bar{\sigma}^2)} \right| \leq C(U)|V|\,|\Delta(\bar{\sigma}^1,\bar{\sigma}^2)| \times \qquad (4.25)$$

$$\exp\{-\gamma(U) \operatorname{dist}(\Delta(\bar{\sigma}^1,\bar{\sigma}^2), \Lambda)\},$$

where $\Delta(\bar{\sigma}^1,\bar{\sigma}^2) = \{s \in \partial V: \bar{\sigma}^1_s \neq \bar{\sigma}^2_s\}$. Indeed, one has to consider any sequence $\bar{\sigma}_1, \bar{\sigma}_2, \ldots, \bar{\sigma}_{|\Delta(\bar{\sigma}^1,\bar{\sigma}^2)|}$, where the first term is $\bar{\sigma}^1$, the last one is $\bar{\sigma}^2$, and $|\Delta(\bar{\sigma}_i,\bar{\sigma}_{i+1})| = 1$, and then apply (4.21), (4.24) to each pair $\bar{\sigma}_i, \bar{\sigma}_{i+1}$. Now, considering again the initial problem, define

$$\tilde{V} = \{s \in V: \operatorname{dist}(s,\Lambda) \leq \operatorname{dist}(t,\Lambda)\}. \qquad (4.26)$$

From r-Markov property it follows that

$$Q^U_{V,\Lambda}(\{\sigma_\Lambda\}|\bar{\sigma}^i) = \sum_{\sigma_{V\setminus\tilde{V}} \in \Omega_{V\setminus\tilde{V}}} Q^U_{\tilde{V},\Lambda}\left(\{\sigma_\Lambda\}|\sigma_{V\setminus\tilde{V}} \cup \bar{\sigma}^j_{(V\setminus\tilde{V})^c}\right) \times \qquad (4.27)$$

$$\times\, Q^U_{V,V\setminus\tilde{V}}(\{\sigma_{V\setminus\tilde{V}}\}|\bar{\sigma}^j), \quad j=1,2.$$

Let us apply (4.25) with $\tilde{V}$ instead of V. For any $\sigma', \sigma'' \in \Omega_{V\setminus\tilde{V}}$

$$\left| \ln \frac{Q^{U}_{\tilde{V},\Lambda}(\{\sigma_\Lambda\}|\bar{\sigma}^{1}_{(V\setminus\tilde{V})^c} \cup \sigma')}{Q^{U}_{\tilde{V},\Lambda}(\{\sigma_\Lambda\}|\bar{\sigma}^{2}_{(V\setminus\tilde{V})^c} \cup \sigma'')} \right| \le C(U)\,|\tilde{V}|\,(|\partial\tilde{V}\cap V|+1) \times \tag{4.28}$$

$$\times \exp\{-\gamma(U)\operatorname{dist}(t,\Lambda)\} .$$

But $|\tilde{V}| \le C_1|\Lambda|(\operatorname{dist}(t,\Lambda))^{\nu}$, $|\partial\tilde{V}| \le C_2|\Lambda|(\operatorname{dist}(t,\Lambda))^{\nu-1}$, where $C_1 = C_1(\nu)$, $C_2 = C_2(\nu,r)$, hence (2.25) follows from (4.27), (4.28) for any $\gamma < \gamma(U)$ and some K , thus the proof is finished.

Let us turn now to semiinvariants.

Proposition 4.4. $U \in M(\mathfrak{A}_{IIa})$ implies that $U \in \mathfrak{A}_{IIc}$.

Proof. Let us apply Theorem 4.1 to the function

$$g(\tilde{U}) = \langle \psi_1^{,k_1}, \ldots, \psi_m^{,k_m} | \tilde{U}, V, \bar{\sigma} \rangle . \tag{4.29}$$

An à priori bound is (2.15):

$$M = C^{k_1+\ldots+k_m} k_1!\ldots k_m!$$

Statement II of Proposition 4.2 implies that the derivative (4.8) can be nonzero only when the family $\{A_1,\ldots,A_m, B_1,\ldots,B_q\}$ is connected. But the bound $\operatorname{diam} B_j \le r$, and the definition of the quantity $d(A_1,\ldots,A_m)$ (see IIc) result in the estimate

$$q \ge c\, d\,(A_1,\ldots,A_m)$$

for some $c = c(r,\nu)$, which proves IIc.

We mentioned in Comment 2.7 that the convergence of the semiinvariants is exponential when $V\to\infty$. This follows from the following estimate (which is, in fact, quite exact):

Proposition 4.5. Let U be a completely analytical interaction. Then for some $C(U) < \infty$, $\alpha = \alpha(U) > 0$, for all finite $V_1, V_2 \subset \mathbf{Z}^{\nu}$, $\bar{\sigma}^1, \bar{\sigma}^2 \in \Omega$, $\psi_1,\ldots,\psi_m$ with $|\psi_i| \le 1$ and $A_i \in V_1 \cap V_2$, $k_1,\ldots,k_m$ the bound

$$|\langle \psi_1^{,k_1},\dots,\psi_m^{,k_m}|U,V_1\bar\sigma^1\rangle - \langle \psi_1^{,k_1},\dots,\psi_m^{,k_m}|U,V_2,\bar\sigma^2\rangle| \le$$
$$\le C^{k_1+\dots+k_m}(k_1!\dots k_m!)\exp\{-\alpha(\mathrm{dist}(A_1\cup\dots\cup A_m,\Delta)+d(A_1,\dots,A_m))\} \tag{4.30}$$

holds with

$$\Delta = \Delta(V_1,\bar\sigma^1,V_2,\bar\sigma^2) = (V_1 \triangle V_2)\cup\{t\in \partial V_1\cap\partial V_2:\ \bar\sigma_t^1 \neq \bar\sigma_t^2\}\ . \tag{4.31}$$

Proof. We apply Theorem 4.1 to the function

$$g(\tilde U) = \langle \psi_1^{,k_1},\dots,\psi_m^{,k_m}|\tilde U,V_1,\bar\sigma^1 - \psi_1^{,k_1},\dots,\psi_m^{,k_m}|\tilde U,V_2,\bar\sigma^2\rangle \tag{4.32}$$

Condition IIa implies the à priori bound:

$$|g(\tilde U)| \le 2\,C^{k_1+\dots+k_m}(k_1!\dots k_m!)\ . \tag{4.33}$$

From statements I, II of Proposition 4.3 follows that the non-vanishing of the derivative implies that the family $\{A_1,\dots,A_m,\ B_1,\dots,B_q\}$ is connected with $B_i\cap\Delta\neq\emptyset$ for some i. Hence for some $c=c(\nu,r)$

$$q \ge c(d(A_1,\dots,A_m)+\mathrm{dist}(A_1\cup\dots\cup A_m,\ \Delta))\ ,$$

which proves (4.30).

The only remaining thing to prove is the following statement.

Proposition 4.6. $U\in\mathcal{U}_{IIb}$ implies that $U\in\mathcal{U}_{Ia}$.

Proof. From (2.13), (2.14) the following formal expansion follows:

$$\begin{aligned}\ln Z_V(\tilde U|\bar\sigma) &= \ln Z_V(U|\bar\sigma) + \\ &+\sum \left(\ell_{1,1}!\,\ell_{1,2}!\dots\ell_{s,u_s}!\right)^{-1}\langle \chi_{1,1}^{,\ell_{1,1}},\chi_{1,2}^{,\ell_{1,2}},\dots,\chi_{s,u_s}^{,\ell_{s,u_s}}|U,V,\bar\sigma\rangle\times \\ &\times\left(\tilde U_{B_1}(\sigma^{1,1})-U_{B_1}(\sigma^{1,1})\right)^{\ell_{1,1}}\left(\tilde U_{B_1}(\sigma^{1,2})-U_{B_1}(\sigma^{1,2})\right)^{\ell_{1,2}}\times\dots \\ &\dots\times\left(\tilde U_{B_s}(\sigma^{s,u_s})-U_{B_s}(\sigma^{s,u_s})\right)^{\ell_{s,u_s}}.\end{aligned} \tag{4.34}$$

Here the functions $\chi_{i,j}$ are defined by (4.19), and the summation goes over all $s>0$, all sequences $(B_1,\dots,B_s)$ with different elements, such that $B_i \cap V \neq \emptyset$, all integers $\ell_{1,1},\ell_{1,2},\dots,\ell_{s,u_s}$, such that $\ell_{i,1}+\ell_{i,2}+\dots+\ell_{i,u_i} > 0$, with $\sigma^{i,j}$ running over Ω_{B_i}, $j=1,\dots,u_i=|\Omega_{B_i}|$.

We are going to show that the series (4.34) is absolutely convergent and then to estimate its sum. For this purpose we shall use the bound (2.16) and we suppose that $\tilde{U}\in \mathcal{O}_\varepsilon(U)$. Then the sum is estimated by

$$\sum (C\varepsilon)^{\ell_{1,1}+\ell_{1,2}+\dots+\ell_{s,u_s}}\left(\sum_{\Gamma\in G(B_1,\dots,B_s)}\prod_{\gamma\in\Gamma}\phi(|\gamma|)\right), \quad (4.35)$$

where the range of summation is the same. But $u_i=|\Omega_{B_i}|\leq$ $\leq C(r,\nu)$, and performing the summation over all $\ell_{i,j}$, we bound the sum (4.35) by

$$\sum_s (C_1\varepsilon)^s \sum_{B_1,\dots,B_s:B_i\cap V\neq\emptyset}\left(\sum_{\Gamma\in G(B_1,\dots,B_s)}\prod_{\gamma\in\Gamma}\phi(|\gamma|)\right), \quad (4.36)$$

if ε is small enough. But the sum (4.36) is finite, because the sets B_i have to be different. So the series (4.34) defines a holomorphic function on $\mathcal{O}_\varepsilon(U)$, provided ε is small. The bound (2.7) follows now from (4.36), (2.17) and the following combinatorics:

Lemma 4.1. Let $\mathbb{G}$ be a finite set. Suppose that the function $\phi(g_1,g_2) > 0$ is such that for any $g_1\in\mathbb{G}$

$$\sum_{g_2\in\mathbb{G}}\phi(g_1,g_2)\leq K<\infty . \quad (4.37)$$

Let $G(g_0)$ be the set of all trees Γ with vertices from $\mathbb{G}$, such that the point $g_0\in\mathbb{G}$ is one of them. Let $|\Gamma|$ be the number of bounds in Γ. Then there exists a value $\varepsilon_0(K)$ such that if $\varepsilon<\varepsilon_0$, for any $g_0\in\mathbb{G}$

$$\sum_{\Gamma\in G(g_0)}\varepsilon^{|\Gamma|}\prod_{(g',g'')\in\Gamma}\phi(g',g'')\leq C(\varepsilon,K)<\infty . \quad (4.38)$$

Proof of the lemma. Let $R_{g_0}(\Gamma)$ be the maximal number of the links of a path on $\Gamma\in G(g_0)$ beginning in g_0. Let $G_R(g_0)=\{\Gamma\in G(g_0):\ R_{g_0}(\Gamma)\leq R\}$. Let

$$C_R = \max_{g\in\mathbb{G}} \sum_{\Gamma\in G_R(g)} \varepsilon^{|\Gamma|} \prod_{(g_1,g_2)\in\Gamma} \phi(g_1,g_2) . \tag{4.39}$$

The tree $\Gamma \in G_R(g_0)$ can be specified by the set $g_1,\ldots$ $\ldots,g_\ell \in \mathbb{G}\setminus\{g_0\}$ of vertices, which are nearest neighbours g_0 in the tree Γ, together with the trees $\Gamma_j \in G_{R-1}(g_j)$, $j=1,\ldots,\ell$ (which can be empty). Now,

$$\begin{aligned}
&\sum_{\Gamma\in G_R(g_0)} \varepsilon^{|\Gamma|} \prod_{(g',g'')\in\Gamma} \phi(g',g'') \le \\
&\le \sum_{\ell=1}^{\infty} (\varepsilon(1+C_{R-1}))^{\ell} \sum_{\{g_1,\ldots,g_\ell\}\subset\mathbb{G}\setminus\{g_0\}} \phi(g_0,g_1)\ldots\phi(g_0,g_\ell) = \\
&= \prod_{g\in\mathbb{G}\setminus\{g_0\}} (1+\varepsilon(1+C_{R-1})\phi(g_0,g))-1 \le \\
&\le \exp\left\{\varepsilon(1+C_{R-1}) \sum_{g\in\mathbb{G}\setminus\{g_0\}} \phi(g_0,g)\right\} -1 \le \exp\{\varepsilon K(1+C_{R-1})\}-1 .
\end{aligned} \tag{4.40}$$

As g_0 is arbitrary, we have

$$C_R \le \exp\{\varepsilon K(1+C_{R-1})\}-1 , \tag{4.41}$$

C_0 being equal to zero. Let y_a be the unique positive solution of the equation

$$y = e^{a(y+1)} - 1 , \tag{4.42}$$

which exists for $0<a<e^{-1}$. Supposing $C_{R-1} < y_{\varepsilon K}$ one has

$$C_R \le \exp\{\varepsilon K(1+C_{R-1})\}-1 \le \exp\{\varepsilon K(1+y_{\varepsilon K})\}-1 = y_{\varepsilon K} ,$$

hence $C_R \le y_{\varepsilon K}$ for all K. This means that the bound (4.38) holds with $\varepsilon_0(K) = (Ke)^{-1}$, $C(\varepsilon,K) = y_{\varepsilon K}$.

References

[1] Dobrushin R. L., Shlosman S. B. Constructive criterion for the uniqueness of Gibbs field, this volume.

[2] Dobrushin R. L. A new method of the study of the Gibbs perturbations of Gaussian fields (in preparation).

[3] Shlosman S. B. Mediation on Ĉzech models: uniqueness, half-space non-uniqueness and analyticity properties (in preparation).

[4] Lebowitz J. L., Penrose O. Analytic and clustering properties of thermondinamic functions and distribution functions for classical lattice and continuum systems. Comm. Math. Phys. 11, 99-124 (1968).

[5] Gallavotti G., Miracle-Sole S. On the cluster property above the critical temperature in lattice gase. Comm. Math. Phys. 12, 269-274 (1969).

[6] Duneau M., Jagolnitzer D., Souillard B. Decrease properties of truncated correlation functions and analyticity properties for classical lattice and continuous systems. Comm. Math. Phys. 31, 191-208 (1973).

[7] Duneau M., Souillard B., Jagolnitzer D. Analyticity and strong cluster properties for classical gases with finite range interactions. Comm. Math. Phys. 35, 307-320 (1974).

[8] Duneau M., Souillard B. Cluster properties of lattice and continuum systems. Comm. Math. Phys. 47, 155-166 (1976).

[9] Dobrushin R. L. Analyticity of correlation functions in one-dimensional classical systems with slowly decreasing potentials. Comm. Math. Phys. 23, 269-289 (1973).

[10] Dobrushin R. L. Analyticity of correlation functions in one-dimensional classical systems with power decay of the potential. Matem. Sbornik 94, No. 1, 16-48 (1974).

[11] Malyshev V. A. Cluster expansions in lattice models of statistical physics and quantum field theory. Uspehi Mat. Nauk 35, No. 2, 3-53 (1980).

[12] Malyshev V. A., Minlos K. A. Gibbsian random fields (the method of cluster expansion). Moscow, 1985 (in print).

[13] Dobrushin K. L., Pechersky E. A. Uniqueness conditions for finitely dependent random fields. In "Random Fields" Vol. 1, North-Holland, Amsterdam-Oxford-N-Y, 232-262 (1981).

[14] Simon B. Correlation inequalities and the decay of correlation in ferromagnets. Comm. Math. Phys. 77, 111-126 (1980).

[15] Lieb E. M. A refinement of Simon's correlation inequality. Comm. Math. Phys. 77, 127-136 (1980).

[16] Statulevičus V. A. On large deviations. Z. Wahrsch. verw. Geb. 6, 123-144 (1966).

[17] Dobrushin R. L., Nahapetian B. S. Strong convexity of the pressure for lattice systems of classical statistical physics. Teor. Mat. Fiz. 20, No. 2, 223-234 (1974).

[18] Dobrushin R. L., Tirozzi B. The central limit theorem and the problem of equivalence of ensembles. Comm. Math. Phys. 54, 173-192 (1977).

[19] Gruber S., Hintermann A., Merlini D. Analyticity and uniqueness of the invariant equilibrium states for general spin classical systems. Comm. Math. Phys. 40, 83-99 (1975).

[20] Van Hove L., Sur l'integrale de configuration pour les systems de perticules à une dimension. Physica 16, 137-143 (1950).

DIFFUSION, MOBILITY AND THE EINSTEIN RELATION

Pablo A. Ferrari , Sheldon Goldstein

and Joel L. Lebowitz

Abstract. We investigate the Einstein relation $\sigma = \beta D$ between the diffusion constant D and the "mobility" , of a "test particle" interacting with its environment : β^{-1} is the temperature of the system where D is measured and σE is the drift in a constant external field E . The relation is found to be satisfied for all model systems in which we can find a unique stationary non-equilibrium state of the environment, as seen from the test particle in the presence of the field. For some systems, e.g. infinite systems of hard rods in one dimension, we find non unique stationary states which do not satisfy the Einstein relation. For some models in a periodic box the Einstein relation is the most direct way of obtaining D . A precise macroscopic formulation of the Einstein relation which makes it mathematically very plausible is given.

I. Introduction.

We investigate stationary states of various classical model systems in which a charged "test particle" (tp) is subject to a constant external field E in the x-direction . The suitably defined drift or mean velocity of this tp , $u(E)$, is generally expected, for small fields to be proportional to the diffusion constant D of the tp at $E = 0$, when the system is in equilibrium. More precisely,

$$\sigma \equiv \lim_{E\to 0} \frac{u(E)}{E} = \beta D \tag{1.1}$$

where β the reciprocal of the temperature characterizing the equilibrium state of the system at $E = 0$. (We set Boltzmann's constant and the charge of the tp equal to unity. We use the usual physicist's normalization for D : for standard Brownian motion, $(dW)^2 = dt$, $D = 1/2$, and for the general one dimensional Brownian motion $\langle W_D(t)^2 \rangle = 2Dt)$.

Equation (1.1) is the first example of a class of general relations between linear transport coefficients and equilibrium fluctuations e.g. σ and D . It was derived by Einstein for Brownian particles in a fluid using physically intuitive quasi-equilibrium arguments [1]. Their most general formulation, as Einstein-Green-Kubo (EGK) relations, is usually derived via formal perburbation arguments around the equilibrium state, see below and refs.[2].

The validity of the EGK relations, or at least of some of their experimental consequences appear well established in many cases. A convincing mathematical derivation, and in some cases even a precise formulation, is however lacking at present [3] . The purpose of this presentation is to discuss the meaning and status of equation (1.1) for various model systems

Formulation of Problem.

We shall call a system for which the tp has differentiable spatial trajectories, i.e. in which the microscopic velocity $\underline{v}$, $\underline{v} \in \mathbb{R}^d$ of the tp is well defined a mechanical system. For such a system

$$u(E) = \langle v_x \rangle_E \ , \ \beta^{-1} = M \langle v_x^2 \rangle_o \tag{1.2}$$

where M is the mass and v_x the x-component of $\underline{v}$. The subscripts E and 0 refer to expectation values in the appropriate stationary measures with and without the field. The microscopic action of the field in such a mechanical system is an acceleration $M^{-1}E$ of the test particle in the x-direction - while leaving all other interactions unchanged. We shall later discuss different models of such mechanical systems. Note that according to our terminology a particle whose velocity undergoes an Ornstein-Uhlenbeck process is a mechanical system. A system which evolves in a deterministic manner according to Newton's equations of motion will be called Newtonian.

In addition to mechanical models we shall also consider systems where the microscopic dynamics is modeled by Brownian motion, a continuous spatial process whose velocity is not well defined. In these models the electric field acts by adding a drift term proportional to $\dot{E}dt$ to the displacement of the tp . The definition of β appearing in (1.1) is now relatcd to the behavior of the tp in equilibrium with an external "confining potential" . A similar situation occurs for models in which the position of the tp takes values on a lattice, $X \in \mathbb{Z}^d$. Here the electric field acts to "suitably" bias jumps in the x-direction .

In general, for mechanical and non-mechanical systems in which the tp is free to move in all of $\mathbb{R}^d$ or $\mathbb{Z}^d$, there will be no stationary probability measure completely describing the entire systems, environment plus tp , since the position of the latter will not be localized in the "steady state". However, if we ignore the position of the tp and consider only the remaining coordinates, describing the environment relative to the tp (including the velocity of the tp in the case of mechanical systems), this problem disappears. It is to stationary probability measures for this relative description - the environment seen from the test particle- to which $\langle\ \rangle_o$ and $\langle\ \rangle_E$ refer.

For mechanical systems it is natural to define $u(E)$ as $\langle v_x \rangle_E$, but this makes no sense for non-mechanical systems. However, in all cases $X(t)$, (the x-coordinate of) the position of the tp at time t $(X(0)=0)$ can be naturally defined as a random variable on the path space for the evolution of the environment seen from the tp , equipped with the invariant path measure P_E arising from $\langle\ \rangle_E$. We then have that $\langle X(t) \rangle_E / t$ is independent of t and we define $u(E)$ by (we abuse notation and write $\langle\ \rangle_E$ for the expectation with respect to P_E) ,

$$u(E) \equiv \langle X(1) \rangle_E \tag{1.3}$$

This definition agrees with the previous one in the case of mechanical systems, since in this case $X(t) = \int_0^t v(s)ds$. We also note that if the process describing the evolution of the environment seen by the tp starting from the state $\langle\ \rangle_E$ is ergodic, then

$$\lim_{t\to\infty} \frac{X(t)}{t} = u(E) \tag{1.4}$$

a.s. with respect to P_E .

Observe also that while for mechanical systems the diffusion constant D is given by $\int_0^\infty < v_x(0)\, v_x(t) >_o dt$, an expression which makes no sense for non-mechanical systems, in all cases (under consideration) we have

$$D = \lim_{t\to\infty} (2t)^{-1} < X(t)^2 >_o \tag{1.5}$$

Note that equation (1.1) implies in particular that σ is zero whenever D vanishes. This is trivial in the case where the tp is confined to a compact spatial region by an external potential, e.g. by walls. It is more interesting in the one dimensional system of Brownian particles with hard cores or lattice gases with jumps restricted to nearest neighbor empty sites. In these cases it is known [4] that $< X^2(t) > \sim t^{1/2}$ so that D is zero. We show indeed that in these cases $u(E) = 0$: In fact the infinite volume stationary measures $< >_E$ for $E \neq 0$ turn out to be limits of finite volume $E \neq 0$, stationary states containing N particles with either periodic or rigid wall boundary conditions. In the latter case the finite volume states are generalized Gibbsian for which of course $u(E) = 0$, while in the periodic case we verify (1.1) for finite N , with $u_N(E)$, $D_N \sim N^{-1}$.

The situation is less clear for interacting Newtonian particles with hard cores. The diffusion constant for a tp , having the same mass M as the other "fluid" particles, is non-zero [5] . However, just as for Brownian particles, we explicitly find stationary measures as limits of finite volume (generalized) Gibbs states. These states, which are spatially identical to those found for Brownian particles, have Maxwellian velocities, so that $u(E) = 0$. (They are however no longer limits of stationary states with periodic boundary conditions). Moreover, by Galilean invariance, these states are imbedded in a family, members of which can be found assigning $u(E)$ any value whatsoever. Presumably, none of these states have anything to do with the Einstein relation, not even the one(s) for which it is satisfied. We expect, but do not prove, that there are also other stationary states, arising in the limit $t \to \infty$ (under the evolution with $E \neq 0$) from the initial state $< >_o$, for which we expect that the Einstein relation is satisfied. Indeed, in cases where there is more than one stationary state $< >_E$ (for given β and density), it is only for those arising in the above way that we expect the Einstein relation to be satisfied and hence we believe that $u(E)$

should be defined in these states. In fact this is essentially what the usual derivation of the Kubo formulas (EGK relations), via perturbation of the equilibrium state in effect does [2]. We present it here as heuristics.

Perturbation Argument.

Let the time evolution of the appropriate probability measure be given by the forward generator of the process $L_o + EL_1$

$$\frac{\partial\mu}{\partial t} = (L_o + EL_1)\mu \tag{1.6}$$

where L_o and L_1 are independent of E. We consider now the state at time $t > 0$ when $\mu(0) = \mu_o$, the equilibrium state, when $E = 0$; $L_o \mu_o = 0$. Then by the standard Dyson formula

$$\mu(t) = \mu_o + E \int_0^t \exp[(t-t')L_o]L_1\mu(t')dt' \tag{1.7}$$

Let now $U = U_o + U_1 E$ be the function whose expectation value in the correct stationary state is $u(E)$. (For a mechanical system $U_o = v_x$, $U_1 = 0$; for a diffusion system U_1 is a constant). We then write formally, assuming that the average of U converges as $t \to \infty$ to the correct $u(E)$ and that the limit (1.1) exists and is given by letting $t \to \infty$ after expanding $\langle U \rangle$ in E,

$$\sigma = \langle U_1 \rangle_o + \int_0^\infty \langle U_o^{(0)}(t')A \rangle_o \, dt' \tag{1.8}$$

where

$$A = \mu_o^{-1} L_1 \mu_o = L_1 \log \mu_o \quad . \tag{1.9}$$

The subsript $\langle\ \rangle_o$ means that the average is with respect to μ_o and the superscript $U_o^{(o)}(t')$ indicates that the time evolution is taken with the generator L_o. For a mechanical system (1.8) is simply

$$\sigma = \beta \int_0^\infty \langle v_x(t')v_x \rangle_o \, dt' = \beta D \tag{1.10}$$

A similar formula [6] is obtained for a diffusive system with the appropriate choice of U_o.

We have avoided making precise statement about the limits $t \to \infty$, $E \to 0$ leading to (1.8). This is deliberate; we have no rigorous (or even very convincing) arguments about the validity of (1.8) for general systems [3]. It is precisely this lack of knowledge which led us to the present work. We believe that in addition to the specific models considered explicitly here the formulation in the last section has some promise of leading to rigorous results for general systems. In particular that formulation makes Einstein's original argument precise and very convincing.

We proceed to give some examples of models with "static" environments for which the Einstein relation holds. In later sections we will investigate models with "dynamic" environments in which the tp is one of the particles in an interacting system and give a new formulation of the Einstein relation.

II. Static Environments.

a) Markovian Mechanical Models.

In this widely used class of models the velocity of the tp undergoes either i) a Markov jump process specified by a transition rate $K(v,v')dv$ or ii) an Ornstein-Uhlenbeck process. (We take here $v \in \mathbb{R}$ since extra dimensions do not introduce any essential new elements.) Setting $x(0) = 0$ the position at time t is given by $x(t) = \int_0^t v(t')dt'$. The position and the velocity distribution of the tp satisfies a "linear Boltzman equation"

$$\frac{\partial f(x,v,t)}{\partial t} + v\frac{\partial f}{\partial x} + \frac{1}{M}\left[-\frac{\partial U(x)}{\partial x} + E\right]\frac{\partial f}{\partial v} = (Kf)(x,v,t) \quad (2.1)$$

We have included here also a force term coming from an external potential $U(x)$ (which we generally take to be zero) in order to clarify the role of E . The operator K on the right side of (2.1) represents the effect of the environment : it is the forward generator of the Markov velocity process and is independent of U, E and x . K is assumed to have a non-degenerate eigenvalue zero, corresponding to the Maxwellian velocity distribution $h_\beta(v)$

$$Kh_\beta = 0 \tag{2.2}$$

$$h_\beta(v) = (2\pi/\beta M)^{-1/2} \exp[-(1/2)\beta Mv^2] \quad . \tag{2.3}$$

The stationary solution of (2.1) for $E = 0$ is the equilibrium distribution

$$f_o(x,v) = C \exp[-\beta U(x)]h_\beta(v) \; . \tag{2.4}$$

C is a normalization constant whose meaning is clear when $\exp[-\beta U(x)]$ is integrable - or the particle is in a rigid box. When $U(x) = 0$ or is bounded periodic with period L it is simplest to interpret (2.4) as holding for x in $\mathbb{R}/L$, i.e. we can think of the tp defined in the periodic box of length L . Alternatively f_o is the stationary Poisson density of independent particles in the x,v plane.

The diffusion constant D for the tp undergoing the process defined by (2.1) with U zero or bounded periodic can be shown to exist under mild assumptions on K , and to be given, as usual, by the integral of the velocity autocorellation function as in Equation (1.10) with $\langle \; \rangle_o$ referring to the normalized equilibrium distribution (2.4). It might appear that, in this case at least, very few additional assumptions are required to jusfity (1.8) and thus prove (1.1) (see however eq. (2.9) and the subsequent discussion). We do not attempt to investigate this here - instead we limit ourselves to two examples where $h(v;E)$ can be computed explicitly and (1.1) can thus be shown to hold.

For the OU process [9]

$$Kh = M^{-1}\gamma\frac{d}{dv}\,[vh + (\beta M)^{-1}\,\frac{d}{dv}\,h] \tag{2.5}$$

and $D = (\gamma\beta)^{-1}$. It is easy to check explicitely (or by performing a Galilean transformation) that for $U = 0$ the stationary velocity distribution is given by

$$h(v;E) = h_{\beta}(v - E/\gamma) \tag{2.6}$$

and so $\sigma = \gamma^{-1} = \beta D$.

A problem may arise, however, when K is an integral operator (case i).

$$(Kh)(v) = \int K(v,v')h(v')dv' - \int K(v',v)h(v)dv' \tag{2.7}$$

so that EL_1 becomes a singular perturbation : E multiplying the highest derivative. To see this let us consider the simplest jump process, one without memory,

$$K(v,v') = \tau^{-1}h_{\beta}(v) \tag{2.8}$$

with τ a constant.

The stationary solution of (2.1) can now be easily obtained for the case $U = 0$;

$$h(v;E) = \int_{-\infty}^{v} \alpha \exp -\alpha(v-v')\; h_{\beta}(v')dv' , \quad \alpha = M/\tau E \tag{2.9}$$

Clearly $h(v;E)$ is not differentiable at $E = 0$. The Einstein relation is nevertheless satisfied,

$$\langle v \rangle_E = \tau E/M = \beta DE . \tag{2.10}$$

The first part of (2.10) is obtained most easily by multiplying (2.1) by v and integrating over v , while the second equality follows readily from (1.10) (In fact the same method can be applied to the case $K(v,v') = W(v-\lambda v')$, $W(\eta)$ even and integrable, $\lambda < 1$. It gives $\sigma = \tau/(1-\lambda)M = \beta D$, $\tau^{-1} = \int W(v)dv$) . Thus the existence of an expansion of

$h(v;E)$ about $E=0$, is certainly not necessary for (1.1) to hold, what is then? Our formulation in section 5 suggests that (1.1) is valid in essentially all cases where the motion of the tp, for $E=0$, converges (in the limit of macroscopic length and time scales) to Brownian motion. The problem however remains open.

b. Diffusion in a Random Environment.

We continue the investigation of static environments by considering the motion of a test particle diffusing in a potential U with a "conductivity" $a(x)$. The distribution function $\rho(x,t)$ for this model (there is no velocity variable) satisfies the equation

$$\frac{\partial\rho(x,t)}{\partial t} + \frac{\partial}{\partial x}\left(\gamma^{-1}\left[-\frac{\partial U}{\partial x} + E\right]\rho\right) = \frac{\partial}{\partial x}\, a\, \frac{\partial}{\partial x}\,\rho \tag{2.11}$$

with $a\gamma \equiv \beta^{-1}$, independent of x, required to make the stationary measure, for $E=0$, assume its equilibrium value, c.f. (2.4). For $U=0$ and a constant, the Einstein relation is then satisfied by assumption. The question about (1.1) now occurs when U and/or a is periodic or more generally themselves form a spatially stationary random process. We shall consider this case in one dimension where one can obtain explicit expressions for σ and D and show that they satisfy (1.1).

Consider ergodic, translation invariant random fields $U(x)$ and/or $a(x)$, $x \in \mathbb{R}$, defined on some probability space $<\Omega, A, P>$. (We assume that w may be identified with U_w, a_w, or (U_w, a_w).) We write $<.>$ for the expectation w.r.t. P. For each $w \in \Omega$ the forward generator of the corresponding process $X(w,t)$ is

$$L_w\rho = -\frac{\partial}{\partial x}\left[\gamma_w^{-1}\left(-\frac{\partial U_w}{\partial x} + E\right)\rho\right] + \frac{\partial}{\partial x}\left[a_w\frac{\partial}{\partial x}\rho\right] = L_o\rho - E\frac{\partial}{\partial x}(\beta a\rho) \tag{2.12}$$

and the stationary solutions to (2.11) satisfy

$$L_w\rho_w(x) = 0 \quad . \tag{2.13}$$

According to (2.12) $L_w\rho = \text{div}\, J_w(\rho)$, where J_w is the current (operator).

Thus the first integration constant of (2.13) is the current $J = J(\rho_w)$, which can be interpreted as

$$J(\rho_w) = \lim_{t\to\infty} \frac{N_0(t)}{t}$$

where $N_0(t)$ is the (expected) signed number of crossings of the origin (total flux) up to time t of independent particles, each performing $X(w,t)$ and distributed initially according to a non homogeneous Poisson process of density $\rho_w(x)$.

If P is ergodic one can prove (using ergodicity of the process induced in the space of the environments) that the $\lim_{t\to\infty} \frac{X(w,t)}{t}$ exists and is independent of w . Thus the effective velocity of the system is given by

$$u(E) = \lim_{t\to\infty} \frac{X(w,t)}{t} \quad , \; P \quad \text{a.s.} \tag{2.14}$$

and we have also that $< \rho_w(0) > u(E) = J(\rho_w)$, so

$$u(E) = \frac{J(\rho_w)}{<\rho_w(0)>} \tag{2.15}$$

Now we compute $u(E)$ in two cases :

a) random potential (a is constant [= 1] in eq. (2.11)) and

b) random conductivities (U is zero in eq. 2.11).

For the case of a random potential the stationary density ρ_w is the solution of the equation

$$\frac{\partial}{\partial x}(\gamma^{-1} [\frac{\partial U_w}{\partial x} - E] \rho) + \frac{\partial^2}{\partial x^2} \rho = 0 \; . \tag{2.16}$$

Putting the first integration constant $J(\rho_w) = 1$ we obtain

$$\rho_w(x) = \int_x^\infty e^{-\beta E(x'-x)} \, e^{\beta[U(x')-U(x)]} \, dx' \tag{2.17}$$

Taking $x = 0$ and using eq. (2.15) we get

$$u(E) = \frac{1}{\langle\rho_w(0)\rangle} = \beta E\left(\int_0^\infty dx\, \beta E\, e^{-\beta Ex} \langle e^{-\beta U(0)} e^{\beta U(x)} \rangle\right)^{-1} \tag{2.18}$$

Thus

$$\sigma \equiv \lim_{E\to 0} \frac{u(E)}{E} = \beta(\langle e^{-\beta U(0)} \rangle \langle e^{\beta U(0)} \rangle)^{-1} \tag{2.19}$$

The rightmost factor is the diffusion constant D of the tp moving in the random potential (provided e.g. $e^{-\alpha U(0)} \in L^2(p)$ and P is mixing under translations), so we have that $\sigma = \beta D$.

For the case of random conductivities, stationary densities ρ_w satisfy the equation

$$\frac{\partial}{\partial x}[a_w \frac{\partial}{\partial x} \rho] - \beta E \frac{\partial}{\partial x}(a_w \rho) = 0 .$$

Taking again the first integration constant $J(\rho_w) = 1$ we obtain, for $x = 0$, $\rho_w(0) = \int_0^\infty \frac{e^{-\beta Ex}}{a(x)} dx$.

Thus $\sigma \equiv u(E) = \frac{1}{\langle\rho_w(0)\rangle} = \beta E \langle 1/a \rangle^{-1}$ which implies that

$$\lim_{E\to 0} \frac{u(E)}{E} = \beta \langle 1/a \rangle^{-1}$$

It is known [7,6] that $\langle 1/a \rangle^{-1}$ equals the diffusion coefficient in this case, so relation (1.1) has been verified for the cases of random potentials and random conductivities: see also Appendix B.

III. Dynamic Environments.

1. Non-crossing Particles in one dimension.

a. Ornstein-Uhlenbeck Dynamics.

We consider first the case of mechanical particles on the line which interact via a finite range even pair potential $\phi(r)$, smooth for $r > 0$. In addition there is a hard core, i.e. they cannot cross. This is irrelevant when $\phi(r) \nearrow \infty$ as $r \searrow 0$ but we shall keep it anyway. When two particles meet, or collide, they exchange velocities. All particles have the same mass and between collisions they undergo independent Ornstein-Uhlenbeck (O.U.) processes with friction constants γ/M and reciprocal temperature β, i.e.

$$dv_i = -(\gamma/M)[v_i\, dt - \sqrt{2/\beta\gamma}\, dW_i] - (1/M) \sum_{j\neq i} \frac{\partial}{\partial x_i} \phi(x_i - x_j) dt \qquad (3.1)$$

where the W_i are independent standard Wiener processes. For a given density, the, presumably unique, stationary state of this system is a Gibbs state at temperature β^{-1} for the Hamiltonian $H = (1/2)M\, v_i^2 + \Sigma_{j\leq i} \phi(x_i - x_j)$.

It is known rigorously for the case when $\phi(r) = 0$ but presumably true also for the general non-crossing case that the mean square displacement of a particle in an infinite system in equilibrium with particle density ρ behaves asymptotically like $t^{1/2}$ [8] rather than t , and so $D = 0$.

The Einstein relation then predicts for this system (and other one dimensional non-crossing systems we discuss later) that σ should be zero. As we shall see later this is true in a very strong sense with $u(E) = 0$ for all values of E acting on a test particle.

We shall, however, first consider a system consisting of N O.U. particles on a circle of length L . We show that the diffusion constant D_N and mobility σ_N are of $O(N^{-1})$ and satisfy the Einstein relation. The displacement $x(t)$ of the tp is now of course to be interpreted as $\int_0^t v(t')dt'$, i.e. like an angular variable.

To investigate the diffusion of a tp in such a system, we simply note two facts : a) since there is no crossing D_N must equal $\overline{D}_N$, the diffusion constant of the "center-of-mass" $\overline{x}$ of the system. b) The motion of $\overline{x}$ is entirely independent of the forces between the particles (as long as they satisfy Newton's law of action and reaction). Thus, setting

$$\bar{x} = N^{-1} \sum_{i=1}^{N} x_i \ , \ x_1 \leq x_i \leq x_1 + L$$

$$\bar{v} = N^{-1} \sum_{i=1}^{N} v_i \ . \tag{3.2}$$

We have from (3.1)

$$d\bar{v} = \frac{-\gamma}{M} [\bar{v}\, dt - \sqrt{2/\beta\gamma}\, N^{-1} \sum_{i=1}^{N} dW_i] \ . \tag{3.3}$$

This gives immediately

$$D_N = \bar{D}_N = (\beta\gamma N)^{-1} \tag{3.4}$$

When there is a field E acting on a tp, say the first particle, then (3.1) is modified for $i = 1$ by the addition of a term $(E/M)dt$ on the right side. The center of mass velocity now satisfies the equation

$$d\bar{v} = \frac{-\gamma}{M} [\bar{v}\, dt - \sqrt{2/\beta\gamma}\, N^{-1} \sum_{i=1}^{N} dW_i] + \frac{E}{MN}\, dt \tag{3.5}$$

It follows that in the stationary state, which we shall discuss next

$$u(E) = \langle \bar{v} \rangle_E = \frac{1}{N\gamma}\, E = \beta^{-1} D_N\, E \tag{3.6}$$

Stationary States.

The stationary non-equilibrium state in the presence of a field on the tp is most easily obtained by considering the problem in the frame of reference moving with the average velocity of the system $E/\gamma N$. In this frame the stationary distribution of the fluid particles relative to the test particle is just the equilibrium measure in the presence of an electrostatic potential $E(x_i - x_1)/N$. More precisely, setting

$$y_i = (x_i - x_1) \in (0,L) \ , \ i = 2,\ldots,N \tag{3.7}$$

the stationary ensemble density will have the form

$$\mu(x_1;\underline{y},\underline{v}) = L^{-1}[\exp[-\beta\hat{U}(\underline{y})] \prod_{i=1}^{N} h_\beta(\hat{v}_i)]/\hat{Z} \tag{3.8}$$

$$= P(\underline{y},\underline{v})/\ L$$

where

$$(\underline{y},\underline{v}) = (y_2,\ldots y_N,v_1,v_2,\ldots,v_N) \ ,$$

$$\hat{v}_i = v_i - E/\gamma N \quad ,$$

$$\hat{U}(y) = \sum_{i=2}^{N} \{ \sum_{j>i}^{N} \phi(y_i-y_j) + \phi(y_i) + Ey_i/N\} \ , \tag{3.9}$$

P is the density for the environment measure and $\hat{Z}$ is a normalization constant. The stationarity of μ follows easily from the fact that μ is a canonical Gibbs state periodic in x_1, of period L (in fact, independent of x_1).

N.B. In choosing the domain of the y_i in (3.7) we have used strongly the fact that there is a hard core between the tp and the fluid particles. This permits the discontinuity in the electrostatic potential and in $P(\underline{y},\underline{v})$ between $y_i = 0$ and $y_i = L$ which correspond simply to the right and left side of the tp .

To see more clearly what is happening, let us consider the case $\phi(r) = 0$, i.e. hard point particles. In this case P is a product measure

$$P(\underline{y},\underline{v}) = h_\beta(\hat{v}_1) \prod_{i=2}^{N} [\lambda(y_i)h_\beta(\hat{v}_i)] \tag{3.10}$$

where (changing the domain of y_i to $(-1/2\,L,1/2\,L)$ for future convenience) $\lambda(y)$ has the form

$$\lambda(y) = \begin{cases} C\ \exp\left[-\beta Ey/N\right] , & 0 < y \leq (1/2)L \\ C\ \exp\left[-\beta E(y+L)/N\right] , & (-1/2)L \leq y < 0 \end{cases} \tag{3.11}$$

$$C = \beta E[1-\exp(-\beta E/\rho)]^{-1}/N \ , \ \rho = N/L$$

Put differently, the distribution of the fluid particles relative to the position of the tp is Poisson with a density $\rho(y)h_\beta(v_i - E/\gamma N)$, $\rho(y) \equiv (N-1)\lambda(y)$.

Note that there is a discontinuity in $\rho(y)$ at the position of the tp . Putting $\rho(0+) = \rho_R$ and $\rho(0-) = \rho_L$, the densities to the right and left of the tp , we immediately find

$$\beta^{-1}[\rho_R - \rho_L] = (\frac{N-1}{N})E \tag{3.12}$$

$$\rho_L/\rho_R = \exp[-\beta E/\rho] \tag{3.13}$$

The left side of (3.12) represents the difference in pressure Δp , exerted on the tp by the fluid particles – the net remaining force on it

$$E - \Delta p = E/N = \gamma\ u(E) \tag{3.14}$$

is just enough to produce the average motion.

Infinite Volume Stationary States.

Taking now the thermodynamic limit $N \to \infty$, $L \to \infty$, $N/L \to \rho$ of this stationary state, we obtain a Poisson field (for the environment measure) with constant densities ρ_R (ρ_L) for $y > 0 (< 0)$ with $\rho_R = \rho_L + \beta E$ and on either side of the tp a Maxwellian velocity distributions $h_\beta(v)$. In this "blocked" state the electric force on the tp is entirely balanced by the difference in pressure exerted by the fluid on its opposite sides and $u(E) = 0$ as it should be.

It is useful to note here, and this will be important later when we deal with other non-crossing systems, that this stationary measure can also be obtained directly as a limit of constrained equilibrium states in a box with rigid walls situated at $\pm L$. Suppose we want to obtain the infinite volume state with some specified density ρ_L to the left of the tp . We then put $N_L = \rho_L L$ particles to the left of the tp and $N_R = [\rho_L + \beta E]L$ particles to the right of it. Because of the hard core interaction between the tp and the other particles N_L and N_R are conserved quantities. The corresponding equilibrium canonical ensemble density will therefore be given by (with now the particles labelled so that the o-th particle is the tp)

$$\mu(x_{-N_L}, v_{-N_L}, \ldots, x_1, v_1, \ldots, x_{N_R}, V_{N_R}) = Z^{-1} \prod_{i=-N_L}^{N_R} h_\beta(v_i) \cdot \exp[\beta E x_o] \chi(\underline{x}; N_L, N_R) \tag{3.15}$$

where χ is the characteristic function specifying that there be exactly N_L , N_R particles on the left, right of x_o . It is now an easy exercise to show that this state leads to the blocked state when $L \to \infty$. (See section 4 for the lattice version). The above construction will work also when $\phi(r) \neq 0$. We will always end up with two semi-infinite Gibbs states which, except near the boundary, will be uniform with densities ρ_L and ρ_R determined by $p(\rho_R, \beta) - p(\rho_L, \beta) = E$ where $p(\rho, \beta)$ is the pressure in the (infinite one dimensional) Gibbs state with density ρ at temperature β .

Remark. Going back to the stationary non-equilibrium state in the periodic box we can also take there the limit $L \to \infty$ with N fixed. We obtain then a stationary environment measure in which there are no particles on the left and N independent particles with exponentially decaying density on the right. The whole entity is moving to the right with a velocity $E/N\gamma$. Such a stationary state cannot be produced by starting with a rigid box and does not exist when $E = 0$; the particles would then disperse to infinity. To the extent that one can define a diffusion constant here it would correspond presumably to the center-of-mass of the block and would satisfy the Einstein relation.

b. Brownian and Jump Dynamics.

The above analysis can be carried out almost verbatim for the (positional part of) the distribution of interacting Brownian particles.

The results are also essentially identical. The problem is a bit more complicated for particles on the 1-dimensional lattice which can only jump to unoccupied nearest neighbor sites. The absence of complete translational symmetry makes the computations awkward, particularly for the periodic case. They are given in section 4.

c. Hamiltonian Dynamics.

When γ is set equal to zero in (3.1), we are dealing with a conservative system whose evolution is governed by classical mechanics. It is clear that for such a system the velocity of the center of mass is a constant of the motion and thus D_N is infinite. Similarly, under the action of a uniform (non-potential) field on the circle the finite systems center of mass keeps on accelerating, no stationary state is possible and σ_N is also infinite.

The situation is different for the infinite system. Here it was shown by many authors [5] for the case $\phi = 0$, i.e. for hard point particles that $D = \langle|v|\rangle/\rho > 0$, where ρ is the uniform density and $\langle|v|\rangle$ the expectation of the speed of the tp , which has the same mass as the other particles, is given in equilibrium by $(2/\pi M\beta)^{1/2}$. Einstein's relation would now appear to say that when the field is put on the tp there should result a non-equilibrium stationary state in which $u(E)/E \to \beta D$ as $E \to 0$. On the other hand the constrained Gibbs state constructed in (3.15) is stationary also for this system and as before, gives rise in the limit $L \to \infty$ to the "blocked" state in which $u(E) = 0$.

We believe but cannot prove that the resolution of this problem lies in the fact that the blocked state is not the appropriate stationary state for this system. Unlike the case of OU or Brownian particles, the blocked state is not unique here. In particular we expect that there exists an entirely different stationary state, one which would be obtained, as $t \to \infty$, if we turned on the field at $t = 0$ when the system is in equilibrium. Such a state should have quite a different velocity distribution for "out going" particles to the right and the left of the tp . If such a state exists it is presumably given by the construction (1.8) and satisfies the Einstein relation as is expected from the considerations in section 5.

2. Higher dimensions, Crossing particles.

The existence of a diffusion constant D , $0 < D < \infty$, for a tp in a general interacting system of particles has been proven so far only for : a) B-particles interacting via sufficiently soft superstable potentials [6,9].b) Particles on a lattice at infinite temperatures in dimension greater than one (also in one dimension when the jumps extend beyond nearest neighbor sites [10]) and for OU particles - with bounded potentials [11].

In none of these systems is there any rigorous information about the existence of a stationary state in the presence of an electric field E acting on the tp . In fact, for a mechanical system with "soft" interactions, the resistance of the fluid to the motion of the tp might be expected to decrease at large speeds of the latter, as the cross section does, and there will presumably not be any stationary state for $E > 0$. The Einstein relation might still hold however for some kind of "metastable" state in such a system or when E is sealed properly : c.f. section 5 . We shall therefore confine our discussion here to the case where the tp is OU or Brownian while the fluid particles are Newtonian. It will be seen that for these "mixed dynamics" the Einstein relation provides some interesting insights.

We begin with the O.U. mixed dynamics system, i.e. we set $\gamma = 0$ in (3.1) for all $i \neq 1$. It is quite easy to see that the stationary state of the system of N particles in a periodic box with a field E on the tp is simply the Galilean transform of the canonical Gibbs state : the positional part remains canonical Gibbs while the velocity part is transformed, $h_\beta(v_i) \to h_\beta(v_i - E/\gamma)$, <u>independent</u> of N . It follows furthermore from general arguments about spreading Markov processes,[12], that for non-degenerate interactions, i.e. when the phase space cannot be decomposed into separate components, this stationary state is unique.

The perturbation argument for the EGK relations given in section 1, which there seems no reason to doubt, then leads to the result that the diffusion constant for the tp in the equilibrium state, $E = 0$, of this periodic system is $(\beta\gamma)^{-1}$; the same as if there were no other particles present.

This result, while a little surprising at first sight, seems not too unreasonable for the finite periodic system. After all the interactions between the particles conserve momentum and energy and the only dissipation occurs via the tp . In fact, we shall now prove this explicitely for the case where the tp is a B-particle.

For the mixed dynamics in which the tp is Brownian, the tp has no velocity and in place of (3.1) for $i = 1$ we have

$$dx_1 = \gamma^{-1}\, E - \frac{\partial}{\partial x_1} \sum_{i=2}^{N} \phi(x_i - x_1)\, dt + 2/\beta\gamma\, dW$$

We can now compute the diffusion constant for $E = 0$. Let

$$V = \sum_{i=2}^{N} v_i\ .$$

Then

$$dx_1 + \gamma^{-1} M dV = \sqrt{2/\beta\gamma}\, dW\ .$$

and

$$\frac{x_1(t) - x_1(0)}{\sqrt{t}} = \sqrt{2/\beta\gamma}\,\frac{W(t)}{\sqrt{t}} - \gamma^{-1} M\, \frac{V(t) - V(0)}{\sqrt{t}}$$

Since $V(t)$ forms a stationary stochastic process (with $\langle V^2 \rangle_o < \infty$) the last term on the right does not contribute in the limit $t \to \infty$. We therefore obtain that the diffusion constant of the tp is the same as if no other particles were present. (We also (more or less) immediately obtain the invariance principle for the motion of the tp) .

Consider now the passage to the thermodynamic limit $N \to \infty$, $L \to \infty$, $N/L^d \to \rho$. We obtain then (for both the B and O.U tp) a uniform Gibbs state in the frame of reference moving with velocity E/γ .

It is clear however, that this stationary state is not the relevant one for the Einstein relation. For starting with an infinite system in equilibrium and putting the field on the tp will surely lead to a state in which the velocity of the fluid particles "far away" will remain essentially unaffected by the electric field in higher dimensions. Also in one dimension for crossing particles, the fluid far away will be moving relative to the tp . It is this state in which $u(E)$ should be computed. It will then presumably satisfy the Einstein relation, with the correct equilibrium D for the infinite system.

We can say a little more about this model if we consider again the non-crossing case. For the finite system in a periodic box the fluid particles must have the same diffusion constant as the test particle, $D = (\beta\gamma)^{-1}$. In the thermodynamic limit we will have the additional stationary states obtained as the Galilean transforms of the blocked states discussed earlier. This leads to the family of stationary states moving with the velocity α and having a pressure jump Δp connected by the relation

$$\alpha = (E-\Delta p)/\gamma \ .$$

For $\gamma = 0$, $E = \Delta p$ and α is arbitrary as we found for the purely mechanical system.

The question now arises as to what is the diffusion constant of the tp , again the same as that of the fluid particles because of non-crossing, in the infinite equilibrium system with $E = \Delta p = 0$. For the O.U. tp, it is clearly not just $(\beta\gamma)^{-1}$ since when $\gamma \to 0$ it should go to $\langle |v| \rangle/\rho$.

The origin of the problem with all these stationary states appears to be the interchange of limits $t \to \infty$, necessary to obtain the stationary state for $E \neq 0$ and diffusion constant for $E = 0$ in the finite periodic system, with the thermodynamic limit $L \to \infty$. We get the "wrong" stationary state and diffusion constant. What then is the right answer ?

IV. Jump Processes on the Lattice.

a) Infinite one dimensional lattice gas.

We consider a one dimensional lattice gas in which all the particles but one have a symmetric rate of jump (i.e. the rate of jumping to the right = the rate of jumping to the left = 1/2). The test particle is subjected to an external field : it jumps with the rate p(resp. $q = 1-p$) to the right (left), $p \geq q$. The relation between p, β and E is given by $q/p = e^{-\beta E}$. (For this choice the (formal) Gibbs state satisfies detailed balance). The interaction of the particles is merely simple exclusion, so when a particle attempts to jump to an occupied site the jump is suppressed.

We describe the system directly as it is seen from the tagged particle ("environment process"). The generator acting on cylindric functions $f : \{0,1\}^{\mathbb{Z}} \cap \{\eta : \eta(0) = 1\}$ is given by

$$L_p f(\eta) = \sum_{\substack{x,y\neq 0 \\ x-y=1}} (1/2)[f(\eta_{xy})-f(\eta)] +$$

$$p[1-\eta(1)]\,[f(\tau_1\eta_{01})-f(\eta)] + q[1-\eta(-1)][f(\tau_{-1}\eta_{0-1})-f(\eta)] \quad (4.1)$$

where $(\tau_x\eta)(z) = \eta(x+z)$ and

$$\eta_{xy}(z) = \begin{cases} \eta(z) & \text{if } z \neq x,y \\ \eta(x) & \text{if } z = y \\ \eta(y) & \text{if } z = x \end{cases}$$

The semigroup $S_p(t)$ corresponding to the generator L_p determines a unique strong Markov process η_t on $\{0,1\}^{\mathbb{Z}}$, in such a way that $S_p(t)f(\eta) = E\, f(\eta_t)$ where E denotes the expectation with respect to the process with initial configuration η .

The set of extremal invariant measures J_e is given by (see theorem A1 in the appendix)

$$J_e = \{\mu_\rho : 0 \leq \rho \leq 1\} \cup \{\tilde{\mu}_n : n \geq 0\} \quad (4.2)$$

where μ_ρ is a Bernoulli measure with parameters $\rho_\ell = \rho$ to the left of the origin and $\rho_r = (1-q/p) + \rho(q/p)$ to the right of the origin. Thus $q/p = \frac{1-\rho_r}{1-\rho_\ell}$. μ_n is concentrated on configurations with no particles to the left of the origin and n particles to its right (cf. eq. (A.11) below).

The position $X(t)$ of the tagged particle is given by the algebraic number of shifts of the system (corresponding to the last two terms in the generator (4.1)) in the interval $[0,t]$. Consider that the initial configuration $\eta(0)$ is distributed accordingly with $\mu_\rho\{X(t)\} \equiv 0$.

The Einstein relation in this case follows trivially : It is known (cf. [4]) that when $E = 0$, $D = 0$ for all $\rho > 0$. On the other hand $U(E) = \frac{EX(t)}{t} = \lim_{t\to 0} \frac{EX(t)}{t} = p(1-\rho_r) - q(1-\rho_\ell) = 0$ for all E and all $\rho > 0$.

The Einstein relation is, in fact, satisfied for this model in a somewhat stronger sense. In section c) below we show that for a sequence of periodic approximations the Einstein relation is satisfied with non-zero diffusion constant D_N and mobility σ_N . Moreover, these quantities converge to their infinite volume values (0) . Finally, we will see that the stationary states we have described in this section arise as limits of the stationary states for the periodic approximations.

In the next section, we show that the same thing is true for box approximations except that here, of course, the diffusion constant and drift are 0 .

b) Finite lattice gas.

We consider now finite approximations to the preceding model : the particles now move on a finite lattice of length $2L$ with reflecting walls at $-L$ and L . Note first that the Einstein relation is now trivial. Moreover, stationary states for this model are easy to find, even for $E \neq 0$, e.g. Gibbs states. The Gibbs states, however, don't have a good limit as $L \to \infty$. But if we condition the Gibbs states on the number of particles to the left and right of the tp , in the appropriate way as $L \to \infty$, we obtain a sequence of stationary states (stationary because particles cannot cross) converging to the measure μ_ρ of the infinite case. This is based on the fact that for these constrained Gibbs states, given the position x of the tp , the particles on its left (right) are uniformly independently distributed with density $\rho_\ell(\rho_r)$ - which depend upon **x**. What must do is show that x is well localized as $L \to \infty$ at a position which gives the correct value for ρ_ℓ/ρ_r. This we proceed to do.

In the box $-L,L$ put the tp and N additional particles, M of them to the left of the tp. The position of the test particle we denote by x. The (average) density to the left is $\rho_\ell = M/(L+x)$;

the density to the right of the tp is $\rho_r = 1-\rho_\ell = N-M/(L-x)$

Let $M/L \sim a$ and $N-M/L \sim b$ as $L \to \infty$. Then, writing $y = x/L$, $\rho_\ell \sim a/1+y \equiv \rho_\ell(y)$ and $\rho_r \sim b/1-y \equiv \rho_r(y)$.

The distribution of the tp corresponding to the constrained Gibbs state (i.e. conditioned on there being M(N-M) particles to the left (right)) when the field acts as before $(q/p = e^{-\beta E})$ is given by

$$f(x) \sim e^{\beta E x}\binom{L+x}{M}\binom{L-x}{N-M} \quad .$$

Using the approximation $\ln n! \sim n \ln n$, we obtain

$$f(x) = e^{F(x)}$$

where

$$\begin{aligned} f(x) &= \beta E x + (L+x) \ln(L+x) + \\ &+ (L-x) \ln (L-x) - (L+x-M) \ln(L+x-M) \\ &- M \ln M - (N-M)\ln(M-N) - \big(L-x-(N-M)\big) \ln(L-x-(N-M)) \quad . \end{aligned}$$

Under the change of variables $y = x/L$ we obtain $F(x) \sim L\phi(y)$

where $\phi(y) = \beta E y + (1+y)\ln(1+y)+(1-y)\ln(1-y)-(1+y-a)\ln(1+y-a)$

$$-(1-y-b)\ln(1-y-b)+g(N,M)$$

where $g(N,M)$ arises from the terms of $F(x)$ which don't depend upon x.

We thus have that the distribution of $y \sim e^{L\phi(y)}$ so that for large L y is near the maximum y_o of $\phi(y)$. Setting $\phi'(y_o) = 0$ we find that

$$e^{-\beta E} = \frac{(1+y_o)(1-b-y_o)}{(1-y_o)(1-a+y_o)}$$

i.e.,

$$q/p = \frac{1-\rho_r(y_o)}{1-\rho_\ell(y_o)} \quad .$$

Moreover $e^{L\phi(y)} \sim e^{-L/2\phi''(y_o)(y-y_o)^2}$, so that $|y-y_o| \sim 1/\sqrt{L}$. Thus in the thermodynamic limit y is localized at y_o and hence we obtain the state described in section 4a).

c) EGK relation for a periodic lattice model.

We here consider the symmetric lattice gas with the tp subjected to an electric field as in section 4a but now with the particles moving in a one dimensional periodic box of length $L+1$, $L > 1$. Let $X(t)$ be the position of the tp in $\mathbb{Z}$ induced by our process : $X(t)$ is the algebraic number of jumps performed by the tp up to time t (jumps to the left make a negative contribution). Let $0 = X_o(t) < X_1(t) < \ldots < X_N(t) \leq L$ be the positions of the particles relative to the tp and let $Y_i(t) = X_i(t) + X(t)$, $i = 0,\ldots,N$, define the motion of the ith particle in $\mathbb{Z}$ ($Y_o(t) = X(t)$.) Then by considering the motion $\overline{Y}(t)$ of the center of mass

$$\overline{Y}(t) = (1/N+1) \sum_{i=0}^{N} Y_i(t) \tag{4.3}$$

we easily compute the diffusion constant for $E = 0$, at least in the limit $L \to \infty$: It is easy to check that $\overline{Y}(t)$ is a martingale (with respect to the σ-algebra generated by the motion of the entire system up to time t). Therefore, since $|Y_o(t) - \overline{Y}(t)| < L$ for all $t \geq 0$, we have that

$$D = \frac{E(\overline{Y}(t)^2)}{2t} = \lim_{t\to 0} \frac{E(\overline{Y}(t)^2)}{2t}$$

$$= < \frac{\sum_{i=0}^{n} 1(X_{i+1}-X_i>1)}{(N+1)^2} >_0 \tag{4.4}$$

Here $1(\ )$ is the indicator function and $X_{N+1} \equiv L+1$. The last equation follows easily from the fact that each particle jumps to each unoccupied neighboring site with rate 1/2 .

We are primarily interested in the situation in which N and L are fixed. In this case the RHS of (4.4) is not as easy to compute as it is when $< \ >_o$ is Bernoulli with density ρ (grand canonical ensemble). With this slight modification we find that

$$D = < \frac{\sum_{x=0}^{L} \eta(x)(1-\eta(x+1))}{[\Sigma\eta(x)]^2} >_o$$

$$\sim \frac{\rho(1-\rho)L}{(L\rho)^2} = \frac{1-\rho}{\rho}\frac{1}{L}$$

since for large $L\rho$ we have that $\Sigma\eta(x) \sim L\rho$

We don't wish to make the approximation above more precise, since we will compute D and u explicitly after again slightly modifying the model. The real problem is with the computation of u(E): in order to compute u(E) we need detailed information on the stationary measure $< \ >_E$ which is not so easy to obtain.

Nevertheless, if instead of fixing the length L+1 of the box and allowing N to be random we fix N and allow L to be random in an appropriate way the computations become much easier. Consider the process $\xi_t^N(y) \in \mathbb{N}^{N+1}$, $y = 0,\ldots,N$ where $\xi_t(y)$ is the number of empty sites to the right of the y-th particle. Let $\nu = \nu_{\rho,E,N}$ be a probability measure on $\mathbb{N}^{N+1}$ satisfying

a) $< L > = < \sum_{y=0}^{N} (\xi(y)+1) > = \frac{(N+1)}{\rho}$.

b) ν is a product measure (i.e. $\xi(y)$ are independent random variables) of geometric distributions with parameters a_y (i.e

$\nu\{\xi(y) = k\} = a_y^k\ (1-a_y))$

c) a_y satisfies the balance equations (4.5) below. One can prove by direct computation that ν is stationary for the process $\xi_t^N(y)$, and hence defines a stationary state $\langle\ \rangle_E$ for our periodic system (with random length).

We want to compute $\lim_{E\to 0} \frac{u(E)}{E}$. We find it convenient to consider the quantity

$$c \equiv c(\rho,E,N) = \frac{u(E)}{p-q}$$

since $\beta E \sim p-q$ as $E \to 0$,

$$\sigma = \lim_{(p-q)\to 0} \beta\cdot c\ .$$

Since $a_y = \nu\{\xi:\xi(y) > 0\}$ the average velocity of the y-th particle, $y \neq 0$, is given by $(1/2)a_y - (1/2)a_{y-1}$ while that of particle zero is $pa_o - qa_N$. (We are identifying N with -1). We thus have that a_y, $0 \leq y \leq N$ must satisfy the following equations $(q/p = e^{-\beta E})$.

$$\begin{aligned} &(1/2)a_y - (1/2)\,a_{y-1} = c(p-q) \\ &pa_o - qa_N = c(p-q) \end{aligned} \tag{4.5}$$

from which it follows that (for $E > 0$)

$$a_o = c(2Nq+1)$$

$$a_N = c(2Np+1)$$

$$a_y = (1-y/N)a_o + (y/N)\,a_N\ ;\quad 0 < y < N$$

The relationship between ρ and c is then given by (cf. a) above)

$$1/\rho = \nu\,\{\frac{\sum_{y=0}^{N}(\xi(y)+1)}{N+1}\} = (1/N+1)\sum_{y=0}^{N}\frac{1}{1-a_y} =$$

$$\frac{1}{\rho} = \frac{N}{N+1} \sum_{y=0}^{N} \frac{1}{N-(N-y)c(2Nq+1)-yc(2Np+1)}$$

and taking the limit $(p-q) \to 0$ we obtain

$$\frac{1}{\rho} = \frac{N}{N+1} \sum_{y=0}^{N} (N-N(N+1) \lim_{(p-q)\to 0} c)^{-1} = (1-(N+1) \lim_{p-q\to 0} c)^{-1}$$

Thus

$$\sigma = \lim_{(p-q)\to 0} \beta c = \frac{\beta(1-\rho)}{N+1} \quad .$$

On the other hand, (see eq. 4.4)

$$D = \frac{\sum_{y=0}^{N} \langle 1(x_{y+1}-x_y \quad >1)\rangle_o}{(N+1)^2}$$

$$= \sum_{y=0}^{N} \left\langle \frac{1(\xi(y)>0)}{(N+1)^2} \right\rangle_o = \frac{(N+1)(1-\rho)}{(N+1)^2} = \frac{1-\rho}{N+1}$$

This proves relation (1.1) for this model.

V. Macroscopic Formulation of Einstein Relation.

In our discussions so far the tp has been treated entirely on a microscopic level - asking for a description of the stationary states of the tp on the spatial and temporal scale on which the basic dynamics of the model is prescribed. While this level clearly gives the most detailed information the Einstein relation concerns quantities, D and σ which are measured on very long, i.e. macroscopic, spatial and time scales. Furthermore being a linear transport coefficient, σ is calculated in the limit $E \to 0$ so the system is really very close to equilibrium - hence, of course, the EGK relations. It seems therefore sensible to formulate the Einstein relation in a more macroscopic way. In fact this turns out to be possible and has the (not so incidental) advantage that it is not necessary to first find $u(E)$ at finite E and then take the limit $E \to 0$. Instead one goes to the appropriate macroscopic length and time scales by setting

$$x' = \varepsilon x \ , \ t' = \varepsilon^2 t$$

where ε is a small parameter related to the electric field E which is assumed to change as $E = \varepsilon E'$. A form of (1.1) is then the following : A) . If for $E' = 0$, when the system is in equilibrium, the rescaled trajectory of the test particle converges (weakly) to Brownian motion with diffusion constant D , i.e.

$$x_\varepsilon(t) \equiv \varepsilon x(t/\varepsilon^2) \xrightarrow[\varepsilon \to 0]{} W_D(t) \tag{5.1}$$

then in the presence of a field εE

$$x_\varepsilon(t; E) \xrightarrow[\varepsilon \to 0]{} \sigma E\, t + W_D(t) \ , \ \sigma = \beta D \ . \tag{5.2}$$

We believe that A can in fact be proven for all non-mechanical cases considered here and in particular for a B-particle in a random environment or interacting with other B-particles [13] . It also seems to hold in the case of an O.U. particle in a periodic potential, recently shown by Rodenhausen to satisfy (1.1) [14].

It should be noted that the scaling of the electric field is just such that it remains effective, neither zero nor infinite, on the macroscopic scale σ is then the mobility, as in Stoke's law, for a dilute concentration of B-particles in a fluid. It is presumably this situation which Einstein had in mind. We are now in a position to present what we regard as the best argument for the Einstein relation, a rigorous reformulation of Einstein's original argument. We first observe that if we drop from equation (5.2) the relation $\sigma = \beta D$ and replace σE in (5.2) by a more or less arbitrary function $u(E)$ of the field, what we obtain is a sort of regularity condition for the macroscopic behavior of the tp . Moreover, if the field is allowed to vary with x on a macroscopic scale

$$E_\varepsilon(x) = \varepsilon F(\varepsilon x) = \nabla_x U(\varepsilon x)$$

the regularity condition (5.2) should become

$$x_\varepsilon(t;F) \xrightarrow[\varepsilon\to 0]{} \int_0^t u(F(x(s)))ds + W_D(t) \tag{5.3}$$

(where the function $u \to u(F)$ does not depend on the particular field $x \to F(x)$ under consideration, and the environment has the same initial distribution as for $F = 0$, i.e. equilibrium).

Now suppose we have macroscopic regularity (5.3). Consider a potential $U(x) \to \infty$, $|x| \to \infty$, sufficiently rapidly. If the system is at temperature β^{-1}, so that it has a stationary state for which the marginal distribution of $x_\varepsilon \equiv x' \sim e^{-\beta U(x')}$, then the distribution $\rho_\beta \sim e^{-\beta U(x)}$ must also be stationary for the limiting diffusion. Since the current in this state must be zero, we have that

$$\begin{aligned} 0 = J(\rho_\beta) &= u(F(x))\,\rho_\beta(x) - D\nabla_x \rho_\beta \\ &= u(F(x))\rho_\beta(x) - D\beta\, F(x)\rho_\beta(x) \end{aligned}$$

so that $u(F)/F \equiv \sigma$ does not in fact depend on F and $\sigma = \beta D$. (It does not much matter here whether we regard x as in $\mathbb{R}$ or $\mathbb{R}^d$. Moreover for $d > 1$, it is sufficient to assume that the drift is a vector valued function $\underline{u}(F)$ of the local field. It then follows from the (Einstein) argument that $\underline{u}(F) = \sigma\underline{F} = \beta D\underline{F}$, where σ and D may be tensors).

Put somewhat differently, if $\rho_\beta \sim e^{-\beta U(x)}$ is to be stationary for the limiting diffusion, whose (forward) generator is

$$L\rho = -\nabla\cdot(u(F)\rho(x)) + D\Delta\rho$$

then σ and D must be related by $\sigma = \beta D$. Thus the Einstein relation is more or less an immediate consequence of macroscopic regularity (5.3) and the very meaning of a system's being at temperature β^{-1}, as Einstein originally argued. (Interestingly enough this appears to be the case in the Newtonian system of hard points when the initial state, before the field is turned on, is a Gibbs state but <u>not</u> when the initial state is one in which all the particles move with velocities ± 1[17].)

APPENDIX A.

In this appendix we sketch the proof of equation (4.2) for the process of section 4 in the infinite lattice. More precisely we prove the following :

Theorem A.1. Let η_t be the simple exclusion process as seen from the test particle ("environment process"), for which the motion of all the particles but the test particle moves with rates p(resp. q) to the right (left), $p > q$, i.e. the infinitesimal generator is given in equation (4.1). Then, the set J_e of extremal invariant measures for the process is given by

$$J_e = \{\mu_\alpha : 0 \leq \alpha \leq 1\} \cup \{\tilde{\mu}_n : n \geq 0\} \tag{A.1}$$

where μ_α is a Bernoulli measure with parameters $\alpha_- = \alpha$ to the left of the origin and $\alpha_+ = (1+q/p) + \alpha(q/p)$ to the right of the origin, and μ_n is concentrated on configurations with no particles to the left of the origin and n-particles to its right (see equation A.11) below for a formal description).

In order to prove theorem A.1 we consider the following partition of x :

$$\begin{aligned} X_\infty &= \{\eta : \sum_{x>0} \eta(x) = \sum_{x<0} \eta(x) = \infty \\ X_\infty^+ &= \{\eta : \sum_{x>0} \eta(x) = \infty , \sum_{x<0} \eta(x) < \infty\} \\ X_\infty^- &= \{\eta : \sum_{x>0} \eta(x) < \infty , \sum_{x<0} \eta(x) = \infty\} \\ X_n^+ &= \{\eta : \sum_{x>0} \eta(x) = n , \sum_{x<0} \eta(x) < \infty\} \end{aligned} \tag{A.2}$$

Now one can prove [15] that

$$\begin{aligned} J_e = [J \cap M(X_\infty)]_e \cup [J \cap M(X_\infty^+)]_e \cup [J \cap M(X_\infty^-)] \cup \\ \cup [J \cap \bigcup_{n\geq 0} M(X_n^+)] \end{aligned} \tag{A.3}$$

where $M(y)$ is the set of probability measures concentrated on y . We prove theorem A.1 by showing :

$$[J \cap M(X_\infty)]_e = \{\mu_\alpha .\ 0 < \alpha \leq 1\} \tag{A.4}$$

$$[J \cap M(X_\infty^+)]_e = \mu_0 \tag{A.5}$$

$$[J \cap M(X^-)]_e = \phi \tag{A.6}$$

$$[J \cap M(X_n^+)\ ] = \tilde{\mu}_n \quad , n \geq 0 \tag{A.7}$$

Proof of (A.4). Let us introduce the zero range process naturally related to the lattice gas as seen from the test particle we are discussing. A given configuration $\eta \in X_\infty$ can be represented by a doubly infinite sequence $x_i : i \in \mathbb{Z}$, $x_i < x_{i+1}$, $x_0 = 0$ where x_i denotes the site occupied by the i-th particle ($i \in \mathbb{Z}$) . (In the same way configurations belonging to X_∞^+ and X_∞^- can be represented by semi-infinite sequences and those belonging to X_n by finite sets of sites).

Let $x_u(t)$ be the position of the particle initially at site x_u . The process $\xi_t = \{\xi_t(u)\}_{u \in \mathbb{Z}}$, $\xi_t(u)$ describing the evolution of the number of sucessive empty sites to the right of the u-th particle, is the so called zero range process :

$$\xi_t(u) = x_{u+1}(t) - x_u(t) \quad . \tag{A.8}$$

It is a process with state space $y = \mathbb{N}^{\mathbb{Z}}$ and infinitesimal generator L_{zr} given by (on h-cylindric)

$$\begin{aligned} L_{zr}h(\xi) = {} & p[h(\xi_{0\,-1}) - h(\xi)] + q[h(\xi_{-10}) - h(\xi)] \\ & +(1/2) \sum_{x \neq 0} [h(\xi_{x,x-1}) - h(\xi)] + (1/2) \sum_{x \neq -1} [h(\xi_{x,x+1}) - h(\xi)] \end{aligned} \tag{A.9}$$

where

$$\xi_{x,y}(z) = \begin{cases} \xi(z)-1 & \text{if} \quad z = x \quad \text{and} \quad \xi(x) > 0 \\ \xi(z)+1 & \text{if} \quad z = y \quad \text{and} \quad \xi(x) > 0 \\ \xi(z) & \text{if either} \quad z \neq y \ , \ z \neq x \quad \text{or} \quad \xi(x) = 0 \end{cases}$$

A set of invariant measures for this process is described in [16]:

$$\mathcal{J}(L_{zr}) = \{\gamma_a\}_{0 \leq a < 1}$$

where ν_a is the geometric product measure defined by

$$\gamma_a\{\eta(x) = k\} = a_x^k(1-a_x) \tag{A.10}$$

where $a_x = a$ if $x < 0$ and $a_x = (q/p)a$ if $x \geq 0$. We find $\{a_x\}$ by solving the detailed balance equation :

$$a_o(p+(1/2)) = a_{-1}q + a_1(1/2)$$

$$a_{-1}(q+(1/2)) = a_o p + a_{-2}(1/2) \tag{A.10.b}$$

$$a_k = (1/2)(a_{k-1}+a_{k+1}) \qquad k \neq 0,1$$

$a_x = \gamma_a \{\eta(x) > 0\}$, so $0 \leq a_x < 1$.

From equation (A.10) one proves, using coupling techniques as in [15, 16], that the only extremal invariant measures for the η_t-process on $M(X_\infty)$ are those described in equation (4.4).

<u>Proof of (A.5) (A.6) (A.7).</u>

In the semi-infinite and finite case one has to look for solutions of equation (A.10b) allowing one (or two) of the a_k to be one. The state space for the corresponding zero range model is semi-infinite (or finite). For X_∞^+ for instance, we consider $a_k = 1$ for a fixed $k < 0$. This implies that in the corresponding zero range model at site κ "there are infinitely many particles" so, from site k to site $k+1$ particles enter with intensity 1/2 (respect q) if $k < -1$ (resp. $k = -1$).] the simple exclusion picture $a_k = 1$ must be read : "there are no particles to the left of the $(k+1)$-th particle", so the rate of jumping to the left for the $(k+1)$-th particle equals $1/2$ if $k < -1$ and p if $k = -1$.

In the semi infinite case we find solutions of the equations analogous to equation (A.10.b) when $k = -1$ and which imply that the density of the invariant the measures must be that described in equation (A.5) for the simple exclusion model. To prove that μ_o is the only invariant measure requires coupling technicalities which we omit.

Equation (A.6) is studied similarly and the counterpart of equation (A.10.b) gives us that there are no solutions with $a_n < 1$ for $n < k$ and $a_k = 1$ for a fixed $k \geq 0$.

Equation (A.7) is proven following the standard methods of finite state markov chains, which imply that

$$\widetilde{\mu}_\eta\{\eta : x_{i_\ell} - x_{i_\ell - 1} = b_{i_\ell} \ , \ \ell = 1,\dots,j \leq \eta\} = \prod_{\ell=1}^{j} a_n(i_\ell)[1-a_n(i\ )b_i]$$

where $$a_n(k) = \frac{z(p-q)}{(n+1)zp+1}\,k + \frac{(\eta+1)zq+1}{(\eta+1)2p+1} \quad . \tag{A.11}$$

APPENDIX B.

In this appendix, we indicate how from ρ_w, the invariant non normalizable measure of section IIb, we may easily obtain an invariant probability measure for the "environment process". This is the process induced by $X(w,t)$ in the space Ω of environments by the relation

$$U_w(t) = \tau_{X(w,t)}\ U_w$$

(or imply $w(t) = \tau_{X(w,t)}w)$ where U_w is the configuration of environments at time $t = 0$, and τ_x denotes translation by x .

In fact, the probability measure

$$d\widetilde{P} = \frac{\rho_w(0)dP}{\langle\rho_w(0)\rangle}$$

is invariant for the environment process. The key to this is the fact that $\rho_w(x)$ depends on x only through the environment seen from x, i.e. $\rho_w(x) = \rho_{\tau_x w}(0)$. To see this note first that $\rho_w(x)dx\ P(dw)$ is invariant for the process $(X(t,w),w)$, in which the environment does not change. The second component of this process, $w_t \equiv w$, is of course not the environment process. However, after the change of variables $(x,w) \to (x,\tau_x w)$ we obtain the process $(X(t,w),w(t))$, with invariant measure $d\mu = \rho_w(0)dx\ P(dw)$. The w marginal for this measure should be an invariant (probability) measure for the environment process, but unfortunately this is not well defined, since μ is not normalizable. But we may regard x as a variable defined modulo L (for any $L > 0$) both for the measure μ and for the process $(X(t,w),w(t))$ since both μ and the process $(X(t,w))$ don't essentially depend upon x , $X(t,w) - X(0,w)$ depending only upon the autonomous process $w(t)$. In this way we obtain a normalizable measure μ_L which is stationary for the process $(X(t,w),w(t))$ so that the w marginal $\rho_w(0)P(dw)$ is stationary for $w(t)$.

The stationarity of $\tilde{P}$ may also be seen directly : We may assume without loss of generality that $< \rho_w(0) > = 1$. Then, denoting the transition probability for the environment process by Q^t , and the transition probability for the position x of the tp by P_w^t , we have, for f a "nice" function of the environment, that

$$P(Q^t f) = \iint dx\ P(dw)\ \rho_w(0)\ P_w^t(0,x)\ f(\tau_x w)$$

$$= \iint dx\ P(dw)\ \rho_{\tau_{-x} w}(0) P^t_{\tau_{-x} w}(0,x)\ f(w)$$

$$= \iint dx\ P(dw)\ \rho_w(-x)\ P_w^t(-x,0) f(w)$$

$$= \int P(dw)\ f(w) \int dx\ \rho_w(x)\ P_w^t(x,0)$$

$$= \int P(dw)\ f(w)\ \rho_w(0) = \tilde{P}(f)$$

where we have used the translation invariance of P, the fact that $\rho_w(x) = \rho_{\tau_x w}(0)$, the homogeneity of the x-process : $P^t_{\tau_y w}(0,x) = P^t_w(y,y+x)$, and the stationarity of $\rho_w(x)$ under P^t_w.

Note that we didn't need to explicitly refer to this invariant probability measure to obtain $u(E)$. In general, assuming ergodicity $u(E) = \lim_{t\to\infty} \frac{X(t,w)}{t}$ $\tilde{P}$ a.s. where $\tilde{P}$ is the stationary probability measure for the environment process. But here $\tilde{P}$ is equivalent to P.

ACKNOWLEDGEMENTS :

We thank E.G.D. Cohen, D. Dürr, C. Kipnis, E. Presutti, G. Papanicolaou and H. Spohn for useful discussions. We also thank the organizers of the Random Fields Conference in Köszeg, where a talk on the subject of this paper was given by one of us (J.L.L.), for their patience in waiting for the manuscript. P.A.F. thanks the Math. Dept. of Rutgers and J.L.L. for their very friendly hospitality. Finally, we thank the IHES for their hospitality while this was being written and (hopefully) clarified, and Ms. F. Breiner, for typing it.

The work was supported in part by NSF grants DMR81-14726 and PHY 8201708, at Rutgers University. P.A.F. was supported by FAPESP Grant 021719-9 and CNPq Grant 201682-83.

REFERENCES

1. A. Einstein, Ann. of Physik 17 549 (1905); also 19, 371 (1906). - E. Nelson, Dynamical Theories of Brownian Motion, Princeton University 1967. - Selected Papers on noise and stochastic processes, J.L. Doob, L.S. Ornstein, G. E. Uhlenbeck, S.O. Rice, M. Kac, S. Chandrasekhar. Edited by Nelson Wax, Dover Publications New York (1954).

2. M.S. Green, J. Chem. Phys., 20 (1952) 1281 - Kubo J. Phys. Soc. Jap. 12 (1957) 570 - R. Zwanzig, Ann. Rev. Phys. Chem. 16, 67 (1965) - D. Forster, Hydrodynamic fluctuations, Broken symmetries and Correlation Functions, Reading Mass : W.A. Benjamin Inc. 1975.

3 N. van Kampen, Phys. Norv. 5 (1971) 279 - E.G.D. Cohen, Physica 118 A (1983) 17-42.

4 T.E. Harris, J. Appl. Probab. 2 (1965) 323-338 - R. Arratia Ann. Prob. 11 (1983) 362-373 - H. Rost, M.E. Vares, Oral communication.

5. D.W. Jepsen, J. Math. Phys. 6 (1965) 405 - J.L. Lebowitz and J.K. Percus, Phys. Rev. 155 (1967) 122 - F. Spitzer, J. Math. Mech. 18 (1969) 973 - F. Spitzer, Lecture Notes in Math. 89 (1969) 201, Springer, Berlin 1969.

6. A. De Masi, P.A. Ferrari, S. Goldstein and D. Wick, An Invariance Principle for Reversible Markov Processes, with Applications to Random Motions in Random Environments, in preparation.

7 G. Papanicolaou and S.R.S. Varadhan, Statistics and Probability : Essays in honor of C.R. Rao, North Holland 1982 547-552.

8. D. Durr, S. Goldstein, J.L. Lebowitz, in preparation.

9. M. Guo, Limit Theorems for Interacting Particle systems, Phd. Thesis, Courant Institute 1984.

10. C. Kipnis, SRS Varadhan, to appear in Comm. Math. Phys.

11. S.R.S. Varadhan, private communication

12. S.R. Foguel, The Ergodic Theory of Markov Processes, Van Nostrand, NY 1969.

13. To be written by ...

14. Rodenhausen, Communication at the Random Fields Conference Köszeg (1984).

15. P.A. Ferrari, preprint 1984.

16. E.Andjel, Ann. Prob. 10 , 525-547 (1982)

17. P. Calderoni, D.Durr, Preprint.

INSTITUT DES HAUTES ETUDES SCIENTIFIQUES, 91440, Bures sur Yvette, France.

Permanent addresses :

For S.G. and J. L.L. : Dept. of Math. Rutgers University, New Brunswick, N.J. 08903, U.S.A.

For P.A.F. : Instituto de Matematica e Estatistica, Universidade de Sao Paulo, Cx Postal 20570, Sao Paulo, BRASIL.

TRANSITION FROM PURE POINT TO CONTINUOUS SPECTRUM FOR RANDOM SCHRÖDINGER EQUATIONS : SOME EXAMPLES

B. Souillard

Abstract

I review some recent results on Schrödinger equations with random potentials, and specially discuss the known examples where a transition in the nature of the spectrum occurs when varying the coupling constant or the energy : a transition from pure point spectrum with power decaying eigenfunctions to purely continuous spectrum has been proven recently for two classes of disordered systems, a transition which differs from the Mott-Anderson transition proven in certain Anderson models.

I will also discuss relevance of localization theory to hydrodynamics and plasma physics.

The week before the present meeting in Köszeg was held in Braunschweig (Germany) a conference on "localization in impure metals". At this conference were present the theoreticians of localization and experimentalists from many domains : 3-d doped semiconductors, metallic glasses, 2-d accumulation layers such as field effect transitors or heterojunctions with modulated doping, very narrow wires... What is this subject called "localization" which has been worth the Nobel prize for Anderson and for Mott and can attract the interest of people from so different fields ?

Let us consider a particle, an electron, in a potential $\lambda V(x)$. This potential will be taken random in order to model the disorder present in the sample. If we treat this electron as a classical particle, it will remain localized in a bounded region if its energy E is smaller than the percolation threshold E_c of the random field $\lambda V(x)$; in contrast if $E > E_c$ it will move in an infinitely extended region and propagate to infinity.

If we now treat this electron as a quantum particle, which we must do for example in order to discuss the electric conductivity of disordered systems at low temperatures, we are led in the common approximation of independant electrons to study a Schrödinger Hamiltonian

$$H = -\Delta + \lambda V \tag{1}$$

acting on $L^2(\mathbb{R}^d)$ where V is a random potential : if we take into account the well known tight-binding approximation we are in contrast led to study the finite difference version of this Hamiltonian defined as

$$(H\psi)(x) = \sum_{y:\,|y-x|=1} \psi(y) + \lambda V(x)\ \psi(x) \tag{2}$$

acting on $\ell^2(\mathbb{Z}^d)$. The continuous and the discrete cases are expected to yield the same results, besides the obvious differences. The random potential in equation (2) is often considered as given by independant and identically distributed random variables $\{V_{(x)}\}_{x\in\mathbb{Z}^d}$: we will also suppose here for simplicity, that these random variables have zero expectation and posses a density which is L^∞ and with moments of order $2+\varepsilon$; these conditions in many of the results quoted are not necessary and we refer to the papers for complete results.

What do we want to know ? One of the basic questions is to determine for almost-all realization of the potential V(x) the nature of the spectrum of H near some energy E : is it pure point, singular continuous, absolutely continuous or a superposition of these types. If the spectrum is pure point near some energy E, then a wave packet built up with energies near E will basically remain in a finite region during time evolution and a solid with Fermi level at E should be an insulator. In contrast if the spectrum is continuous, the particle will go to infinity as $t\to\infty$ and a solid with Fermi level at E may be a conductor [1][2]

Now when $E \ll E_c$, it is clear on the physical ground that quantum tunneling will not spoil the localization of the classical particle : this is because randomness of the potential does not allow constructive tunneling interferences. This fact is however not necessarily easy to prove! Some progress have been done recently going in the good direction : it has been proven that almost-surely the diffusion constant vanishes [3][4], and that there is no absolutely continuous spectrum [5] . One expects that in such situations the spectrum is only pure point.

From the pioneering works of Anderson and Mott, two remarquable facts have been discovered which are responsible for the emergence of localization as a full domain of physics :

1. The quantum localized regime persists much farther than the classical one.
2. There exists a sharp transition from extended states to localized states when varying the coupling constant or the energy at least for $d \geqslant 3$: this Mott-Anderson transition explains a certain number of metal-insulator transitions observed in various materials at low temperatures when varying the concentration of impurities.

The most spectacular result connected with point 1 is the complete exponential localization of all states for one-dimensional systems, for arbitrary small disorder. As for mathematical results, it was proven that for any λ and almost-all realization of the potential, the spectrum of H is pure point and the wave functions are exponentially decaying. Other proofs applying to various cases have been also given since [7-11] . We do not discuss these results in greater detail, since our main purpose here is to discuss the transition from pure point to continuous spectrum. Let us just mention a fact which is useful in the proof of one of the results below : among the present various proofs of localization for one-dimensional systems, the one of reference [11] is, up to now the strongest one in the sense that it is the only one which allows to study non homogeneous random potentials, that is random potentials which do not posess translation invariance properties.

Let us come now to the second point mentionned above : the theory predicts for $d \geqslant 3$ a transition from exponentially localized to extended states in appropriate domains of E and λ; this is expected to correspond to a transition from a situation with only pure point spectrum and exponentially decaying eigenfunctions to a situation with only continuous spectrum. As far as mathematical results are concerned the Mott-Anderson transition is only proven on the Bethe lattice, and we discuss the corresponding results below. On the other hand, a new type of transition, namely a transition from pure point spectrum with power law localized states to purely continuous spectrum, has been discovered recently in other situations, and we will discuss it later on.

So let us come now to the discussion of the results on the Mott-Anderson transition on the Bethe lattice. A Bethe lattice is a connected graph with no closed loops and a constant coordination number at each lattice site; we can consider a model of type (2) associate to this graph. Why are we interested in such a model ? In the statistical mechanics of phase transitions, one knows the crucial role played by the mean field theory : it was the basis of our understanding of phase transitions and it allows to compute the critical exponents for systems in large enough dimension d ; unfortunately no mean field theory is known for the localization problem. More precisely the models which would be natural candidates for a mean field theory of localization do not exhibit a transition. It is thus natural to look for the next simple model which should exhibit a transition : it should be the model on the Bethe lattice. Unfortunately this model for localization is still very hard. Abou-Chacra, Anderson and Thouless [12] could establish a limit of stability of localized states and numerically investigate the region of extended states. Recently we have studied this problem again [13] . Although we have not obtained a complete solution of it, we could at least get a certain number of exact results : for a large class of distributions of random potentials we prove that almost surely

1. The spectrum of H is pure point for $\lambda > \lambda_1$, and is pure point for $|E| > E_1(\lambda)$.
2. The spectrum of H is purely absolutely continuous for $\lambda < \lambda_2$ in $|E| < E_2(\lambda)$.

Thus we have here the first model where the Mott-Anderson transition is proven. The next results are concerned with the critical properties of the model, and for them we need to require stronger assumption on the distributions of random variables V_x and in particular we need that their density probability r is an analytic function. Then

3. the density of states is an analytic function of E and λ ; in particular it is analytic at the transition points.

4. the localization length, which governs the rate of decay of the eigenfunctions in the pure point spectrum, diverges as $|E - E_c|^{-1}$ at the threshold, yielding a critical exponent $\nu = 1$ for the localization problem on the Bethe lattice.

This is the first critical exponent exactly computed for the localization problem. This result implies that the critical exponent ν is equal to 1/2 for dimension d large enough. This fact, and possible consequences on the question of finding the upper critical dimension for the localization transition are discussed in [14] .

Let us turn now to the other type of transition which has been found recently and which we alluded to above. Let us first consider a continuous one-dimensional Schrödinger equation in an electric field : we have thus to study the Hamiltonian

$$H = \frac{d^2}{dx^2} - Fx + V(x) \tag{3}$$

When $F = 0$ and $V(x)$ is a random potential, then almost-surely H has only pure point spectrum and exponentially decaying eigenfunctions as mentionned previously. What happens for $F \neq 0$?

1. if the random potential in $V(x)$ is almost-surely sufficiently smooth (e.g. has two bounded derivatives) then for any $F \neq 0$, H has only absolutely continuous spectrum and in fact an initial state is uniformly accelerated [15].

2. if in contrast the random potential $V(x)$ is given as a Krönig-Penney potential $V(x) = \sum_{n \in \mathbb{Z}} V(n)\,\delta(x-n)$, $V(n)$ independant random variables, a model which is the natural one for several physical situations, then [16]

if $0 < F < F_1$, H has almost-surely only a pure point spectrum with power low decaying wave functions at $+\infty$

if $F > F_2$, H has almost-surely only a continuous spectrum.

Some of these results were anticipated in the physical literature by perturbation treatment or numerical calculations [17][18].

The previous transition is strongly related to the following one [16] : let us consider a one-dimensional discrete Schrödinger equation of the type

$$(H\psi)(n) = \psi(n+1) + \psi(n-1) + \lambda|n|^{-\alpha} V(n)\psi(n) \tag{4}$$

The $V(n)$ are again random, and the only difference with equation (2) is that here the random potential $V(n)$ is multiplied by $|n|^{-\alpha}$ and thus the disorder decreases as $|n|$ increases. For $\alpha = 0$, we have the usual one-dimensional case and the spectrum is pure point with exponentially decaying eigenfunctions. What happens for $\alpha > 0$? Then almost surely [16] the essential spectrum of H is $[-2,2]$ and

1. if $0 < \alpha < 1/2$, it is pure point and the eigenfunctions decay as $\exp\{-|n|^{(1-2\alpha)}\}$.

2. if $\alpha = 1/2$, it is pure point with power low decaying eigenfunctions if $\lambda > \lambda_1$, and it is purely continuous if $\lambda < \lambda_2$ in $|E| < E_1(\lambda) < 2$.

3. if $\alpha > 1/2$, it is purely continuous. In fact, S. Kotani [19] has proven it is purely absolutely continuous (scattering theory shows it is purely absolutely continuous only for $\alpha > 1$).

In all these statements we have always supposed that

V(n) has a distribution with a finite moment of order $2+\varepsilon$ It is curious to note that if we choose instead for the V(n) a Cauchy distribution (the so called Lloyd model), then we can prove [16] that, in the electric field case with a Krönig Penney potential, the spectrum is almost-surely pure point for all F, and in the discrete model described above the essential spectrum is almost-surely pure point now for any $0<\alpha<1$; in both cases the eigenfunctions decay as $\exp\{-|n|^{\alpha}\}$, $0<\alpha<1$. It is the first case where one can rigorously prove a difference of behaviour between Cauchy and distributions with moments of order $2+\varepsilon$.

As a conclusion, I would like to stress that these phenomenons are not specific to quantum problems. The basic mechanism at work is interference : is it constructive or destructive ? So localization and Anderson-Mott transition can be exhibited by other wave propagation problems. Let us consider for example the propagation of shallow water waves on a rough bottom [20] : the waves will be partly neglected and partly transmitted by the fluctuations of the bottom, and localization will arise in various situations; this can be observed experimentaly. Let us also consider the propagation of electromagnetic waves in a plasma [21] : here also they will be partly reflected and partly transmitted, this time because of the density fluctuations inside the plasma; we have shown that localization is relevant for real plasmas in various circumstances and moreover that localization phenomenon has consequences on the instabilities inside the plasma, yielding a new mechanism to turn the convective ones into absolute ones.

Other aspects of disordered systems

We have focuss here mainly on the transitions from pure point to continuous spectrum. However disordered systems present many other interesting aspects and some progress have been also achieved in these directions. Many aspects complementary to this talk can be found in the talk of Pastur [22] or in [23] . A very large bibliography on the topic can be found in [24].

References

[1] D. Ruelle, Nuovo Cimento A61 (1969) 655

[2] W. Amrein, V. Georgescu, Helv. Phys. Acta 46 (1973) 633

[3] J. Fröhlich, T. Spencer, Commun. Math. Phys. 88 (1983) 151.

[4] H. Holden, F. Martinelli, Commun. Math. Phys. (in press).

[5] F. Martinelli, E. Scoppola, "A remark on the absence of absolutely continuous spectrum in the Anderson model for large disorder or low energy". Preprint Bielefeld.

[6] Y. Gold'sheid, S. Molchanov, L. Pastur , Funct. Anal. i Pril. 11 (1977) 1; S. Molchanov, Math. USSR Izv. 12 (1987) 69.

[7] H. Kunz, B. Souillard, Commun. Math. Phys. 78 (1980) 201.

[8] G. Royer, Bull. Soc. Math. France 110 (1982) 27.

[9] R. Carmona, Duke Math. J. 49 (1982) 191.

[10] J. Lacroix, Ann. Inst. Elie Cartan N°7, Nancy (1982).

[11] F. Delyon, H. Kunz, B. Souillard, J. Phys. A16 (1983) 25.

[12] R. Abou-Chacra, P.W. Anderson, D.J. Thouless, J. Phys. C6 (1973) 1734.

[13] H. Kunz, B. Souillard, J. Physique (Paris) Lett. 44 (1983) 411, and to be submitted to J. Stat. Phys.

[14] H. Kunz, B. Souillard, J. Physique (Paris) Lett. 44 (1983) 503.

[15] F. Bentosela, R. Carmona, P. Duclos, B. Simon, B. Souillard, R. Weder, Commun. Math. Phys. 88 (1983) 387.

[16] F. Delyon, B. Simon, B. Souillard, Phys. Rev. Lett. 52, (1984) 2187 and "From power pure point to continuous spectrum in disordered systems", Preprint Ecole Polytechnique (1984)

[17] V.N. Prigodin, ZH. Eksp. Teor. Fiz. 79 (1980) 2338.

[18] C.M. Soukoulis, J.V. Jose, E.N. Economou, Ping Sheng, Phys. Rev. Lett. 50 (1983) 754.

[19] S. Kotani, Private communication.

[20] E. Guazzelli, E. Guyon, B. Souillard, J. Phys.(Paris) Lett. 44 (1983) 837.

[21] D. Escande, B. Souillard, Phys. Rev. Lett. 52 (1984) 1296.

[22] L. Pastur, in these Proceedings.

[23] B. Souillard, in Proceeding of the "Les Houches Workshop on Common Trends in Particle and Condensed Matter Physics, Phys. Rep. 103 (1984) 41.

[24] B. Simon, B. Souillard, Franco-American Meeting on the Mathematics of Random and Almost-Periodic Potentials, to appear in J. Stat. Phys.

B. Souillard

Centre de Physique Théorique
Ecole Polytechnique
F-91128 Palaiseau Cedex - France.

RIGOROUS STUDIES OF CRITICAL BEHAVIOR II

Michael Aizenman*

The Institute for Advanced Study
Princeton, New Jersey 08540, USA

1. Introduction

The task addressed in the talk is the analysis of the critical behavior in statistical mechanical systems, and related problems in the theory of random fields. While it would have been very useful to have mathematical means of approaching this subject at the level of generality presented here in Professor Dobrushin's analysis of the high temperature phases, we find that the richness of the behavior exhibited at low temperatures calls for more restricted treatments of special models. Nevertheless, it was found that the somewhat delicate issues of the phase structure and the critical behavior may be approached by methods which apply to a number of systems. In particular, the picture offered by random walk representations led to very useful insights for the study of ferromagnetic systems, ϕ^4_d field theory, and percolation models. (It is hoped that random surface expansions might play a somewhat similar role for gauge fields and some models of spin glasses.)

Since most of the talk given at the Conference is covered by a paper which appears in the Proceedings of the Sitges 1984 meeting, I shall focus here on the newest results which are not discussed there. (Their presentation here is self-contained.)

A brief outline of [1] is given in Section 2. The results which are derived in this sequel to it deal with the phase structure of ferromagnetic systems of one-component spin variables with the Hamiltonian

$$H = -\sum J_{x,y}\sigma_x\sigma_y - h\sum\sigma_x \quad ; \quad J_z \geq 0 \quad . \tag{1}$$

Our methods apply to systems with single spin distributions, $\rho(\sigma)\,d\sigma$, which belong to the Griffiths-Simon class. This includes the following standard examples: i) Ising spins $\rho(\sigma) = \delta(\sigma^2 - 1)$, ii) "ϕ^4" variables $\rho(\sigma) = \exp(-\lambda\sigma^4 + b\sigma^2)$ $\lambda > 0$, iii) the "classical n spin" $\rho(\sigma) = \sum_{-n}^{n} \delta(\sigma - k)$, and iv) the uniformly distributed $\rho(\sigma) = I[-1, 1]$. Since at one point we need the "Gaussian domination bounds" whose derivation required reflection positivity, the strongest results are restricted to the nearest-neighbor interactions. For such systems we prove here the following:

a) Absence of an intermediate phase. The above one-component systems, in $d > 2$ dimensions, exhibit a direct transition from the high-temperature phase, at which the correlations decay exponentially, to the low-temperature phase, characterized by spontaneous magnetization. (It is known that the mass vanishes as $T \to T_c + 0$). What is ruled out is an intermediate regime, in which the correlation functions could, for example, behave like in the Kosterlitz-Thouless phase of the two-dimensional, two-component, plane rotor system.

b) Mean-field bound on the critical exponent of the spontaneous magnetization. It is shown here that for $T < T_c$ the spontaneous magnetization satisfies:

$$M(T) \geq \text{const.}(T_c - T)^{1/2} \qquad (1.2)$$

i.e., quite generally $\beta \leq 1/2$, where β stands for the critical exponent.

c) Convergent upper and lower bounds on the critical temperature. A method is derived for the use of the values of the magnetic susceptibility of finite systems (which are in principle calculable) for rigorous upper and lower bounds on the critical temperature, whose discrepancy can be made arbitrarily small by performing the calculation for a system of a large, yet predetermined, size. This complements the previous upper bounds of Simon (based on the Simon-Lieb inequality).

While the convergence is only by a fractional power law, such bounds can be used for theoretical results, such as:

d) "Restricted Continuity" of T_c -- as a function of the interaction. The restriction is to regimes in which the systems are uniformly regular, in a sense introduced below. Thus, for the nearest neighbor models in $d > 2$ dimensions we can prove that T_c depends continuously on the spin distribution. Another example is discussed in Section 6.

The above results are derived from a differential inequality which was proven, and can be understood, within the context of a random walk representation. I shall not discuss its derivation here;

however in the next section I recapitulate that general scheme -- and the content of the talk given in Kösheg.

2. A Brief Outline of Part 1.

Random walk representations have turned out to be quite useful for the analysis of statistical mechanical models with the Hamiltonian (1.1) and of the closely related ϕ^4_d field theory. Such representations led to both heuristic insights and to rigorous arguments. For the latter, the R. W. representations were applied as means of derivation of various relations between the physical quantities of the given model. The new relations were used in the form of either direct inequalities, like a bound on the truncated four-point function in terms of the two-point function, or differential inequalities in which the bound is on the derivative of some quantity with respect to a ("bare-") parameter (such an inequality is used in the next section). Analogies with random walk problems suggest relations which may be both valid and useful.

Some of the results which were arrived at by this approach (by a number of authors) are referred to in [1]. Current ideas for further progress call for the derivation of inequalities involving partial derivatives of one dynamical quantity with respect to another (e.g., the correlation length). Such an approach has been implemented in the analysis of an analogous Brownian motion problem mentioned later in this section.

Two random walk representations for the Ising model and the ϕ^4_d field theory were presented in the works of the author [2], and Brydges, Fröhlich and Spencer [3]. While the two are different, one offering some advantages for the strongly-coupled, and the other for the perturbative, regimes they share a number of basic properties (introduced in [2,4]). These are quoted in the first part of this report, which includes also a description of a random walk representation for Ising spin systems, whose derivation is simpler than that of [2].

The relevance of our understanding of the intersection properties of simple random walks, or the paths of independent Brownian motions, spurred the reconsideration of some classical results. This led to a new method for the analysis of the probability that the paths of two independent Brownian motions are within a short distance of each other. In recent, parallel, works of Frölich and Felder [7] (which follows an earlier suggestion of Fröhlich [5], and the author [6], a "renormalization group" differential equation was set for this quantity. It was shown there that simple bounds on the corresponding β function are quite effective, leading to a unified treatment of the probability of intersection (which vanishes for $d \geq 4$) above, below, and at the critical dimension $d = 4$. (In fact, since the linear part of the β function was determined exactly, the above mentioned bounds provided new information -- on the sign and the magnitude of the "second order" terms).

In view of the correspondence of the renormalized coupling constant to the probability of intersection of certain (self-interacting) random walks, these works suggest a new approach to the study of the ϕ^4_d field theory, and of issues like the hyperscaling in the three-dimensional Ising model. The sights are now set on non-perturbative bounds on β functions; however, this still requires a considerable sharpening of our methods.

3. Absence of an Intermediate Phase

We turn now to the results stated in the Introduction. The main result is stated in this section as *Theorem 3.1*. Its proof is given in Section 4.

We consider here one-component spin systems with the translation invariant Hamiltonian (1.1), for which $|J| = \sum_x J_{0,x} < \infty$. The single site spin distribution $\rho(\sigma)$ is assumed to belong to the Griffiths-Simon class [8,9], and to decay (in σ) faster than the Gaussians, i.e., $\int \rho(d\sigma)\, e^{b\sigma^2} < \infty$ for any $b \in \mathbb{R}$.

The G.-S. class consists of spin distributions for which the variable σ can be represented by means of a sum of ferromagnetically coupled Ising spins, or a distributional limit of such sums. Included are some of the most frequently referred to one-component spin variables (some of them continuous) which are listed in the Introduction.

In particular, the following analysis applies to the standard Ising models. We shall later require another, dynamical, hypothesis, and prove it for the nearest neighbor models in $d > 2$ dimensions.

i) Preliminaries

Let me first describe the problem and some preliminary results, most of them well known. In general, the spins described above (which may be represented as "block variables") inherit most of the properties which were initially derived for Ising spins; like various "dimensionally balanced" correlation inequalities. Likewise, the Lee-Yang theory [10] applies and shows that phase transitions can occur only at zero magnetic field ($h=0$).

For a finite system, in a region $\Lambda_L = [-L,L]^d \subset \mathbb{Z}^d$ with periodic boundary conditions, we denote by $\chi_L = \chi_L(\beta)$ the magnetic susceptibility at $h=0$:

$$\chi_L = \sum_{x\in\Lambda_L} \langle\sigma_0\,\sigma_x\rangle_L \tag{3.1}$$

where $\langle f(\sigma)\rangle_L = \mathrm{tr}\, e^{-\beta H} f(\sigma)/\mathrm{tr}\, e^{-\beta H}$. β is the inverse temperature, "tr" stands for $\int \prod_{x\in\Lambda_L} \rho(\sigma_x)\, d\sigma_x$, and "periodic boundary conditions" means that the interaction J is replaced by $J^L_{x,y} = \sum_{n\in\mathbb{Z}^d} J_{x,y+2Ln}$.

The corresponding quantities for the free boundary conditions are denoted by $\langle\sigma_0\,\sigma_x\rangle_L^{f.b.}$ and $\chi_L^{f.b.} (\le \chi_L)$. At the infinite volume limit, $L\to\infty$, we drop the subscript L .

The f.b. correlation functions are monotone non-decreasing as a function of L , and thus converge to a limit. At high temperatures (= small β) the magnetic susceptibility at $h=0$, $\chi^{f.b.} = \sum_x \langle\sigma_0\,\sigma_x\rangle$, is finite and the following result applies.

Proposition 3.1: If $\chi_{f.b.} < \infty$, for an interaction with $h=0$ and $|J| = \sum J_{0,y} < \infty$, then

$$\lim_{L\to\infty} \langle\sigma_0\,\sigma_x\rangle_L = \lim_{L\to\infty} \langle\sigma_0\,\sigma_x\rangle_L^{f.b.} \quad ; \quad \lim_{L\to\infty} \chi_L = \chi_{f.b.} \tag{3.2}$$

and if, furthermore, the interaction decays exponentially, then the limiting correlation function satisfies:

$$\langle\sigma_0\,\sigma_x\rangle \le \text{const.}\ e^{-m|x|} \tag{3.3}$$

with a strictly positive "mass" $m(\beta)$.

For finite range models this follows easily from Simon's analysis [11]. The proof of the generalization stated here is in the Appendix.

The spontaneous magnetization $M(\beta)$ is defined as:

$$M(\beta) = \lim_{h\to 0+} \langle\sigma_0\rangle(\beta,h) \tag{3.4}$$

The limit exists and is independent of the boundary conditions since $\beta\langle\sigma_0\rangle = \partial P(\beta,h)/\partial h$ where $P(\beta,h)$ (the bulk-free energy) is a convex function of h .

The non-vanishing of $M(\beta)$ at $h=0$ is the manifestation of a spontaneous symmetry breaking. This phenomenon occurs at low temperatures (for $d\geq 2$), and is implied by the long-range order, which refers to the non-vanishing of $\langle\sigma_0\,\sigma_x\rangle$ for $|x|\to\infty$. The latter does not occur in the high temperature phase described in Proposition 2.1, however, the stronger statement is also true.

Proposition 3.2: If $\chi<\infty$ *, as in Proposition 1, then*

$$M(\beta) = 0 \quad . \tag{3.5}$$

The proof (which may be well known) is given in Appendix II. Following is a general criterion for the opposite phase.

Proposition 3.3: If for some β

$$\limsup_{L\to\infty} \chi_L(\beta)/|\Lambda_L| = D > 0 \quad , \tag{3.6}$$

then there is both a non-vanishing spontaneous magnetization:

$$M(\beta) > D^{1/2} \quad , \tag{3.7}$$

and long-range-order, in the sense that any limiting function defined with p.b.c., and a subsequence of volumes, satisfies

$$\sum_{x\in\Lambda_k} \langle\sigma_0\,\sigma_x\rangle/|\Lambda_k| \geq D \qquad \underline{\textit{for any } k} \quad . \tag{3.8}$$

The above criterion for the spontaneous magnetization was derived by Griffiths [12], and (3.8) is obtained by the Schwarz inequality. For the understanding of the above statements it is useful to express them in terms of the quantities $S_\Lambda = \sum_{x\in\Lambda} \sigma_x/|\Lambda|$.

In the absence of more information, one could define three critical temperatures (we regard T and β^{-1} as synonymous):

$T_c^{h.t.} = \inf\{T|\chi(\beta)<\infty\}$ (h.t. -- for the high temperature phase; if $J_{0,x}$ decays exponentially, then so does $\langle\sigma_0\sigma_x\rangle$ for $T>T_c^{h.t.}$.)

$$T_c^{\text{s.m.}} = \sup\{T \mid M(\beta) > 0\}$$

$$T_c^{\text{l.r.o.}} = \sup\{T \mid (3.8) \text{ holds with some } D > 0\} \tag{3.9}$$

From what is said above

$$T_c^{\text{l.r.o.}} \leq T_c^{\text{s.m.}} \leq T_c^{\text{h.t.}} \quad . \tag{3.10}$$

Examples where the last is a <u>strict</u> inequality, are provided by the two-component plane rotor and its approximating discrete clock models in two dimensions. In particular, the latter have an intermediate Kosterlitz-Thouless-type phase at which $M = 0$ -- yet the correlations decay only by a power law [13,14,15].

Our main result is that for a general class of one-component systems in $d > 2$ dimensions, all the critical temperatures in (3.10) are equal.

<u>ii) The Regularity Condition</u>.

Since our analysis leads also to a re-derivation of the existence of a phase transition, one might expect that it should require some additional hypothesis which, in particular, should rule out the one-dimensional finite-range systems. We shall now introduce such a condition.

Let us denote by B_L the sum:

$$B_L = \sum_{x \in \Lambda_L} \langle \sigma_0 \sigma_x \rangle_L^2 \quad , \tag{3.11}$$

where $\langle -- \rangle$ is with the periodic boundary conditions. Removing from this "bubble diagram" its "zero momentum component" we obtain the quantity

$$B_L^T = B_L - \chi_L^2 / |\Lambda_L| = \sum_{x \in \Lambda_L} \langle \sigma_0 ; \sigma_x \rangle_L^2 \quad , \tag{3.12}$$

where the truncated function is defined as

$$\langle \sigma_0 ; \sigma_x \rangle_L = \langle \sigma_0 \sigma_x \rangle_L - \chi_L / |\Lambda_L| \quad . \tag{3.13}$$

Since σ is not restricted to be an Ising spin, its scale, as well as those of χ and B, are quite arbitrary at the level of generality considered here. However, the combinations

$$\hat{\chi}_L = \beta|J|\chi_L \qquad \hat{B}_L^{(T)} = (\beta|J|)^2 B_L^{(T)} \quad , \tag{3.14}$$

where $|J| = \sum_x J_{0,x}$, are invariant under the simple "field strength renormalization."

<u>*Definition*</u>*: We say that the system is* <u>*regular*</u> *if for all* β *and* L

$$1 + \hat{B}_L^T \leq \hat{\chi}_L^2 f(\hat{\chi}_L) \tag{3.15}$$

with some function f *for which*

$$\int_1^\infty f(s)\, ds < \infty \tag{3.16}$$

<u>Remarks</u>:

i) It is easy to see that $\hat{B}_L^T \leq \hat{\chi}_L^2$. Hence, it may always be assumed that

$$1/s^2 \leq f(s) \leq 1 + 1/s^2 \quad . \tag{3.17}$$

ii) In the one-dimensional Ising model at the infinite volume limit (where $\langle\sigma_0 \sigma_x\rangle = e^{-m|x|}$), $B \approx \chi/2$. Thus (3.15) can hold there only with $f(s) \geq (2s)^{-1}$. In this case (3.16) fails, but barely so. One may expect that above its lower critical dimension, the model <u>is</u> regular. We shall see that this is the case in dimensions $d > 2$.

iii) For another insight into the notion of regularity, let us note that if (3.16) holds, and $f(\cdot)$ is monotone, then $f(s) = s^{-1} o(1)$. An immediate implication is that when a regular system exhibits "long-range order," i.e., (3.6) holds, then in the L^2 sense the correlation function strongly condenses into its zero momentum component. The ratio $B_L^T / (\chi_L^2/|\Lambda_L|)$ is bounded by $|\Lambda_L|/\chi_L\, o(1) = D^{-1} o(1)$ which vanishes as $L \to \infty$.

iv) For systems endowed with reflection positivity, regularity is essentially implied, up to considerations of discretization effects, by the familiar condition (which has already played a key role in proofs of the existence of a phase transition [16, 17]):

$$\int_{[-\pi,\pi]^d} dp \, \frac{1}{E(p)} < \infty \tag{3.18}$$

where

$$E(p) = - \sum_{x \in \mathbb{Z}^d} e^{ipx} \left[J_{0,x} - |J| \delta_{x,0} \right] . \tag{3.19}$$

(3.18) and (3.16) are related via the "Gaussian domination" (or "equipartition") bound of Fröhlich, Simon and Spencer [16], which states that in RP systems

$$G_L(p) = \sum_{x \in \Lambda_L} e^{ipx} \langle \sigma_0 \sigma_x \rangle_L \tag{3.20}$$

obeys

$$G_L(p) \leq \left[2\beta \, E(p) \right]^{-1} , \tag{3.21}$$

at the dual momenta $p \in (\Pi/L)^{\mathbb{Z}^d} \cap [-\pi,\pi)^d$.

By the Plancherel identity:

$$B_L^T = (2\pi)^{-d} \text{ "} \int^{(L)}_{\substack{[-\pi,\pi)^d \\ p \neq 0}} dp \text{ "} G_L(p)^2 \tag{3.22}$$

where " $\int^{(L)} dp$ " represents the "Riemann sum" over the discrete set of momenta specified above, with the weights $(\pi/L)^d$, excluding $p = 0$.

Combining (3.21) with the obvious inequality $G_L(p) \leq \chi_L$, and using (3.22), one gets

$$\left(1 + \hat{B}_L^T \right) / \hat{\chi}_L^{\,2} \leq f^{(L)}(\hat{\chi}_L) \tag{3.23}$$

with the functions:

$$f^{(L)}(s) = \frac{1}{s^2} + (2\pi)^{-d} \text{"}\int_{\substack{[-\pi,\pi)^d \\ p\neq 0}}^{(L)} dp\text{"} \min\left\{\left(\frac{|J|}{2E(p)s}\right)^2, 1\right\} , \qquad (3.24)$$

for which

$$\int_1^\infty f^{(L)}(s)\, ds \leq 1 + (2\pi)^{-d} \text{"}\int_{\substack{[-\pi,\pi)^d \\ p\neq 0}}^{(L)} dp\text{"} \frac{|J|}{E(p)} \qquad (3.25)$$

In particular, we can establish regularity for the following important class of models.

Proposition 3.4: In $d > 2$ dimensions, non-degenerate nearest neighbor models (for which $J_{x,y} \neq 0$ iff $|x-y| = 1$), are regular.

Proof: These models are reflection positive [16], and the above discussion applies, with

$$E(p) = 4 \sum_{i=1}^{d} J_i \sin^2(p_i/2)$$

$$\approx |p|^2 \quad \text{if} \quad J_i \equiv 1 \quad . \qquad (3.27)$$

Furthermore, the discretization effects in (3.24) and (3.25) are easy to control here. (3.24) -- for whose integration it is convenient to use: $\min\{a^{-2},1\} \leq 2/[a^2+1]$, implies that (3.15) is satisfied with

$$f(s) = \begin{cases} c\ s^{-\min\{d/2,2\}} & d \neq 4 \\ c\ s^{-2}[1+\ln_+ s] & d = 4 \end{cases} \qquad (3.28)$$

This function is integrable for $d > 2$, as suggested already by (3.25). ■

Other examples of reflection positive interactions are [17]

$$J_{x,y} = 1/|x-y|^\alpha \quad , \quad \text{and} \quad \left[1+|x-y|^\alpha\right]^{-1} \qquad (3.29)$$

in $d = 1$ dimension. Using the analysis of Ref. [17] one can show that these are regular for $1 < \alpha < 2$.

It should be emphasized that (3.18) is only a sufficient condition (for RP interactions). I would conjecture that finite-range one-component systems, of the type considered here, are regular even in $d = 2$ dimensions -- where (3.18) fails.

Statement of the Main Result

Theorem 3.1: In any regular system, of the type described at the beginning of this section (with ferromagnetic pair interaction $|J| < \infty$, *and spins in the Griffiths-Simon class),*

$$T_c^{\text{l.r.o.}} = T_c^{\text{s.m.}} = T_c^{\text{h.t.}} \tag{3.30}$$

and that critical temperature $(T_c \equiv \beta_c^{-1})$ *satisfies*

$$\beta_c^{\text{m.f.}} \leq \beta_c \leq \beta_c^{\text{m.f.}} \exp\left(\int_1^\infty f(s)\, ds\right) \tag{3.31}$$

where $\beta_c^{\text{m.f.}} = (|J|\langle\sigma^2\rangle_0)^{-1}$ *is the critical value in a mean-field approximation.*

Furthermore, at any $T < T_c$ *the spontaneous magnetization obeys*

$$M(\beta) \geq (t/|J|)^{1/2}[1 - O(t)] \quad ; \quad t = (\beta - \beta_c)/\beta_c \tag{3.32}$$

and, moreover, there is long-range-order and (3.8) holds with $D = (\beta|J|)^{-1} \ln(\beta/\beta_c)$ *(whose square root is described in the R.H.S. of (3.32)).*

For $T > T_c$ *the condition in Proposition 1 is satisfied, and*

$$\beta/(\beta_c - \beta) \leq \hat{\chi}(\beta) \leq g((\beta_c - \beta)/\beta_c) \tag{3.33}$$

with the function $g(t)$ *defined by the following equation*

$$t = \int_{g(t)}^\infty f(s)\, ds \quad . \tag{3.34}$$

4. Proof of the Main Result

Our analysis is based on the following inequality, derived in Aizenman and Graham [18] (on p. 273) for the class of systems described at the beginning of the preceding section, at zero magnetic field $(h = 0)$.

$$\frac{\partial \chi_L}{\partial \beta} \geq \frac{|J|\ \chi_L^2}{1 + (\beta|J|)^2 \sum\limits_{x \in \Lambda_L} \langle \sigma_0 \sigma_x \rangle^2} \left[1 - \frac{\chi_L}{(\beta|J|) \sum\limits_{x \in \Lambda_L} \langle \sigma_0 \sigma_x \rangle^2} \right] \tag{4.1}$$

In terms of the dimensionless quantities introduced in (3.14), we obtain simply:

$$\frac{\partial \hat{\chi}_L}{\partial \ln \beta} \geq \frac{\hat{\chi}_L^2}{1 + \hat{B}_L} \quad . \tag{4.2}$$

(4.2) supplements the upper bound

$$|J| \chi_L^2 \geq \partial \chi_L / \partial \beta \tag{4.3}$$

which follows from the Lebowitz inequality [19].

The bounds (4.1) and (4.3) have already been used for the study of the high temperature (= low β) phase. The new idea which is applied here is to analyze the transition from the high to the low temperature phases (defined within the infinite volume limit) without first removing the "finite size cutoff." Since $\chi_L(\beta)$ are real analytic functions of β , one may wonder how is the phase transition manifested in a finite system. We shall see a clear answer to that question.

Proof of Theorem 3.1

Combining (4.2) with (3.12) and (3.15) we get:

$$\frac{\partial \hat{\chi}_L}{\partial \ln \beta} \geq \frac{1}{1/|\Lambda_L| + f(\hat{\chi}_L(\beta))} \tag{4.4}$$

For an intuitive grasp of the situation, let us note that if $\hat{\beta}_L$ is defined by the condition

$$f(\hat{\chi}_L(\hat{\beta}_L)) = 1/|\Lambda_L| \tag{4.5}$$

then

$$\frac{\partial\hat{\chi}_L}{\partial \ln\beta} \geq \begin{cases} |\Lambda_L|/2 & \text{for } \beta > \hat{\beta}_L \\ [2f(\hat{\chi}_L)]^{-1} & \text{for } \beta < \hat{\beta}_L \end{cases} \tag{4.6}$$

Integrating this inequality in both directions from $\hat{\beta}_L$ one can see that for $\beta > \hat{\beta}_L$, $\hat{\chi}_L(\beta - \hat{\beta}_L) \geq (\beta - \hat{\beta}_L)|\Lambda_L|/2$ i.e., $\hat{\chi}$ acquires the size of the volume; whereas for $\beta < \hat{\beta}_L$, $\hat{\chi}_L(\beta)$ is less than a finite valued function of the difference $(\beta - \hat{\beta}_L)$, which is independent of L. (For the integration of (4.6) into the regime $\beta < \hat{\beta}_L$ it is convenient to view β as a function of $\hat{\chi}_L$.) We shall now make this argument in a slightly more efficient way, and also establish some uniform bounds on the value of $\hat{\beta}_L$ (and other finite-volume approximants of β_c).

i) Upper and Lower Bounds on $\hat{\chi}_L(\beta)$.

For each L, $\hat{\chi}_L$ is a smooth and strictly increasing function of β . This allows us to reverse their functional dependence, and express (4.6) in the following form:

$$\frac{\partial \ln\beta}{\partial\hat{\chi}_L} \leq |\Lambda_L|^{-1} + f(\hat{\chi}_L) \quad . \tag{4.7}$$

Integrating it we learn that for each pair $0 < \beta_1 < \beta_2$:

$$\ln\frac{\beta_2}{\beta_1} \leq \frac{\hat{\chi}_L(\beta_2) - \hat{\chi}_L(\beta_1)}{|\Lambda_L|} + \int_{\hat{\chi}_L(\beta_1)}^{\hat{\chi}_L(\beta_2)} f(s)\, ds$$

$$\leq \hat{\chi}_L(\beta_2)/|\Lambda_L| + \int_{\hat{\chi}_L(\beta_1)}^{\infty} f(s)\, ds \tag{4.8}$$

Let β_L be some value chosen according to a rule which will be specified below. Regardless of the choice of β_L, (4.8) implies the following pair of bounds:

$$\frac{\hat{\chi}_L(\beta)}{|\Lambda_L|} \geq \ln \frac{\beta}{\beta_L} - \varepsilon_L \quad , \quad \text{for} \quad \beta > \beta_L \quad , \tag{4.9}$$

and

$$\hat{\chi}_L(\beta) \leq g\left(\ln \frac{\beta_L}{\beta} - \delta_L\right) \quad , \quad \text{for} \quad \beta < \beta_L \quad ; \tag{4.10}$$

where $g(\cdot)$ is the function defined by (3.24), the ε and δ are:

$$\varepsilon_L = \int_{\hat{\chi}_L(\beta_L)}^{\infty} f(s)\, ds \quad , \quad \delta_L = \frac{\hat{\chi}_L(\beta_L)}{|\Lambda_L|} \quad , \tag{4.11}$$

and (4.10) is proven via the inequality

$$\int_{\hat{\chi}_L(\beta_L)}^{\infty} f(s)\, ds \geq \ln\left(\frac{\beta_L}{\beta}\right) - \delta_L \quad . \tag{4.12}$$

<u>ii) The Choice of</u> β_L

We shall now choose β_L so that as $L \to \infty$ both

$$\varepsilon_L \to 0 \quad \text{and} \quad \delta_L \to 0 \quad . \tag{4.13}$$

By the integrability of $f(\cdot)$, the former is equivalent to:

$$\hat{\chi}_L(\beta_L) \to \infty \quad , \tag{4.14}$$

whereas the second requirement is

$$\hat{\chi}_L(\beta_L) \, / \, |\Lambda_L| \to 0 \quad . \tag{4.15}$$

To accomplish both, we choose β_L by the condition

$$\hat{\chi}_L(\beta_L) = |\Lambda_L|^{1/2} \quad . \tag{4.16}$$

(One can also take $\hat{\beta}_L$, defined by (4.5).)

iii) Bounds on β_L

Let us now establish some uniform upper and lower bounds on β_L , which are needed for a compactness argument.

For a lower bound we use a familiar argument (see references in [20]) based on (4.2). Re-writing it as $|\partial\chi_L^{-1} / \partial\beta| \leq |J|$, and integrating from $\beta = 0$, one gets

$$\chi_L^{-1}(\beta) \geq \chi_L^{-1}(0) - \beta|J| \quad . \tag{4.17}$$

Since

$$\chi_L(\beta = 0) = \int \sigma^2 \ \rho(\sigma)\, d\sigma = \langle\sigma^2\rangle_0 = (\beta_L^{m.f.}|J|)^{-1} \quad , \tag{4.18}$$

we have

$$\beta_L \geq \beta_c^{m.f.} - [|J|\hat{\chi}_L(\beta_L)]^{-1} \quad . \tag{4.19}$$

To get an upper bound, let us assume $\beta_L \geq \beta_c^{m.f.}$. Then (4.8), with $\beta_1 = \beta_2$ and $\beta_2 = \beta_c^{m.f.}$, implies

$$\beta_L \leq \beta_c^{m.f.} \exp\left[\hat{\chi}_L(\beta_L)/|\Lambda_L| + \int_{\hat{\chi}_L(\beta_c^{m.f.})}^{\infty} f(s)\, ds\right] \tag{4.20}$$

However, using the Griffiths inequality in the obvious way,

$$\hat{\chi}_L(\beta_c^{m.f.}) \geq \beta_c^{m.f.} \ |J| \ \langle\sigma_0^2\rangle_L \geq \beta_c^{m.f.} \ |J| \ \langle\sigma_0^2\rangle_0 = 1 \quad . \tag{4.21}$$

Substituting this in (4.20), we get

$$\beta_L \leq \beta_c^{m.f.} \exp\left[\delta_L + \int_1^\infty f(s)\, ds\right] \quad . \tag{4.22}$$

iv) Conclusion

The uniform bounds (4.19) and (4.22) imply that there is a subsequence of volumes for which β_L converges to a limit, which clearly obeys (3.31). Restricting L to such a subsequence, we define

$$\beta_c = \lim_{L\to\infty} \beta_L \quad . \tag{4.23}$$

(Our conclusions imply that β_c is, in fact, independent of the subsequence, i.e., the limit exists in the usual sense.)

Since β_L were chosen so that $\varepsilon_L, \delta_L \to 0$, and the function $g(\cdot)$ is continuous, (4.9) and (4.10) imply that

a) for $\beta > \beta_c$

$$\limsup_{L\to\infty} \frac{\hat{\chi}_L(\beta)}{|\Lambda_L|} \geq \ln \frac{\beta}{\beta_c} = \frac{\beta-\beta_c}{\beta_c}\left[1 + 0\left(\frac{\beta-\beta_c}{\beta_c}\right)\right] \tag{4.24}$$

b) for $\beta < \beta_c$

$$\liminf_{L\to\infty} \hat{\chi}_L(\beta) \leq g(\ln \frac{\beta_c}{\beta}) \leq g\left(\frac{\beta_c-\beta}{\beta_c}\right) \tag{4.25}$$

Applying now Propositions 2.1 and 2.3, we see that for $\beta > \beta_c$ there is spontaneous magnetization and long-range order, whereas for $\beta > \beta_c$ the correlations decay exponentially. Combined with Proposition 2.2, this leads to the claimed equality of the critical points, (3.30).

The bounds established above imply all the inequalities stated in Theorem 2.1, with the exception of the lower bound in (3.33). That however is a known consequence of (4.2), derived by integrating it <u>down</u> from β_c (as opposed to (4.17)). A careful argument is presented in Ref. [21].■

Notice that the only characteristics of β_L which were used are those described by (4.14) and (4.15). Since the critical point to which β_L converge has an independent characterization, we can make the following observation:

Corollary 4.1: Any sequence of inverse temperatures, $\tilde{\beta}_L$, for which both

$$\lim_{L\to\infty} \hat{\chi}_L(\tilde{\beta}_L) = \infty \tag{4.26}$$

and

$$\lim_{L\to\infty} \hat{\chi}_L(\tilde{\beta}_L)/|\Lambda_L| = 0 \tag{4.27}$$

converges to β_c.

An interesting question which has some bearing on the theory of finite-size scaling is to determine the actual behavior of $\chi_L(\beta_c)$. For example, the inequality (6.3) of Ref. [2] implies that if Ref. [22] is correct, then

$$\hat{\chi}_L(\beta_c) \geq \sqrt{|\Lambda_L|} \quad , \tag{4.28}$$

i.e., our β_L are lower bounds:

$$\beta_L \leq \beta_c \quad . \tag{4.29}$$

5. Convergent Upper and Lower Bounds on β_c

In this section it is shown how finite size calculations may be used for rigorous bounds on the critical temperature. The first result of this type was the convergent sequence of lower bounds (on β_c) of Simon [11]. We supplement it here with upper bounds. Furthermore, we complete the result stated in Corollary 4.1, where a sufficiency criterion was established for the convergence of finite volume approximants to β_c, by presenting explicit error bounds on $\beta_c - \tilde{\beta}_L$. Although the convergence is not very fast, such results are useful for theoretical purposes, as is demonstrated in the next section.

The new upper bounds are expressed by the following inequality:

Proposition 5.1: In a regular system of the type described in Section 3, for any β *and* K ,

$$\beta_c \leq \beta \exp \left[\int_{\hat{\chi}_K^{f.b.}(\beta)}^{\infty} f(s)\, ds \right] \quad . \tag{5.1}$$

In order to extract from (5.1) good numerical bounds, one should find, for as large K as practical, values of β for which $\hat{\chi}_K^{f.b.}(\beta)$ is very large, yet small, compared with $|\Lambda_K|$.

Proof: The inequality is obviously true for $\beta \geq \beta_c$. For $\beta \leq \beta$, the upper bound (3.33) -- in its stronger version of (4.25), implies that

$$\ln \frac{\beta_c}{\beta} = \int_{g(\ln \frac{\beta_c}{\beta})}^{\infty} f(s)\, ds$$

$$\leq \int_{\hat{\chi}(\beta)}^{\infty} f(s)\, ds \leq \int_{\hat{\chi}_K^{f.b.}(\beta)}^{\infty} f(s)\, ds \tag{5.2}$$

where in the first step we used the definition of $g(\cdot)$ ((3.34)), and in the last one the Griffiths inequality. Exponentiating (5.2) one gets (5.1). ■

Our lower bounds are based on an extension of Simon's result [11]. To formulate it we define the mean range of an interaction as

$$R(J) = \sum_x \|x\|\, J_{0,x} / \sum_y J_{0,y} \tag{5.3}$$

with $\|x\| = \max_{1 \leq i \leq d} \{|x_i|\}$, for $x = (x_1 , \ldots , x_d)$.

We shall now assume that $R(J) < \infty$.

Proposition 5.2: In a system of the above type (i.e., spins in the S.G. class and the Hamiltonian (1.1) with $h = 0$ *and* $|J| < \infty$ *), if for some* β *and* $L > 0$,

$$\hat{\chi}_L^{f.b.}(\beta) \leq L / R(J) \quad , \tag{5.4}$$

then the condition of Proposition 3.1 is satisfied, i.e., $\hat{\chi}(\beta) < \infty$ *, and*

$$\beta < \beta_c \quad . \tag{5.5}$$

Alternatively stated, at the threshold of the high temperature phase,

$$\hat{\chi}_L^{f.b.}(\beta_c^{h.t.}) \geq L / R(J) \tag{5.6}$$

for each $L > 0$.

The Proof of Proposition 5.2 is given in Appendix I. (The restriction imposed here on the spin distribution may be relaxed.) Combining it with the previous result, we get:

<u>*Corollary 5.1*</u>: *Let* β_L^s *be defined by the following generalization of Simon's condition* [11]

$$\hat{\chi}_L^{f.b.c.}(\beta_L^s) = L / R(J) \quad . \tag{5.7}$$

Then, in a regular system, for each $L > 0$:

$$\beta_L^s \leq \beta_c \leq \beta_L^s \, e^{\int_{L/R(J)}^{\infty} f(s)\, ds} \tag{5.8}$$

Proposition 3.4 shows that for the nearest neighbor models, one may use $f(s) = C\, s^{-\min\{\frac{d}{2},\, 2\}} \left[1 + I[d=4]\,|\ln s|\right]$. In particular, in $d = 3$ dimensions, we have:

$$\beta_L^s \leq \beta_c \leq \beta_L^s \left[1 + O(L^{-1/2})\right] \quad . \tag{5.9}$$

Of course, better <u>upper bounds</u> may be obtained by using (5.1) with higher values of β than β_L^s ; minimizing empirically the right-hand side of (5.1). Combining (5.1) with (4.8), where we take $\beta_1 = \beta \quad \beta_2 = \beta_L^s$, we have the following error bounds for such estimates:

<u>*Corollary 5.2*</u>: *In any regular system of the type described above, for any* β *and* L :

$$\beta \exp\left\{-\left[\frac{\hat{\chi}_L(\beta)}{|\Lambda_L|} + \int_{L/R(J)}^{\infty} f(s)\, ds\right]\right\} \leq \beta_c \leq \beta \exp\left[\int_{\hat{\chi}_L^{f.b.}(\beta)}^{\infty} f(s)\, ds\right] \tag{5.10}$$

Summary: Each set of values of $\left\{\beta \, , \chi_L(\beta) \, , \chi_L^{f.b.c.}(\beta)\right\}$ yields both upper and lower bounds on β_c . However, the above results may be best utilized by a separate use of (5.1) and (5.8). Before engaging in such a calculation, one would like to know the accuracy it would yield with a given L . Our a priori over-estimate for $d=3$ is $O(L^{-1/2})$. This may be slightly too pessimistic, although the inequality (4.4) suggests that the actual discrepancy between the bounds would still not decay faster than a power of L . For a better method, one should have a more complete theory of the finite-size scaling.

In the next section, we shall see examples of theoretical applications of the bounds derived here.

6. Convergence of the Critical Temperatures of Finite-Width Slabs

We shall now consider ferromagnetic systems with a fixed coupling $J_{x,y}$, on the infinite slabs $:[-L\, ,\, L] \times \mathbb{Z}^d$; with periodic boundary conditions in the first component of the position vector.

If $d \geq 2$, then for each L there is a positive critical temperature whose inverse we denote by $\beta_c(d\, ;\, L)$. One may expect that

$$\lim_{L\to\infty} \beta_c(d\, ;\, L) = \beta_c(d+1) \tag{6.1}$$

where the right-hand side is the critical point for the system on $\mathbb{Z}^{d+1}$. However, (6.1) is not totally obvious. In particular, it fails for $d=1$, since $\beta_c(1\, ;\, L) = \infty$ for each L , while $\beta_c(2) < \infty$. Nevertheless, (6.1) "should" be true once d is above the lower critical dimension.

The conjecture that an analog of (6.1) (with $d=2$) holds for the bond percolation model, was used in Ref. [23] for the derivation of rigorous block-variable techniques which allowed one to study a random surface problem up to its critical point.

Proposition 6.1: *Let* $\beta_c(d\,;L_1,\ldots,L_k)$ *denote the critical inverse-temperatures for ferromagnetic systems on the "thick slabs"* $\bigotimes_1^k[-L_i\,,L_i]\otimes\mathbf{Z}^d$, *with the nearest neighbor interaction* $J_{0,z}=\delta_{|z|,1}$ *and a single-spin distribution in the G.S. class. If* $d\geq 3$, *then*

$$\lim_{L_1,\ldots,L_k\to\infty}\beta_c(d\,;L_i\,,\ldots L_k)=\beta_c(d+k) \tag{6.2}$$

for every fixed k .

The proof is by a simple application of the bounds of the preceding section, and the <u>uniform regularity</u> of the slab systems, which is expressed in the next Lemma.

Lemma 6.1: *There exists a constant* $C<\infty$ *such that for each* $d\geq 3$, k , L *and* $L_1,\ldots,L_k\leq L$, *the above system in the finite volume* $\bigotimes_1^k[-L_i\,,L_i]\otimes[-L\,,L]^d$ *satisfies* (3.15) *with* $f(s)=C\,s^{-1/2}$.

The proof is left as an exercise in the technique used in the Proof of Proposition 3.4. It requires also some fairly simple bounds on Riemann sums. The value of C to which we refer above, is the one which (3.28) yields for $d=3$.

<u>Proof of Proposition 6.1</u>

The main idea of the proof is to approximate the critical temperatures of both the thick d-dimensional slabs, and the $d+k$ dimensional system by the temperature $\hat{\beta}_L^{-1}$ defined by (4.16) for the $d+k$ dimensional cube $[-L\,,L]^{d+k}$, where

$$L=\min\{L_1\,,\ldots,L_k\} \tag{6.3}$$

Since the bounds (5.10) apply to all those systems with the same function, described in Lemma 6.1, we learn that

$$A_L^{-1} \leq \beta_c(d\,;L_1\,,\ldots,L_k)\,/\beta_c(d+k) \leq A_L \tag{6.4}$$

with

$$A_L = \exp\left[\int_{(2L)^{(d+k)/2}}^{\infty} f(s)\,ds + (2L)^{-(d+k)/2} + \int_L^{\infty} f(s)\,ds\right] =$$

$$= \exp\left[2C(2L)^{-(d+k)/4} + (2L)^{-(d+k)/2} + 2C\,L^{-1/2}\right] \quad . \tag{6.5}$$

(6.5) not only implies (6.2), but it also shows that the convergence is not slower than $O(L^{-1/2})$. ■

Remark: A similar argument shows that in $d \geq 3$ dimensions, β_c depends continuously on $\rho(\cdot)$.

Appendix I: Extension of Simon's High Temperature Criterion to Interactions of Unbounded Range

In Propositions 3.1 and 5.2, reference was made to an extension of Simon's analysis [11] in which the condition that the interaction be of finite range is replaced by the weaker requirement of exponential decay. In this Appendix we prove the necessary results. An interesting outcome is that the relevant length scale is not the minimal distance beyond which J_x vanishes, but rather the mean range $R(J)$ which is the first moment of $\|x\|$ with respect to J_x -- defined by (5.3).

Our starting point is the following inequality:

$$\langle\sigma_0\sigma_x\rangle - \langle\sigma_0\sigma_x\rangle' \leq \sum_{\{y,z\}\in B} \langle\sigma_0\sigma_y\rangle'\ \beta J_{y,z}\ \langle\sigma_y\sigma_z\rangle \tag{I.1}$$

in which $\langle - \rangle$ refers to the correlation functions in some finite system and $\langle - \rangle'$ denotes the correlations in a weakened system, which is obtained by setting $J_{y,z}$ to zero on a collection of pairs $\{y,z\} \in B$.

For the models considered here, (I.1) is a consequence of the most elementary of the techniques used in Refs. [1, 18]. Such an inequality was derived also for a somewhat different class of spins, for which $\rho(\sigma) = e^{-V(\sigma^2)}$ with a convex function $V(\cdot)$ [24].

The primes in (I.1) represent a significant improvement of the inequality obtained by Simon [11] on the basis of the Lebowitz inequality [19]. (The effect is somewhat similar to that of Lieb's strengthening of the original Simon "site" inequality [11,25]).

We shall now apply (I.1) to compare the correlation function for a finite volume system with the periodic boundary conditions on $\Lambda_L \subset \mathbf{Z}^d$, with the one obtained with the free boundary condition in $\Lambda_K \subset \Lambda_L$. The p.b. interaction, in Λ_L, is given by the couplings $J^{(L)}_{y,z} = \sum_{n\in\mathbf{Z}^d} J_{y,z+2Ln}$; whereas for the f.b. $J_{y,z} = I[y,z \in \Lambda_k]J_{y,z}$. ($I[-]$ denotes here the $\{0,1\}$ valued indicator function.)

An immediate implication of (I.1) is that for each $L \geq K$

$$\langle\sigma_0\sigma_x\rangle_L - \langle\sigma_0\sigma_x\rangle^{f.b.}_K \leq \sum_{\substack{y\in\Lambda_K \\ z\in\Lambda_L\setminus\Lambda_K}} \langle\sigma_0\sigma_y\rangle^{f.b.}_K \beta J_{y,z} \langle\sigma_z\sigma_x\rangle_L +$$

$$+ \sum_{n\in\mathbf{Z}^d\setminus\{0\}} \sum_{y,z\in\Lambda_L} \langle\sigma_0\sigma_y\rangle^{f.b.}_K \beta J_{y,(z+2Ln)} \langle\sigma_z\sigma_x\rangle_L \ . \tag{I.2}$$

For a clearer view of this inequality, let us sum it over $x\in\Lambda_L$. One gets

$$\chi_L - \chi^{f.b.}_K \leq \chi_L \sum_{\substack{y\in\Lambda_K \\ z\in\mathbf{Z}^D\setminus\Lambda_K}} \langle\sigma_0\sigma_y\rangle^{f.b.}_K \beta J_{y,z} \ . \tag{I.3}$$

(I.2) is the basis for the considerations of this section. The results dealing with the exponential decay of the correlation function in the high temperature phase are restricted to exponentially decaying interactions, i.e., those for which there is some $\lambda > 0$ such that

$$\sum_x J_{0,x}\, e^{\lambda|x|} < \infty \ . \tag{I.4}$$

A simple way to obtain exponential decay is by considering the weighted sums,

$$\chi_L^{(\mu)} = \sum_{x \in \Lambda_L} \langle \sigma_0 \sigma_x \rangle_L \, e^{\mu|x|} \quad , \tag{I.5}$$

and $\chi_L^{f.b.(\mu)}$ -- which are defined analogously.

Let us also define

$$\gamma_K^{(\mu)} = \sum_{\substack{y \in \Lambda_K \\ z \in \mathbb{Z}^d \setminus \Lambda_K}} \langle \sigma_0 \sigma_y \rangle_K^{f.b.} \, \beta J_{y,z} \, e^{\mu|z|} \quad . \tag{I.6}$$

Notice that if (I.4) is satisfied, than $\gamma_K^{(\mu)}$ is finite for $\mu \leq \lambda$, and is continuous in the limit $\mu \to 0$.

Lemma I.1: For each $L \geq K$ *and* $\mu \geq 0$,

$$\chi_L^{(\mu)} - \chi_K^{f.b.(\mu)} \leq \chi_L^{(\mu)} \gamma_K^{(\mu)} \quad . \tag{I.7}$$

In particular, if for some K $\gamma_K^{(\mu)} < 1$, *then for all* $L \leq K$

$$\chi_L^{(\mu)} \leq \chi_K^{f.b.(\mu)} \left[1 - \gamma_K^{(\mu)}\right]^{-1} \quad . \tag{I.8}$$

Proof: To prove (I.7) we sum (I.2) over $x \in \Lambda_L$, with the weight $e^{\mu|x|}$, which for the R.H.S. is replaced by $e^{\mu|z|} \, e^{\mu|x-z|}$ ($\geq e^{\mu|x|}$).

(I.8) is an obvious consequence of (I.7), assuming (as an observant reader might note) that $\chi_L < \infty$. That, however, is implied by the stability assumption we made about $\rho(\sigma)$. ∎

We are now ready to prove Proposition 3.1. Its first claim, (3.2), requires only the finiteness of $|J|$. In the corresponding part of the proof we use the above lemma only with $\mu = 0$.

Proof of Proposition 3.1: First let us note that if $\chi^{f.b.} < \infty$, then

$$\gamma_K^{(0)} \to 0 \quad \text{as} \quad K \to \infty \quad . \tag{I.9}$$

That follows, by the monotone convergence theorem, from the bound:

$$\gamma_K^{(0)} \le \sum_{\substack{y,z\in \mathbb{Z}^d \\ z\notin\Lambda_K}} \langle\sigma_0\sigma_y\rangle^{f.b.}\ \beta J_{y,z} \le \hat{\chi}^{f.b.} \quad . \tag{I.10}$$

Assume now that $\chi^{f.b.} < \infty$. Letting K be the first value for which $\gamma_K^{(0)} \le 1/2$, we learn from (I.8) that for each $L \ge K$

$$\chi_L \le 2\chi_K^{f.b.} \le 2\chi^{f.b.} \quad . \tag{I.11}$$

Substituting (I.11) in (I.7) we get

$$\sum_{x\in\Lambda_L} |\langle\sigma_0\sigma_x\rangle_L - \langle\sigma_0\sigma_x\rangle_L^{f.b.}| = \chi_L - \chi_L^{f.b.} \le 2\chi^{f.b.}\ \gamma_L^{(0)} \longrightarrow 0 \ , \quad \text{as } L\to\infty \quad . \tag{I.12}$$

(I.12) clearly implies both of the statements made in (3.2).

Let us now add the assumption that the interaction satisfies (I.4), and prove that the correlation function decays exponentially. First we note that by the continuity of $\gamma_K^{(\mu)}$, there is some $\mu > 0$ for which, with the above value of K :

$$\gamma_K^{(\mu)} < 1 \quad . \tag{I.13}$$

Using it in (I.8), one gets the following uniform bound, which holds for all x , and $L \ge K$.

$$\langle\sigma_0\sigma_x\rangle_L \le e^{-\mu|x|}\ \chi_L^{(\mu)} \le \chi^{f.b.}\left[1 - \gamma_K^{(\mu)}\right]^{-1} e^{-\mu|x|} \quad . \tag{I.14}$$

This clearly proves (3.3). ■

The proof of the following result shows that the value of K used above is not greater than $2\chi^{f.b.}\ R(J)$.

Lemma I.2: If for some M ,

$$\chi_M^{f.b.} < M/R(J) \quad , \tag{I.15}$$

then

$$\gamma_K^{(0)} < 1 \tag{I.16}$$

for some $K \in [1, M]$.

Proof: Averaging over the interval [1, ..., M], we obtain

$$\frac{1}{M}\sum_{K=1}^{M} \gamma_K^{(0)} \leq \sum_{y,z} \langle\sigma_0\sigma_y\rangle_M^{f.b.} \beta J_{y,z} \frac{1}{M}\sum_{K=1}^{M} I\left[y \in \Lambda_K , z \notin \Lambda_K\right]$$

(where use was made of the Griffiths inequality)

$$\leq \sum_{y,z} \langle\sigma_0\sigma_y\rangle_M^{f.b.} \beta J_{y,z} ||z-y|| / M = \chi_M^{f.b.} \beta \sum_x J_{0,x} ||x|| / M =$$

$$= \hat{\chi}_M^{f.b.} R(J) / M \quad . \tag{I.17}$$

(Here, as in (5.4), $||x|| = \max_{1 \leq i \leq d} \{|x_i|\}$.)

If (I.15) holds, then the R.H.S. of (I.17) is less than one, and therefore, so is $\gamma_K^{(0)}$ for at least one value of K in [1 , M]. ∎

Proof of Proposition 5.2: The condition (5.5) is identical to (I.15). If it is satisfied for some L (here denoted by M), then, by Lemma I.2, (I.16) holds for some $K \in [1, L]$, and thus, by (I.8), $\chi^{f.b.} < \infty$. ∎

Appendix II: Vanishing of the Spontaneous Magnetization for $T > T_c^{h.t.}$

The following argument proves that $M(T) = 0$ at the high temperature phase which is characterized by the finiteness of χ .

Proof of Proposition 3.2: The free energy mentioned after (3.4) may be represented as

$$P(\beta,h) = \lim_{L\to\infty} \frac{1}{|\Lambda_L|} \ln\langle e^{h|\Lambda_L|S}\rangle_L + P(\beta,0) \quad , \tag{II.1}$$

where S is the magnetization per site:

$$S = \frac{1}{|\Lambda|} \sum_{x\in\Lambda_L} \sigma_x \quad . \tag{II.2}$$

The Newman inequality [26], applied to the terms obtained by the power expansion of the exponential, leads to the following bound:

$$\langle e^{h|\Lambda_L|S}\rangle_L = \sum_{n=0}^{\infty} h^{2n} \sum_{x_1,\ldots,x_{2n}\in\Lambda_L} \langle\sigma_{x_1}\ldots\sigma_{x_{2n}}\rangle_L / (2n)!$$

$$\leq \sum_{n=0}^{\infty} h^{2n} |\Lambda|^n \chi_L^n (2n-1)(2n-3)\ldots 1/(2n)!$$

$$\leq e^{h^2 \chi_L |\Lambda_L| / 2} \quad . \tag{II.3}$$

Substituting (II.3) in (II.2), we get

$$(0 \leq)\ P(\beta, h) - P(\beta, 0) \leq h^2 \chi/2 \tag{II.4}$$

(where the first bound is by the Jensen inequality.) Therefore, as long as $\chi(\beta) < \infty$,

$$M(\beta) = \lim_{h\to 0+} [P(\beta, h) - P(\beta, 0)]/h = 0 \quad . \blacksquare \tag{II.5}$$

Acknowledgements: I would like to thank here K. Binder for a very stimulating discussion of finite-size scaling theories. It is also a pleasure to thank D. Jasnow and B. Widom for useful comments, the Aspen Center for Physics for providing the uplifting environment in which this work was done, and the organizers of the Conference in Köszeg for a very enjoyable and stimulating meeting.

*J.S. Guggenheim Foundation Fellow. On leave from Departments of Mathematics and Physics, Rutgers University. Supported by N.S.F. Grant PHY-8301493 A01.

References

[1] M. Aizenman, "Rigorous Studies of Critical Behavior," in Applications of Field Theory in Statistical Mechanics, Proceedings , Sitges 1984, L. Garrido (ed.), Springer Verlag (Lecture Notes in Physics), in press.

[2] M. Aizenman, Phys. Rev. Lett. 47, 1 (1981); and Commun. Math. Phys. 86, 1 (1982).

[3] D. Brydges, J. Fröhlich and T. Spencer, Commun. Math. Phys. 83, 123 (1982).

[4] J. Fröhlich, Nucl. Phys. B200 [FS 4], 281 (1982).

[5] J. Fröhlich, in Progress in Gauge Theory, Cargése 1983, G. 't Hooft et al. (eds.), Plenum (New York), in press.

[6] M. Aizenman, "The Intersection of Brownian Paths as a Case Study of a Renormalization Group Method for Quantum Field Theory," to appear in Commun. Math. Phys.

[7] G. Felder and J. Fröhlich, "Intersection Properties of Simple Random Walks: a Renormalization Group Approach," to appear in Commun. Math. Phys.

[8] R. Griffiths, J. Math. Phys. 10, 1559 (1969).

[9] B. Simon and R. Griffiths, Commun. Math. Phys. 33, 145 (1973).

[10] T.D. Lee and C.N. Yang, Phys. Rev. 83, 410 (1952).

[11] B. Simon, Commun. Math. Phys. 77, 111 (1980).

[12] R. Griffiths, Phys. Rev. 152, 240 (1966).

[13] J.M. Kosterlitz and D.J. Thouless, J. Phys. C6, 1181 (1973).

[14] S. Elitzur, R. Pearson and J. Shigemitsu, Phys. Rev. D19, 3698 (1979).

[15] J. Fröhlich and T. Spencer, Commun. Math. Phys. 81, 527 (1981).

[16] J. Fröhlich, B. Simon and T. Spencer, Commun. Math. Phys. 50, 79 (1976).

[17] J. Fröhlich, R. Israel, E.H. Lieb, B. Simon, Commun. Math. Phys. 62, 1 (1978).

[18] M. Aizenman and R. Graham, Nucl. Phys. B225 [FS 9], 261 (1983).

[19] J.L. Lebowitz, Commun. Math. Phys. 35, 87 (1974).

[20] M. Aizenman, in Scaling and Self-Similarity (Renormalization in Statistical Mechanics and Dynamics), J. Fröhlich, ed., Birkhäuser (Boston, 1983).

[21] M. Aizenman and C.M. Newman, J. Stat. Phys. 36, 107 (1984).

[22] K. Binder, M. Nauberg, V. Privman and A.P. Young, "Finite-Size Tests of Hyperscaling," UCSC preprint, 1984.

[23] M. Aizenman, J.T. Chayes, L. Chayes, J. Fröhlich and L. Russo, Commun. Math. Phys. 92, 19 (1983).

[24] D. Brydges, in Gauge Theories: Fundamental Interactions and Rigorous Results, P. Dita, V. Georgescu and R. Purice (eds.), Birkhäuser (Boston, 1982).

[25] E.H. Lieb, Commun. Math. Phys. 77, 127 (1980).

[26] C.M. Newman, Z. Wahr. 33, 75 (1975).

PROGRESS IN PHYSICS

Already Published

PPH 1
Iterated Maps on the Interval as Dynamical Systems
Pierre Collet and Jean-Pierre Eckmann
ISBN 3-7643-3026-0
256 pages, hardcover

PPH 2
Vortices and Monopoles, Structure of Static Gauge Theories
Arthur Jaffe and Clifford Taubes
ISBN 3-7643-3025-2
294 pages, hardcover

PPH 3
Mathematics and Physics
Yu. I. Manin
ISBN 3-7643-3027-9
112 pages, hardcover

PPH 4
Lectures on Lepton Nucleon Scattering and Quantum Chromodynamics
W. B. Atwood, J. D. Bjorken, S. J. Brodsky, and R. Stroynowski
ISBN 3-7643-3079-1
574 pages, hardcover

PPH 5
Gauge Theories: Fundamental Interactions and Rigorous Results
P. Dita, V. Georgescu, R. Purice, editors
ISBN 3-7643-3095-3
406 pages, hardcover

PPH 6
Third Workshop on Grand Unification, 1982
P. H. Frampton, S. L. Glashow, H. van Dam, editors
ISBN 3-7643-3105-4
388 pages, hardcover

PPH 7
Scaling and Self-Similarity in Physics (Renormalization in Statistical Mechanics and Dynamics)
J. Frölich, editor
ISBN 3-7643-3168-2
ISBN 0-8176-3168-2
440 pages, hardcover

PPH 8
Workshop on Non-Perturbative Quantum Chromodynamics
K. A. Milton, M. A. Samuel, editors
ISBN 3-7643-3127-5
ISBN 0-8176-3127-5
284 pages, hardcover

PPH 9
Fourth Workshop on Grand Unification
H. A. Weldon, P. Langacker, P. J. Steinhardt, editors
ISBN 3-7643-3169-0
ISBN 0-8176-3169-0
415 pages, hardcover

DATE DUE

OhioLINK

OCT 13 1997

GAYLORD PRINTED IN U.S.A.